迷你冷水机
MINI CHILLER

- 超强制冷效能。
- 采用日立高效旋转式压缩机。
- 低噪音、小体积、重量轻，移动方便。
- 高精准的智能电子温度控制。
- 纯钛金属蒸发器，耐腐蚀。
- Powerful cooling capacity.
- Quiet with Hitachi high efficiency rotary compressor.
- Compact in size, light in weight, easy to move .
- Accurate intelligent electronic temperature control.
- Evaporator made of titanium, corrosion resistant.

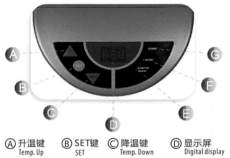

A 升温键 Temp. Up B SET键 SET C 降温键 Temp. Down D 显示屏 Digital display
E 加热指示灯 Heating indicator F 制冷指示灯 Cooling indicator G 电源指示灯 Power supply indicator

- 外接加热棒插孔
  Socket for heater

① 纯钛蒸发器 Titanium evaporator ② 散热系统 Heat dissipation system
③ 高效压缩机 Compressor

Model	Volt	Freq	Power	Recommended aquarium size	Fefrigerating capacity	Weight	Measurement
MINI200	100~120V/220~240V	50/60Hz	1/13HP	≤225L	200W	8kg	200x280x330mm

Shenzhen Xing Risheng Industrial Co., Ltd.
No. 2, Baolong Road Three, Baolong Industrial City,
Longgang, Shenzhen, China

WWW.AKX.CC
400 619 8876

SPIRIT OF AQUASCAPING

Professional Aquarium Syst

1L CO_2 Aluminum Cylinder Set - Pressure Reduced (Face-Up)

- Seamless aluminum CO_2 cylinder, TUV/DOT approved, safety and quality assured.
- Adjustable supporting base prevents cylinder and regulator from falling down.
- Precise pressure reduced regulator provides accurate and reliable CO_2 output.
- Ceramic CO_2 diffuser ensures the best dissolving rate, placing it at the bottom of tank can increase CO_2 diffusing time in the water. Convenient to change the ceramic disc.
- Acrylic intense flow CO_2 bubble counter, transparent material allows accurate CO_2 bubble counting.
- High quality stainless spring check valve ensures no water backflow. Air pipe fastening nuts promise no CO_2 bursting out.
- Air pipe holder included, preventing air pipe from being bended and clogged.

| Air pipe | Air pipe holder | Spring Check Valve | Ceramic CO_2 Diffuser - M | Intense Flow CO_2 Bubble Counter | Precise CO_2 Pressure Regulator | Cylinder Supporting Base | CO_2 Aluminum Cylinder - 1L |

95g CO_2 Disposable Supply Set - Pressure Reduced

- Disposable CO_2 cartridge, easy installation and replacement.
- CO_2 cartridge supporting base, modern design, goes well both f hanging and standing.
- Pressure-Reduced CO_2 Regulator, accurate CO_2 adjusting and dosing promised.
- 2 in 1 CO_2 Diffuser-cone, assures best diffusing effect and clear b counting observation. Easy replacement of diffusing ceramic.
- Spring Check Valve, stainless spring and silicone plug adopted, quality assured, best effect guaranteed.
- Two sets of Air Pipe Holder included, effectively prevents the pip bended and clogged.

| Air pipe | Air pipe holder | Check valve | 2 in 1 CO_2 Diffuser | Vertical Pressure -reduced regulator | Disposable CO_2 Cartridge - 95g | Supporting Base |

1L CO_2 Aluminum Cylinder Set - Professional (Face-Side)

- Seamless aluminum CO_2 cylinder, TÜV/DOT approved, safety and quality assured.
- On-off valve, forged from electroplating copper with plastic steel handle, durable and rust resistance promised.
- Adjustable supporting base prevents cylinder and regulator from falling down.
- Stainless steel twin gauge CO_2 controller provides accurate and reliable CO_2 output.
- 2 in 1 CO_2 diffuser ensures the best dissolving rate, convenient to change the ceramic disc.
- CO_2 indicator monitors precisely CO_2 level, avoid CO_2 waste. Water-proof color chart sticker is reusable.
- High quality stainless spring check valve ensures no water backflow. Air pipe fastening nuts promise no CO_2 bursting out.
- Twin setting timer, provides individual CO_2 and lighting control.
- Air pipe holder included, preventing air pipe from bended and clogged.

| Air pipe | Air pipe holder | Spring check valve | 2 in 1 CO_2 diffuser-M | Solution of CO_2 Indicator | CO_2 Indicator | CO_2 controller twin-gauge pressure -reduced | Twin Setting Timer | Cylinder Supporting Base | CO_2 Aluminum Cylinder - 1L |

95g CO_2 Disposable Supply Set - Easy start up

- Disposable CO_2 cartridge, easy installation and replacement.
- Baking-finished CO_2 cartridge supporting base, slim cut-line design, goes well both for hanging and standing.
- CO_2 regulator with pressure gauge, more accurate CO_2 dosing.
- 3 in 1 compact V CO_2 diffuser, assures best diffusing effect, clear bubble counting, and no back flow! Easy replacement of diffusing ceramic.
- Air pipe holder included, effectively prevents the pipe from bended and clogged.

| Air pipe | Air pipe holder | 3 in 1 compact V CO_2 diffuser (S) | CO_2 flow regulator | Disposable CO_2 Cartridge - 95g | Supporting Base |

CO_2 Diffuser Set   A perfect CO_2 supply set for small tank and beginner!

- So simple in setting up and operation.
- Extendable CO_2 diffusing chamber provides more CO_2 diffusing area.
- Suitable for small water plant tanks.
- Under 25°C 1 ATM condition, disposable CO_2 can contains 5500cc of CO_2 gas. Provide approximately 45 times of CO_2 injection into diffuser chamber.

Check valve

Disposable CO_2 can

CO_2 Diffuser

| Air pipe | Check valve | CO_2 Diffuser | Disposable CO_2 can |

宗洋水族有限公司 TZONG YANG AQUARIUM CO., LTD.

TEL:886-6-230-3818 FAX:886-6-230-6734   www.tzong-yang.com.tw   e-mail:ista@tzong-yang.com.tw

Professional Aquarium System
ISTA

Canister Filter

- Quick start by pressing button continuously
- Energy-saving 20%
- Auto power-off function when pump over heating
- Adjustable water flow by movable hose tape
- High quality filter-Power Material Bio-Ceramic included

By only press the button on the motor stand to quick start. No need to priming water and simplify the operation

The stop cocks can turn 360 degrees to fix any angle of soft pipe.

Equipped with 4 bucket make operation and maintenance safer.

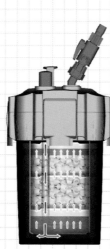

SPONGE

BIO-SPONGE

CERAMIC RINGS

F-1080

F-1240

Hang-on Filter

Ultra thin • Non-return • Auto Priming

New patented structure with non-return valve ensures that no water flows back when power off. After power supply is on, water will circulate quickly to avoid pump racing.

High quality silent design is adopted to prevent the pump from making noise when operating.

Two-stage filtering cartridges enhance filtration efficiency and quality. Extend filtering area to 2 - 3 times wider.

Bio-sponge is attached to water inlet to prevent baby fish from being sucked into the hang-on filter Suitable for both freshwater and saltwater.

Active Carbon | Bio Sponge

Hang-on Filter Adjustable

- No need to add water for re-starting
- Quiet motor operation, adjustable water flow
- Equipped with water inlet sponge, avoid sucking in small fish
- Cartridge filtering sponge, easy for replacement

ADJUSTABLE HANG-ON FILTER

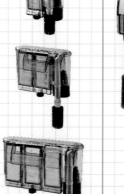

宗洋水族有限公司　TZONG YANG AQUARIUM CO., LTD.

TEL:886-6-230-3818　FAX:886-6-230-6734　www.tzong-yang.com.tw　e-mail:ista@tzong-yang.com.tw

BOYU®

- **National High-Tech Enterprise**
- **Guangdong Top Brand**
- **Research Center of Gardening & Aquarium Equipment Engineering of Guangdong**
- ISO9001:2008   ISO14001:2004   OHSAS18001:1999

Big Flow Rate Air Pump

IPX4

UVC Sterilamp

360°

Ø40mm
Ø32mm
Ø63mm
Ø75mm

ADJUST THE LIGHT TRANSMITTANCE HOLE

UV LIGHT

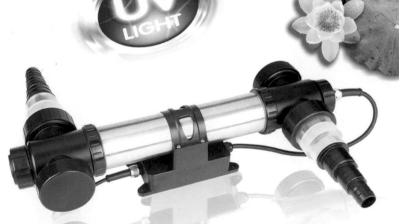

PATENTED

Modular structure designed is allowed to connect multiple UV-C sterilizers to increase the performance result by your purpose. (the connectors for multiple-use are optional parts)

BOYU 博宇

 garden ponds   boyu aquarium   pet products

Http://www.boyuaquarium.com

Natural Life, Natural Enjoyment
--Created by BOYU

External Filter Canister

UV LIGHT

FILTER SPONGE

BIO BALL

ACTIVATED CERBON

BIO SPONGE

UVC Sterilizer

Electronic Heater

QUARTZ BALL

Water Chiller

This appliance can be used for both fresh water and seawater.it can be used inaquariums and other fields which the temperature of water need to be adjusted.

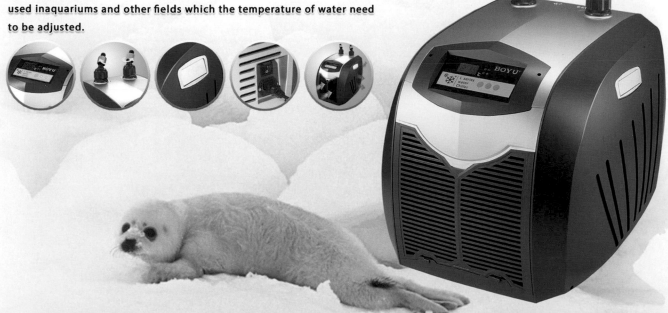

PATENTED

制造商：广东博宇集团
MANUFACTURER: GUANGDONG BOYU GROUP
电话TEL: 86-768-8899328(总机) 8891168
传真FAX: 86-768-8887799 邮编ZIP. 515700
E-mail: boyu@boyuaquarium.com

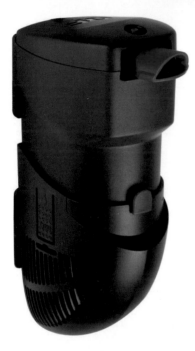

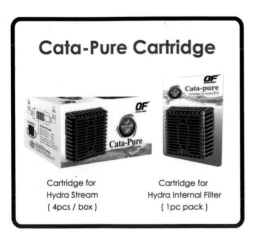

OCEAN NUTRITION™

Preserving Life and Beauty through Nutrition

Betta Products

Betta SPA

- The wild almond leave extract will create a more natural environment for the Betta fish.

- These leaves contain humic acids and tannins and are known to have antibacterial properties.

- Ideal for conditioning the Betta, promotes activity, spawning and colors.

- Contains Yucca extract and calcium.

- Colors the water via natural tannins.

Betta STARTER

- Top performance larval diet.

- To be used from the 3rd day to the end of the 1st month.

- Nutritionally complete food for young Betta fish, which will help promote fast growth and disease resistance.

- Highly attractive for Betta fry.

- With vitamins and a high DHA/EPA ratio.

- Free-flowing floating crumbles.

Betta PRO

- Ideal for the growth and conditioning of the Betta fish.

- To be used from the 2nd month to the end of the 6th month.

- Enhances the natural colors of the Betta.

- Fins and scales grow stronger and last longer.

- Floating pellets (+/- 1mm), does not cloud the water.

Betta FOOD

- High quality daily diet, using the freshest ingredients available.

- To be used for fish from 6 months and older.

- Developed to produce the most nutrient complete food for all Betta species.

- Will enhance the natural colors of the Betta.

- Floating pellets (+/- 1,5mm), does not cloud the water.

Ocean Nutrition Europe

www.ocean nutrition.eu   www.facebook.com/oceannutrition.eu

The Success of Your Aquarium
DYMAX®

REX-AQUARIUM LIGHTING

The Dymax Rex–Led lighting provides a balanced combination of multiple LEDs for optimal photosynthetic activity, coral growth and viewing. These lights provide full spectrum lighting plus essential actinic blue spectral wavelengths that promote and support strong coral growth as well as enhance its colour. This fixture also emits balanced light for accurate viewing of fishes and the aquatic environment in the aquarium.

www.mydymax.com

Chemi-pure and Vitachem are a tag-team of proven power. Combining the industry's longest-lasting and most effective carbon with ion-exchange resins, Chemi-pure is a simple-to-use filter media in a convenient nylon bag that keeps your aquarium water crystal clear. Vitachem is a revolutionary vitamin supplement that boosts your fishes' natural immune system, improving color, vigor, and behavior. Use them together and see the natural wonder in your fish.

Boyd Enterprises
Advanced Aquarist Products
www.chemi-pure.com
1-855-655-2100

CV. MAJU AQUARIUM
CITES REG. A-ID-522

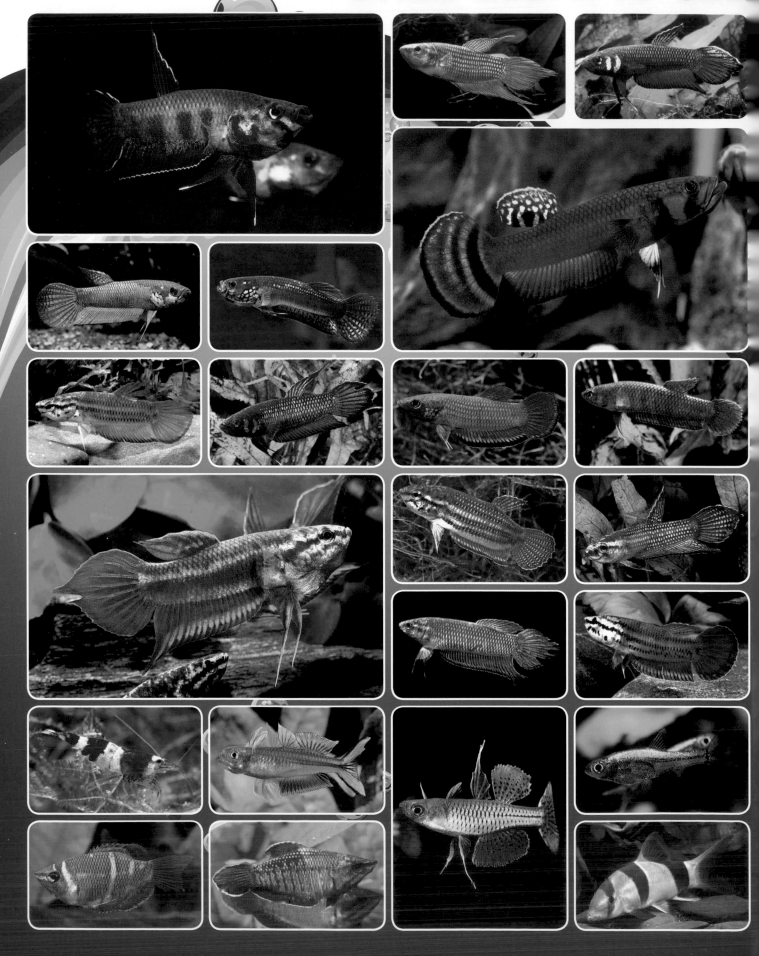

BREEDER AND EXPORTER

Tank bred offering

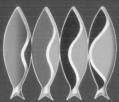

JY LIN TRADING CO., LTD.
The Biggest Ornamental Fish Exporter in Taiwan

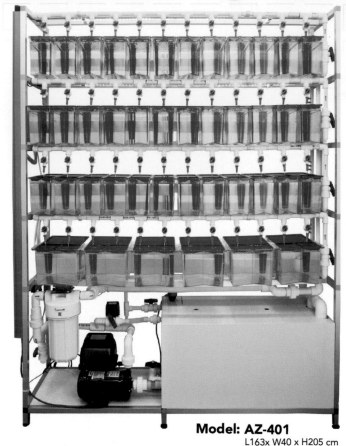

PC Tanks- Configurations and Dimensions

"Aquazoo System Series - General Lab Use

Standard-PC01: Design for Lab Experiment
Dimensions: Length: 163cm, Width: 40cm

Model:	AZ-201	AZ-301	AZ-401	AZ-501
Shelf:	2-shelf	3-shelf	4-shelf	5-shelf
4L-Tanks:	24	36	48	60
Valves:	24	36	48	60
Air Valves:	6	9	12	15
LED Light:	2	3	4	5
Height:	Custom	167cm	198cm	229cm

Standard-PC02: Design for Lab Experiment
Dimensions: Length: 145cm, Width: 40cm

Model:	AZ-201	AZ-302	AZ-402	AZ-502
Shelf:	2-shelf	3-shelf	4-shelf	5-shelf
4L-Tanks:	20	30	40	50
Valves:	20	30	40	50
Air Valves:	6	9	12	15
LED Light:	2	3	4	5
Height:	Custom	167cm	198cm	229cm

Standard-PC03: Design for Lab Experiment
Dimensions: Length: 125cm, Width: 40cm

Model:	AZ-203	AZ-303	AZ-403	AZ-503
Shelf:	2-shelf	3-shelf	4-shelf	5-shelf
4L-Tanks:	16	24	32	40
Valves:	16	24	32	40
Air Valves:	6	9	12	15
LED Light:	2	3	4	5
Height:	Custom	167cm	198cm	229cm

"Aquazoo System Series - Challenge Testing/Screening"

Standard-PC05: For Lab Challenge Testing/Sreening Experiment
Dimensions: Length: 163cm, Width: 40cm

Model:	AZ-205	AZ-305	AZ-405
Shelf:	2-shelf	3-shelf	4-shelf
4L-Tanks:	24	36	48
Valves:	24	36	48
Air Valves:	6	9	12
LED Light:	2	3	4
Height:	Custom	197cm	228cm

Standard-PC06: For Lab Challenge Testing/Sreening Experiment
Dimensions: Length: 145cm, Width: 40cm

Model:	AZ-206	AZ-306	AZ-406
Shelf:	2-shelf	3-shelf	4-shelf
4L-Tanks:	20	30	40
Valves:	20	30	40
Air Valves:	6	9	12
LED Light:	2	3	4
Height:	Custom	197cm	228cm

Standard-PC07: For Lab Challenge Testing/Sreening Experiment
Dimensions: Length: 125cm, Width: 40cm

Model:	AZ-207	AZ-307	AZ-407
Shelf:	2-shelf	3-shelf	4-shelf
4L-Tanks:	16	24	32
Valves:	16	24	32
Air Valves:	6	9	12
LED Light:	2	3	4
Height:	Custom	197cm	228cm

*Dimensions: not include IP65 control box

*Tanks can be changed to 2, 4 and 8 liters.

Please contact us to confirm.

Options:

A: LED Light cycle cabinets
B: Water temperatures(22-32 degree)
C: One Year bio-filter
D: Anti-dust cover
E: Supplemental aeration Tank
F: Water pre-filter system
G: Auto feeding system
H: Alarm systems with remote texting and email notification
S: Slience enhanced

Common:

Auto-cleaning polycarbonate tank(4L, 8L)
40W UV Sterilizers
IP65 control box
High-efficiency bio-fitration systems
Water level detection and changer

Consumables:

Filter pads
Porous foam filters
Activated carbon

Glass Tank: Configurations and Dimensions

Standard-12: Model- ECO Design for Larval Rearing
Dimensions: Length: 125cm, Width: 40cm

Model:	AZ-212-ECO	AZ-312-ECO	AZ-412-ECO
Shelf:	2-shelf	3-shelf	4-shelf
Glass Tank Segment:	6	9	12
Valves:	6	9	12
LED Light:	2	3	4
Height:	Custom	197cm	228cm

Standard-13: Model- for Larval Rearing
Dimensions: Length: 125cm, Width: 40cm

Model:	AZ-213	AZ-313	AZ-413
Shelf:	2-shelf	3-shelf	4-shelf
Glass Tank Segment:	6	9	12
Valves:	6	9	12
LED Light:	2	3	4
Height:	Custom	167cm	198cm

*Standard-13: Model- for Larval Rearing

Dimensions: Length: 125cm, Width: 40cm

IP65 control box

Professional Shrimp Farm & Exporters C-SKY 海天

C-SKY International Trade Co.,Ltd.

海天國際貿易有限公司

www.c-sky.com.tw

service@c-sky.com.tw

Anubias Classical System

阿诺比·经典系列

IP67

WATER PROOF IP67
防水等级

Features:

1. 全新触摸式开关
   Brand new touch-switch

2. 白光、蓝光、RGB三色完美搭配
   Superb collocation of white lighting,
   blue lighting,RGB lighting

3. 颜色模式可随意切换
   Color lighting mode could be switched as wish

4. 带有定时开关功能
   Time switch

A-550
Size:550*550*650mm
Size:550*550*730mm
Lamp:34 LED diodes, 17W

A-800
Size:800*420*500mm
Size:800*420*730mm
Lamp:51 LED diodes, 25W

A-1000
Size:1000*420*530mm
Size:1000*420*730mm
Lamp:66 LED diodes, 33W

A-1000
Size:1000*420*530mm
Size:1000*420*730mm
Lamp:66 LED diodes, 33W

A-1200
Size:1200*450*580mm
Size:1200*450*730mm
Lamp:81 LED diodes, 40W

A-1200
Size:1200*450*580mm
Size:1200*450*730mm
Lamp:81 LED diodes, 40W

A-1500
Size:1500*500*650mm
Size:1500*500*730mm
Lamp:4×54W T5

A-1800
Size:1800*600*650mm
Size:1800*600*730mm
Lamp:4×(24W+39W) T5

专业生产鱼缸木柜及水族用品
Specialized in aquarium tank, cabinet, aquatic equipment production

上海刘波水族宠物用品有限公司
Shanghai Liubo Aquarium & Pet CO.,LTD.

电话/Tel: 0086-21-66131836   网址/Http://www.liubo.cn
电邮/E-mail:liubo@liubo.cn

神龍與皇冠VOL.1

龍魚與魟魚的養殖介紹

神龍與皇冠VOL.2

龍魚與魟魚的養殖介紹

Labyrinth Fish World

600頁，超過1700張產地與繁殖
的圖片，是目前世界上最齊全的
迷鰓魚專書

2009 野生原鬥 展示級鬥魚 辨識年鑑

針對至今所發表的野生原鬥
以及展示級鬥魚品種
飼養、繁殖有詳細的介紹與分類

鰕虎圖典

收錄超過200種
來自台灣與世界各地
淡水至感潮帶的
鰕虎品種！

BRAND NEW

魚蝦疾病根療手冊

飼養的魚兒生病了
該怎麼治療？
本書告訴你，輕鬆治療
愛魚的疾病

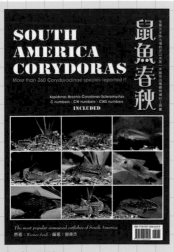

鼠魚春秋

全華文水族市場
有史以來第一本
鼠魚品種最齊全的工具書

神仙世紀

描述所有神仙魚品種
特別是
埃及神仙的繁殖過程及介紹

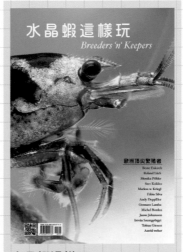

水晶蝦這樣玩

介紹歐洲頂尖繁殖者
養蝦經驗與繁殖的知識跟經驗
藉此更享受玩蝦的樂趣

卵生鱂魚的飼育與賞析

卵生鱂魚的類群與單種介紹外
亦以文字與圖片相互配合
描述卵生鱂魚的飼育與繁殖操作

水族專業書籍出版　　水族海報/DM/包裝設計　　宣傳物印刷承製

BEST VIEW In Chinese Language

魚雜誌 Fish Magazine Taiwan

Mailbox: P.O.BOX 5-85 Muzha New Taipei City 22299 Taiwan (R.O.C.)
Tel: 886-2- 26628587. 26626133　Fax: 886-2-26625595
email: nathanfm@ms22.hinet.net
Website: www.fish168.com

Contents

Labyrinthfishes	7
Labyrinthfishes from Asia	18
The genus Anabas	18
The genus Belontia	24
The genus Betta	30
The genus Colisa	309
The genus Ctenops	339
The genus Helostoma	343
The genus Luciocephalus	347
The genus Macropodus	356
The genus Malpulutta	374
The genus Osphronemus	379
The genus Parasphaerichthys	387
The genus Parosphromenus	392
The genus Pseudosphromenus	499
The genus Sphaerichthys	506
The genus Trichogaster	520
The genus Trichopsis	539
Labyrinthfishes from Africa	555
The genus Ctenopoma	556
The genus Microctenopoma	568
Afterword	574

Publishing House / Fish Magazine Taiwan

Publisher / Nathan Chiang

Author / Horst Linke

Art Supervisor / Lynn Chen

Photographer / Horst Linke

Mail Box /
P.O.BOX 5-85 Muzha New Taipei City 22299 Taiwan (R.O.C.)

Office Number / 886-2-26628587

Fax Number / 886-2-26625595

http://www.fish168.com

E-mail / nathanfm@ms22.hinet.net

Date of Publication Aug. 2014

ISBN 978-986-84527-7-0

LABYRINTH FISH WORLD

with 1768 selected photographs

by
Horst Linke

The Author

Horst Linke has been involved in the aquarium hobby since early childhood. His first study animals included fightingfishes and other labyrinthfishes, and these remain the focus of his interest to the present day. Hence during a trip across the "Dark Continent" in 1963 he took the opportunity to study the natural habitats of the African labyrinthfishes. And this was simultaneously the beginning of a second field of interest: the maintenance and breeding of West African cichlids, as well as the study of their biotopes. Since 1973 he has made numerous research trips to the Ivory Coast, Cameroon, Nigeria, Togo, Ghana, Sierra Leone, Kenya, and Tanzania, as well as to Bangladesh, Borneo, Laos, Malaysia, Myanmar, Hongkong, Sumatra, Java, Singapore, Sri Lanka, Thailand, Vietnam, and China. His interest in South American dwarf cichlids, freshwater angelfishes, and discus has taken him several times to the South American countries of Brazil, Colombia, Peru, and Venezuela. During his travels he has studied natural habitats and collected information that has led to interesting new knowledge regarding maintenance and breeding in the aquarium, and he has also had the opportunity to bring back rare, in most cases previously un- or little-studied fishes for his aquaria back home. Numerous published works have reported his experiences, always very informatively illustrated with his own photos of fishes and biotopes. Because of his good working relationships with scientists at various museums and universities, it has been possible successfully to evaluate the results of his researches.

Horst Linke has for many years been actively involved in the aquarium hobby. To the present day the labyrinthfishes, together with West African cichlids, angelfishes, and discus, remain at the forefront of his interest. Numerous lectures at a wide variety of events have demonstrated the quality of his knowledge and experience. Throughout his fishkeeping career the exchange of information with other aquarists has always been paramount.

The author in the field

Photo by Mike Linke

Foreword

I have been an aquarist ever since my early youth, and my first fishes were Siamese Fighters, *Betta splendens*. That was many decades ago but these fishes have never lost their fascination for me – labyrinthfishes in general, that is, and many of them in particular. Their bright colours and interesting behaviour mean they are always special, and so do their requirements as regards optimal maintenance.

In the interim I have also embraced other groups of fishes, for example the small, colourful West Africans, and Discus with their very interesting brood care, and have been very successful with one of the Kings of the Aquarium, the Altum Angelfish. But I have always continued to keep labyrinthfishes as well. In addition to the numerous different *Betta* species I never cease to be fascinated by the little liquorice gouramis, and that is one reason why in recent years I have visited and studied a multitude of the natural habitats of these small, brightly-coloured *Parosphromenus* species. Unfortunately, in so doing I have established that these habitats – and with them the animal and fish species to which they are home – are increasingly falling victim to the need of humans for living space plus industrial and agricultural usage. Natural disasters such as major flooding and exceptional spates have destroyed or altered some natural habitats, often those limited to a small area. Here too many species have been lost as a result, and sometimes biotopes I had explored in the past were completely changed or no longer existed at all. As a result there are increasing numbers of cases where species that formerly lived at a site can no longer be found, leaving us aquarists with the task of at least keeping these fishes in existence in the aquarium. A good example of this is the worldwide conservation program for *Parosphromenus*-species at www.parosphromenus-project.de as well as efforts to distribute rare but still surviving labyrinthfishes as part of the work of the European Anabantoid Club mit Arbeitskreis Labyrinthfische im VDA (EAC/AKL) at www.aklabyrinthfische-eac.eu under the heading "Have and Search".

During my travels I have discovered numerous new species and often been able to bring them back for the aquarium for the first time. This would often not have been possible without qualified support in the field, so I would like to thank all those who accompanied me during my trips, above all Atison Pumchoosri, Hendra Tommy, Katsuma Kubota, Nathan Chiang, Patrick Yap, and Thomas Sim, without whose help I would not have been able to observe many species in their natural habitats and subsequently in my aquaria back home. I would also like to thank Dr. Jürgen Schmidt for reviewing the manuscript and Mary Bailey for the English translation.

In this book I have tried to present all the species known to and described by science up to 2012, not least because the number has increased enormously over the past decade. Most of the fish names are based on the original descriptions, even though the spelling isn't always correct in the view of various current scientists.

I hope that this book, containing my biotope studies, numerous illustrations, and my suggestions for optimal maintenance, will contribute to the conservation of the host of colourful and extremely interesting labyrinthfishes.

Horst Linke
Schwarzenbach am Wald, Germany, Oktober 2013

3

Contents

Labyrinthfishes	**7**
What is meant by "labyrinthfishes" ?	8
Systematics	9
The morphology of labyrinthfishes	10
The water in the natural habitat	10
Water parameters	10
The pH value	11
British and American measurements	12
The aquarium for labyrinthfishes	**12**
Foods for labyrinthfishes	14
Map of the distribution	**16**
Labyrinthfishes from Asia	**18**
The genus Anabas	**18**
Anabas cobojius	19
Anabas testudineus	21
The genus Belontia	**24**
Belontia hasselti	25
Belontia signata	27
The genus Betta	**30**
Phylogenetic arrangement of the *Betta* species	30
Differences in the reproductive behaviour of *Betta* groups	32
Betta akarensis	35
Betta albimarginata	37
Betta anabatoides	40
Betta antoni	43
Betta apollon	45
Betta aurigans	48
Betta balunga	49
Betta bellica	51
Betta breviobesus	53
Betta brownorum	54
Betta burdigala	56
Betta cf. *imbellis* "Black Imbellis" →*Betta siamorientalis*	185
Betta channoides	59
Betta chini	61
Betta chloropharynx	62
Betta coccina	65
Betta compuncta	68
Betta cracens	69
Betta dennisyongi	71
Betta dimidiata	73
Betta edithae	75
Betta enisae	78
Betta falx	80
Betta ferox	83
Betta foerschi	87
Betta fusca	90
Betta gladiator	92
Betta hendra	93
Betta hipposideros	96
Betta ibanorum	98
Betta ideii	101
Betta imbellis	104
Betta krataios	119
Betta kuehnei	121
Betta lehi	124
Betta livida	127
Betta macrostoma	128
Betta mahachaiensis ←*Betta* sp. Mahachai	131
Betta mandor	136
Betta midas	139
Betta miniopinna	143
Betta obscura	145
Betta ocellata	146
Betta pallida	147
Betta pallifina	150
Betta pardalotus	151
Betta patoti	153
Betta persephone	156
Betta pi	158
Betta picta	160
Betta pinguis	163
Betta prima	166
Betta pugnax	168
Betta pulchra	171
Betta raja	172
Betta renata	175
Betta rubra	178
Betta rutilans	181
Betta schalleri	183
Betta siamorientalis ←*Betta* cf. *imbellis* "Black Imbellis"	185
Betta simorum	189
Betta simplex	191
Betta smaragdina	196

Betta sp. aff. anabatoides→Betta midas 139
Betta sp. Airplaik 202
Betta sp. Bukit Lawang 204
Betta sp. Danau Calak 205
Betta sp. Duc Hua 206
Betta sp. Jantur Gemuruh 208
Betta sp. Kubu 209
Betta sp. Lake Luar 210
Betta sp. Langgam 211
Betta sp. from Mandor→*Betta mandor* 136
Betta sp. Mahachai→*Betta mahachaiensis* 131
Betta sp. from Pampang→*Betta channoides* 59
Betta sp. from Pangkalanbun→*Betta uberis* 301
Betta sp. from Pulau Laut→*Betta ideii* 101
Betta sp. from Sanggau→*Betta antoni* 43
Betta sp. East-Kalimantan 213
Betta sp. Satun 214
Betta sp. Sematan 217
Betta sp. Sengalang 218
Betta sp. Sungai Dareh 220
Betta sp. Tana Merah 221
Betta sp. Tideng Pale 222
Betta sp. Waeng 223
Betta sp. Yao Koi 225
Betta spilotogena 233
Betta splendens 234
Betta stigmosa 281
Betta stiktos 284
Betta strohi 289
Betta taeniata 291
Betta tomi 296
Betta tussyae 297
Betta uberis 301
Betta unimaculata 304
Betta waseri 307

The genus Colisa 309

Colisa bejeus 312
Colisa chuna 314
Colisa fasciata 318
Colisa labiosa 321
Colisa lalia 328
Colisa sp. Lake Inle 337

The genus Ctenops 339

Ctenops nobilis 340

The genus Helostoma 343

Helostoma temminkii 344

The genus Luciocephalus 347

Luciocephalus aura 348
Luciocephalus pulcher 353

The genus Macropodus 356

Macropodus baviensis 357
Macropodus erythropterus→*Macropodus spechti* 370
Macropodus hongkongensis 358
Macropodus lineatus 359
Macropodus ocellatus 360
Macropodus oligolepis 363
Macropodus opercularis 364
Macropodus phongnhaensis 369
Macropodus spechti 370

The genus Malpulutta 374

Malpulutta kretseri 375

The genus Osphronemus 379

Osphronemus exodon 380
Osphronemus goramy 381
Osphronemus laticlavius 385
Osphronemus septemfasciatus 386

The genus Parasphaerichthys 387

Parasphaerichthys lineatus 388
Parasphaerichthys ocellatus 389

The genus Parosphromenus 392

Parosphromenus alfredi 394
Parosphromenus allani 397
Parosphromenus anjunganensis 400
Parosphromenus bintan 402
Parosphromenus cf. *bintan* (blue line) 404
Parosphromenus deissneri 405
Parosphromenus filamentosus 409
Parosphromenus cf. *filamentosus* 412

Parosphromenus gunawani ←*Parosphromenus* sp. Danau Rasau   415
Parosphromenus harveyi   418
Parosphromenus linkei   420
Parosphromenus cf. *linkei*   424
Parosphromenus nagyi   427
Parosphromenus opallios ←*Parosphromenus* sp. "Sukamara"   431
Parosphromenus ornaticauda   433
Parosphromenus pahuensis   437
Parosphromenus paludicola   439
Parosphromenus parvulus   444
Parosphromenus phoenicurus ←*Parosphromenus* sp. Langgam   448
Parosphromenus quindecim ←*Parosphromenus* sp. Manis Mata   451
Parosphromenus rubrimontis ←*Parosphromenus* sp. Bukit Merah   455
Parosphromenus sp. Ampah   459
Parosphromenus sp. Belitong   460
Parosphromenus sp. Calak   463
Parosphromenus sp. Dabo   465
Parosphromenus sp. Danau Rasau →*Parosphromenus gunawani*   415
Parosphromenus sp. Dua   467
Parosphromenus sp. Gawing   469
Parosphromenus sp. Langgam →*Parosphromenus phoenicurus*   448
Parosphromenus sp. Palangan   471
Parosphromenus sp. Pelantaran   473
Parosphromenus sp. Pematanglumut   476
Parosphromenus sp. Pontian Besar   479
Parosphromenus sp. Red   480
Parosphromenus sp. Sentang   481
Parosphromenus sp. Sungaibertam   485
Parosphromenus sp. Sungai Stunggang   489
Parosphromenus sp.Tanjong Malim   491
Parosphromenus sumatranus   493
Parosphromenus tweediei   496

The genus Pseudosphromenus   **499**

Pseudosphromenus cupanus   500
Pseudosphromenus dayi   504

The genus Sphaerichthys   **506**

Sphaerichthys acrostoma   507
Sphaerichthys osphromenoides   510
Sphaerichthys selatanensis   514
Sphaerichthys vaillanti   518

The genus Trichogaster   **520**

Trichogaster leerii   521
Trichogaster microlepis   525
Trichogaster pectoralis   527
Trichogaster trichopterus   530

The genus Trichopsis   **539**

Trichopsis pumila   540
Trichopsis schalleri   544
Trichopsis sp. Hua-Hin   546
Trichopsis vittata   548

Labyrinthfishes from Africa   **555**

The genus Ctenopoma   **556**

Ctenopoma acutirostre   557
Ctenopoma kingsleyae   558
Ctenopoma maculatum   559
Ctenopoma multispinis   560
Ctenopoma muriei   561
Ctenopoma nebulosum   562
Ctenopoma ocellatum   564
Ctenopoma pellegrinii   565
Ctenopoma petherici   566
Ctenopoma weeksii   567

The genus Microctenopoma   **568**

Microctenopoma ansorgii   569
Microctenopoma damasi   570
Microctenopoma fasciolatum   571
Microctenopoma nanum   572
Microctenopoma sp.   573

Labyrinthfishes

Labyrinthfishes are among the most interesting and colourful of aquarium fishes. They are usually easy to maintain, although – depending on the species – they may also have special maintenance requirements. They are subtropical and tropical freshwater fishes whose natural distribution is restricted to Africa and Asia. They include a huge variety of behavioural types. In some individual species the males are so aggressive that encounters between two males frequently lead to battles to the death.

For more than a hundred years the battles of the "Pla-Kat", the biting and tearing fishes, have been known in Thailand and are staged as a competitive sport. The fishes involved are specially bred strains derived from the wild form of *Betta splendens* and occasionally also *Betta imbellis* as well as *Betta smaragdina*.

In order to breed, many labyrinthfishes build nests of air bubbles, which are often reinforced with pieces of plants, at the water's surface or sited in caves lower down. The fishes then lay their eggs on the underside of their nest during a nuptial embrace. A variety of species breed without constructing a nest, but in all species the male embraces his partner with his body. A large number of species practise brood care, with the male devotedly tending and guarding the nest with the developing brood. In the Ceylon Combtail, *Belontia signata*, this applies to both parents and they even shepherd their offspring for several weeks thereafter. In addition, various labyrinthfishes brood their offspring in the mouth.

A number of species make "growling" sounds during courtship or confrontations. Other species can be seen to protrude their fleshy lips as if in a "kiss", both during courtship and when quarrelling.

Some species are even capable of leaving the water during damp or wet weather in order to "crawl" to other habitats. The labyrinthfishes include accomplished jumpers that can accurately target their food out of the water. For this reason the aquarium in which such species are kept must be tightly covered or have a covering of floating plants on the surface of the water. Labyrinthfishes are predominantly warmth-loving, and because they take atmospheric air from the water's surface, this shouldn't be too cold.

Bubblenest of *Betta imbellis*

Labyrinthfishes come in an astonishing range of sizes. Thus the smallest species are barely three centimetres (1.25 in) long when full-grown, while the largest species grow to around 70 cm (27.5 in). Labyrinthfishes are not only aquarium fishes but have other uses as well. The large species are bred as food fishes on fish farms in Asia. They are grown on to splendid specimens in huge open-air expanses of water, and often fed with pig manure. Because dead fishes are prone to be perishable at high temperatures, making lengthy transportation impossible, "air-breathing fishes", which can be kept alive for a long time in damp packaging, are much more valuable. Labyrinthfishes are, however, also tasty delicacies. The Giant Gourami, *Osphronemus goramy*, which can grow to a weight of 7 kg (15.5 lbs), has a very delicate taste. The same applies to the Kissing Gourami, *Helostoma temminckii*, which may only tip the scales at around 1 kg (2.2 lbs) when full-grown, but is nevertheless sold in large numbers in the markets.

But the "useful fishes" also include the many thousands of captive-bred long-finned fightingfishes, *Betta splendens*. These are of commercial importance as many families in Thailand earn a living from their mass production. The multitude of colour and finnage variants seen in *Betta splendens* are not only very popular in Asia, but also make these fishes an export item in demand all over the world.

What is meant by "labyrinthfishes"?

So what does the collective term "labyrinthfishes" mean? Labyrinthfishes are fishes that are equipped with an accessory respiratory organ, a so-called labyrinth, which permits them to take in atmospheric air. This means that they can take in air at the water's surface. If they are prevented from so doing then the majority of species will suffocate in their natural environment, the water. In only a few species is the gill system sufficiently well developed for them to obtain all the oxygen they require by this means alone. The air that is taken in passes through the mouth cavity and is forced into the labyrinth cavity, an upward extension of the head cavity situated above the gill arches on both sides of the head and containing small bony plates arranged in a labyrinthine pattern. These lamellae are covered in skin equipped with numerous blood vessels, and it is via this skin that oxygen is extracted from the air and passes into the bloodstream.

This respiratory organ equips the fishes for inhabiting areas of oxygen-depleted water which attain high temperatures and are often no more than unpleasant waterholes. The oxygen content of the water is often so small that it is only via the uptake of atmospheric air via the labyrinth that the body can be supplied with oxygen and the fishes enabled to exist and survive until the next rainy period. The species that are more demanding in their maintenance live predominantly in flowing woodland watercourses, whose water is oxygen-rich. These species can sometimes even satisfy their oxygen requirement via their gills.

Labyrinth at *Anabas*

Systematics

According to current classification, the labyrinthfishes form the suborder Anabantoidei of the order Perciformes and are divided into three families. These are the family Anabantidae with four genera, the monotypic family Helostomidae with the single genus *Helostoma*, and the family Osphronemidae. The last of these is further divided into four subfamilies with a total of 14 genera.

This gives us the following systematic classification:

Suborder Anabantoidei, labyrinthfishes
- Family Anabantidae, climbing perches
- Genus *Anabas*, true climbing perches
- Genus *Ctenopoma*, bushfishes
- Genus *Microctenopoma*, dwarf bushfishes
- Genus *Sandelia*, Cape bushfishes
- Family Helostomidae, kissing gouramis
- Genus *Helostoma*, kissing gouramis
- Family Osphronemidae, giant gouramis & allies
- Subfamily Belontiinae, combtails
- Genus *Belontia*, combtails
- Subfamily Luciocephalinae, pikeheads & allies
- Genus *Ctenops*, frail gouramis
- Genus *Luciocephalus*, true pikeheads
- Genus *Parasphaerichthys*, Burmese chocolate gouramis
- Genus *Sphaerichthys*, chocolate gouramis
- Genus *Colisa*, western gouramis
- Genus *Trichogaster*, eastern gouramis
- Subfamily Macropodinae, paradisefishes & allies
- Genus *Betta*, fightingfishes
- Genus *Macropodus*, paradisefishes
- Genus *Malpulutta*, spotted gouramis
- Genus *Parosphromenus*, liquorice gouramis
- Genus *Pseudosphromenus*, spiketail paradisefishes
- Genus *Trichopsis*, croaking gouramis
- Subfamily Osphronemidae, giant gouramis
- Genus *Osphronemus*, giant gouramis

This is, perhaps, a good opportunity to mention a second aspect of systematics. Aquarists often get annoyed by the unfamiliar, scientific designations of their fishes and would prefer that only the local trade names were used. But the latter can sometimes be so variable from place to place that there are difficulties identifying a fish as being one and the same species, and in many cases this isn't possible at all on a worldwide or even a national basis.

In 1758 the Swedish naturalist Carl von LINNÉ introduced a universal system for naming animals and plants, the binomial nomenclature that is still in use today. Binomial nomenclature means the giving of a two-part name for the purposes of scientific classification. If, for example, aquarists use the hobby name Leopard Bushfish, the use of this name can lead to misunderstandings, but if they use the designation *Ctenopoma acutirostre* there can be no doubt what fish is meant. While the latter name tells us a lot more as well. To continue with the same example, the designation in full reads *Ctenopoma acutirostre* (PELLEGRIN, 1899). The first part of the name, *Ctenopoma*, identifies the genus and is always written with a capital letter. This tells us what sort of fish we are dealing with as long as we know to what family the genus belongs. The second part of the name, *acutirostre*, identifies the species and is always written in lower case letters. It enjoys priority, and in the event that the genus name is changed, then, come what may, the fish will retain the species name which was assigned to it by the original describer. Because both parts of the name, i.e. the genus and species names, must be Latin or latinised, and the species name must in many cases have the same grammatical gender as the genus name, then if the genus name is changed then at most the final syllable of the species name will change as well. The third part of the name, (PELLEGRIN, 1899), including the date, tells us that the French ichthyologist PELLEGRIN described this fish in the year 1899. However, the fact that the name appears in brackets signifies that the genus name has subsequently been altered. PELLEGRIN described this fish in 1899 as *Anabas acutirostre*. If the species name is followed by a further Latin name – as in the case of the Asian labyrinthfish *Sphaerichthys osphromenoides selatanensis*, for example –then this tells us that we are dealing with the subspecies *selatanensis*.

In this example *Sphaerichthys osphromenoides selatanensis* is thus a subspecies of *Sphaerichthys osphromenoides* and belongs to the genus *Sphaerichthys*, which was erected by CANESTRINI in 1860. Nowadays, however, even though it is used here as an example, this subspecies is regarded as a valid species, *Sphaerichthys selatanensis*.

This treatise on the correct naming of our fishes may appear somewhat complicated at first glance, but demonstrates how the system prevents any misunderstandings. Popular and pseudoscientific names have only a supporting role to play.

The morphology of labyrinthfishes

Next is a short discussion of the morphology of labyrinthfishes. The area from the tip of the snout to the posterior edge of the operculum is termed the head. Thereafter what is termed the body extends to a vertical through the anal opening. The remainder of the fish to the root of the tail – the point where the caudal fin begins – is termed the tail. The caudal peduncle ends at the same point and begins at the posterior end of the base of the anal fin. Labyrinthfishes have a dorsal fin on the back, a pectoral fin on either side of the anterior flank, and beneath the body, usually below the pectoral fins, two ventral fins, often much prolonged, and threadlike in a variety of species. The anal fin is situated beneath the posterior lower body. The caudal fin is attached to the caudal peduncle at the posterior end of the body.

In addition, the distance from the tip of the snout to the end of the caudal fin is termed the total length, or often just the length.

The water in the natural habitat

All biotope data are snapshots of a single moment in time in the natural habitat, and reflect only the state of affairs at the time the research was conducted. The water parameters given here should nevertheless be taken into account for optimal maintenance, and above all for successful breeding. Humic substances are very often important components of the water in the wild, but their significance is very often underestimated and hence they are not taken into consideration when it comes to aquarium maintenance. I regard them as the most important additives for the aquarium water. The natural waters of our labyrinthfishes are often very soft and at the same time extremely mineral-poor. In addition there is often an acidifying hydrogen-ion concentration, i.e. a very acid pH value. This mineral-poor, acid water has a very low "germ" count and hence is a prerequisite for a problem-free existence and successful brood development in a variety of species. Many labyrinthfishes live in water with a high level of humic substances, in so-called black

water. Humic substances in the water not only have an antibacterial effect, but they also protect the epidermis of the fish in that they strengthen it and thereby protect it against outside influences. They cause a type of natural stress that leads to reinforcement of the immune system. Humic substances suppress, prevent, and combat fungal and parasite attacks to a significant degree. They protect the spawn against bacteria, and can also influence the sex ratio in favour of female offspring.

Water parameters

To facilitate problem-free usage and conversion of the water parameters and other data cited in this work, the terms in general use are presented below in tabulated form.

For a better understanding of the data on so-called total hardness, the data are summarised here in the following table:

0° - 4° dGH   = very soft water
4° - 8° dGH   = soft water
8° - 12° dGH = medium-hard water
12° - 18° dGH = hard water
18° - 30° dGH = very hard water
over 30° dGH = extremely hard water

Temperature data are measured in degrees using a thermometer, and may be expressed in degrees Celsius (C), Reaumur (R), or Fahrenheit (F). In Europe temperatures are given predominantly in degrees Celsius (C).

C	R	F
20	16.0	68.0
21	16.8	69.8
22	17.6	71.6
23	18.4	73.4
24	19.2	75.2
25	20.0	77.0
26	20.8	78.8
27	21.6	80.6
28	22.4	82.4
29	23.2	84.2
30	24.0	86.0
31	24.8	87.8
32	25.6	89.6
33	26.4	91.4
34	27.2	93.2
35	28.0	95.0

Enrichment of the aquarium water with humic substances is very important for the optimal maintenance of many species. The active ingredients of black water can be introduced to the aquarium by using a good "tropical water conditioner"

The pH value

The pH value is the most important factor in the aquarium water. Other terms you may hear are the hydrogen-ion concentration, and water with an acid, alkaline (basic), or neutral reaction. Acid water is caused by a preponderance of H (hydrogen) ions, alkaline water by HO (hydroxide) ions. In neutral water the two types of ions are in equilibrium. This means that if the pH value is 7 then we speak of neutral water, i.e. where the ratio of H and HO ions is in equilibrium. A pH value of less than 7 denotes acid water with a higher percentage of H ions, and a pH value of more than 7 is alkaline (or basic) water, with a higher component of HO ions.

The majority of natural waters are mineral-poor and at the same time their hydrogen-ion

Blackwater river in Kalimantan Tengah

concentration means they are neutral to slightly alkaline or even slightly acid. But there are also extremely high HO-ion concentrations, i.e. high pH values; An example of this relevant to labyrinth fishes is the northern regions of Myanmar, habitat of *Parasphaerichthys ocellatus*, *Colisa* sp., *Dario hysginon*, *Badis corycaeus*, and others.

At the other extreme there are the very mineral-poor black water regions. These are often swamp regions and watercourses very rich in humic substances, where the pH value can drop to as low as 4.0 or less. These waters are very low in bacteria and other "germs", and for many fish species such conditions are a prerequisite for living and breeding. The acidity here is determined by the hydrogen-ion concentration. A summary of the hydrogen-ion concentration is given in the following table:

pH value above 8 = very alkaline
pH value above 7 = alkaline
pH value 7 = neutral
pH value below 7 = acid
pH value below 4.5 = very acid

Minerals that cause carbonate hardness – put simply, calcium carbonate and many others – make the water alkaline. By contrast, acids such as hydrochloric acid (HCℓ), phosphoric acid (H_3PO_4), and carbonic acid (H_2CO_3), as well as humic acids (fulvic acids), make the water acid. The same applies to filtration over peat, which has an acid reaction. Thus when it comes to trying to lower the pH value it can be said, put very simply, that "acids eat minerals" (e.g. calcium carbonate), i.e. lower the pH value.

British and American measurements:
1 inch (in., ") = 25.4 mm
1 mm = 0.039 in.
1 foot (ft., ') = 12 in. = 30.48 cm
1 metre = 3.281 ft.
1 yard (yd.) = 36 in.   = 91.44 cm
1 metre = 39.375 in. = 1.094 yds.
1 gallon (USA) = 3.785 litres (l)
1 gallon (Engl.) = 4.546 litres (l)
1 pound (Engl.) = 0.454 kilogram (kg)
1 kilogram = 1000 grams (g) = 2.205 pounds (Engl.)

The aquarium for labyrinthfishes

An aquarium can never be large enough from the viewpoint of its occupants. A large volume of water is more stable and less prone to fluctuations. But a very large maintenance aquarium will also make it more difficult to monitor the fishes to be kept. Hence there are three different aquarium sizes that can be considered for the optimal maintenance of labyrinthfishes.

Firstly, the larger species – such as the gouramis of the genera *Colisa* and *Trichogaster*, the genera *Belontia* and *Macropodus*, the larger mouthbrooding *Betta* species, the African *Ctenopoma* species, the Kissing Gourami *Helostoma temminckii*, and the Giant Gouramis of the genus *Osphronemus* – require tanks with a length of 120 cm (48 in), better 140 cm (55 in), upwards. The tank does not have to be very deep, so that aquaria with a depth of around 40 to 50 cm (16 to 20 in) will be adequate.

For the optimal maintenance of the medium-sized species (such as the bubblenest-building and smaller *Betta*

species, the small *Colisa* species, the chocolate gouramis of the genus *Sphaerichthys*, and the African bubblenesting *Microctenopoma* species), aquaria with a minimum length of 80 cm (31 in) should be used.

Even very small tanks – often with a length of only 25 to 35 cm (10 to 14 in) and containing only 15 to 25 litres (3.3 to 5.5 gallons) – can be optimal for the smallest labyrinthfishes, for example the small red *Betta* species and all *Parosphromenus* and *Parasphaerichthys* species. Such small containers are often not viewed with approval in the aquarium hobby, but more recently have found increasing favour in the form of the currently fashionable so-called "nano aquaria". These small containers offer better opportunities for observing and monitoring the small labyrinthfish species, as they are living in a smaller space. Obviously this requires that population numbers are also small and to some degree these will be tanked in which only one species is maintained.

All maintenance aquaria should have a dense growth of healthy vegetation, while at the same time leaving sufficient open swimming areas. The substrate can consist of light sand. In tanks with decorative plant containers the substrate can be a very thin layer that is siphoned off for cleaning and then poured back in again. The decor should include not only bogwood but also dead beech (*Fagus sylvatica*) leaves or similar. The red-brown colour of these leaves will combine with the green of the plants, the dark brown of the bogwood, and the light sand to create a very decorative effect.

All labyrinthfish aquaria should have an efficient filtration system that purifies the water mechanically and above all biologically, and at the same time creates gentle water movement. Oxygenated, biologically healthy water is an important requirement for keeping labyrinthfishes alive. A partial water change (a quarter to a third of the tank volume) every 8 to 10 days should be regarded as a prerequisite of optimal maintenance.

The maintenance of the smaller aquaria requires a lot of attention and efficient filtration is particularly important here. Integral internal filters operated by an airlift are ideal for the purpose. It is very useful to have small "caves", sited among plants such as Java Fern and Java Moss, and once again a number of the dead leaves of the beech tree can be used for this purpose and will also serve as a source of humic substances. The gaps between the dead leaves and the bottom will be appreciated as shelter and hiding-

Dead, dry leaves from Sea Almond tree, *Terminalia catappa*

The tropical Sea Almond tree, *Terminalia catappa*

places, as in the wild. In this instance a substrate of sand should not be included, for reasons of hygiene. This will permit better monitoring of any excreta, leftover food, and other waste that accumulates on the bottom, and allow it to be removed more efficiently. Water changes of up to a third are important in these small containers and should be performed at least once per week.

Because the smaller labyrinthfish species are predominantly sensitive, often so-called blackwater fishes, the aquarist considering their maintenance in a small aquarium should have prior experience with "mini" or "nano" aquaria. They are usually delicate soft-water fishes that can be maintained optimally only in acid, very soft water. For this reason mature, lime-free tanks should be used for their maintenance. Likewise, water changes should be made using "conditioned" water with its parameters appropriately adjusted prior to use. The addition of humic substances is very important in this regard. The aquarium trade offers a wide range of products for this. Also suitable, however, are the dead leaves of the tropical Sea Almond tree, *Terminalia catappa* (which forms part of the natural landscape in South-East Asia), available in the trade as Sea Almond leaves, as well as the dead leaves of the beech *Fagus sylvatica* and those of various oaks (*Quercus* spp), all trees widespread in northern temperate zones in Europe. All these leaves can be soaked for several weeks in a separate container of water to produce a brown-coloured brew, very rich in humic substances, which can then be used to dose the maintenance tank after water changes. Enrichment of the aquarium water with humic substances can, however, also be achieved using various types of peat. Even the granulated peat available in the trade will produce good enrichment, and the water will also be softened slightly and the pH value lowered in parallel. The use of dead leaves doesn't soften the water, nor does it lower the pH value. The addition of humic substances should form part of the maintenance of all fish species, not only labyrinthfishes. The requisite amount is recommended as 10 mg/l by Professor Christian STEINBERG, the leading German hydrobiologist in this field. Unfortunately the humic-substance content of the water cannot be ascertained via the measuring techniques usual in the aquarium hobby, but it can be estimated from the colour of the water. On this basis the aquarium water should normally be coloured slightly yellowish, but for fishes with a higher humic-substance requirement it should be yellowish to slightly brownish. In the wild, pure blackwaters contain brown to strongly dark red-brown,

clear water and hence have a very high component of humic substances, but this type of heavily-coloured water should not be the goal in the aquarium.

Foods for labyrinthfishes

Labyrinthfishes are for the most part omnivores and are often greedy feeders. They will hunt down anything that moves in the water and at its surface. They will take flake and granulated foods without problem and a wide range of frozen live foods are taken with relish. However, some species are interested only in live food. These include above all the species of the genus *Parosphromenus*. In addition, a few species will take occasional frozen bloodworms (red mosquito larvae). The food offered should always be very varied and live food should be alternated with artificial fare. Vegetable food should also form part of the diet. To keep them in good health our aquarium fishes should undergo one – better two – fast days per week as they are very often overfed out of misplaced kindness.

A huge variety of good organisms are present in the natural habitat. These include water fleas (*Daphnia* and *Moina*) and copepods (*Cyclops* and *Diaptomus*) as well as insect larvae, worms, and non-aquatic creatures that fall into the water. In the wild, shrimps play an especially important role on the menu and probably represent the main food for many species. Shrimps occur in large numbers and in all sizes in the natural biotope, so that this very nutritious food is permanently available to all the fishes.

When it comes to aquarium maintenance then, depending on the time of year and the source, a very wide range of food organisms is available. These include water fleas (*Daphnia* and *Moina*); plus for small and very small fish

Bloodworms (red mosquito larvae)

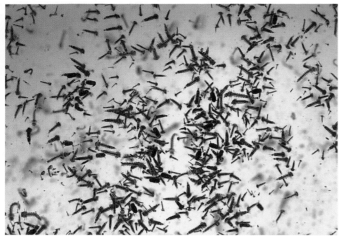
Black mosquito larvae

species there are copepods (*Cyclops* and *Diaptomus*). The larvae of a wide variety of flying insects are a very good food, particularly for the medium-sized and larger species. They include black mosquito larvae (*Culex pipiens*), a food enjoyed greatly by all fishes. At any movement of the water they tumble around the aquarium, but otherwise they hang at the surface. Unfortunately this food is not without its dangers for the aquarist, particularly in tropical countries, as the mosquitoes are vectors for malaria. Glassworms or white mosquito larvae (*Corethra plumicornis*) are not present everywhere or at all times of year. They swim horizontally through the water and are enjoyed by fishes. But the favourite food of many fishes is bloodworms or red mosquito larvae (*Chironomus plumosus*). These remain on the bottom and can crawl very rapidly into the substrate. Unfortunately they often come from polluted waters and hence are not always suitable for our aquarium fishes. For this reason you should ensure they originate from unpolluted sources. Cultivated bloodworms offer major advantages. They are usually free of pollutants and it can (probably) be guaranteed that the fishes will not be affected by pathogens and toxins and contamination with heavy metals from such bloodworms. The same applies to *Tubifex* worms (*Tubifex tubifex*). These too are much enjoyed by fishes but only cultivated stocks can be recommended because of the danger of serious contamination with pollutants in wild-collected supplies.

Freshly-hatched *Artemia* nauplii are a very good food for fry, but they should be genuinely freshly-hatched and not two or three days old. At the time that they become free-swimming *Artemia* nauplii still contain yolk reserves and hence represent a particularly important and very nutritious food for fish fry. Adult brine shrimps of the genus *Artemia* are much enjoyed by all fishes.

Almost all Labyrinthfishes will also feed very enthusiastically on non-aquatic live foods that land on the water's surface. Various species of flies are particularly for this purpose. They can be cultivated without problem and represent a very important source of roughage for the fishes.

Larger *Betta* species also relish earthworms such as Redhead Worms (*Lumbricus rubellus*), Brandlings (*Eisenia foetida*), etc. These should, however, be chopped into smaller pieces before feeding. Another very good type of worm is the small, white, slender Grindal Worm (*Enchytraeus buchholzi*), up to 7 mm (0.25 in long); likewise Whiteworms (*Enchytraeus albidus*), up to 30 mm (1.25 in) long. Unfortunately both species are very nutritious and it is recommended that they are not fed exclusively. Woodlice (*Oniscus asellus* and *Porcellio scaber*) are likewise a very good, readily accepted food.

Almost all the food organisms listed above can be readily cultivated, thereby ensuring a supply of very clean and healthy food. Information on cultivating them is beyond the scope of this book but can be found in the relevant hobby literature.

The above explanations conclude the introductory part of this book, and there follows a catalogue of the individual species within the framework of the genera. The latter are presented in alphabetical order without reference to families and subfamilies, and the species are likewise arranged alphabetically for each genus. Wherever possible a detailed description is given for each species, accompanied by illustrations in colour. My frequent expeditions to Africa and Asia have allowed me to collect a number of rare species and import them myself, and this has in turn permitted a work containing illustrations of many uncommon and very different species. Species for which no photo is available have been excluded now and then, but because these are generally very uncommon species, often not yet imported alive, I hope that these omissions will not be regarded as serious.

This book also discusses the natural habitats of the various labyrinthfishes, with information on the conditions required for maintenance and breeding, based on detailed study in the field. While it would have been preferable to avoid repeating this information, because this book is also intended for reference purposes, an element of repetition seems unavoidable in the case of species with similar behaviour and origins.

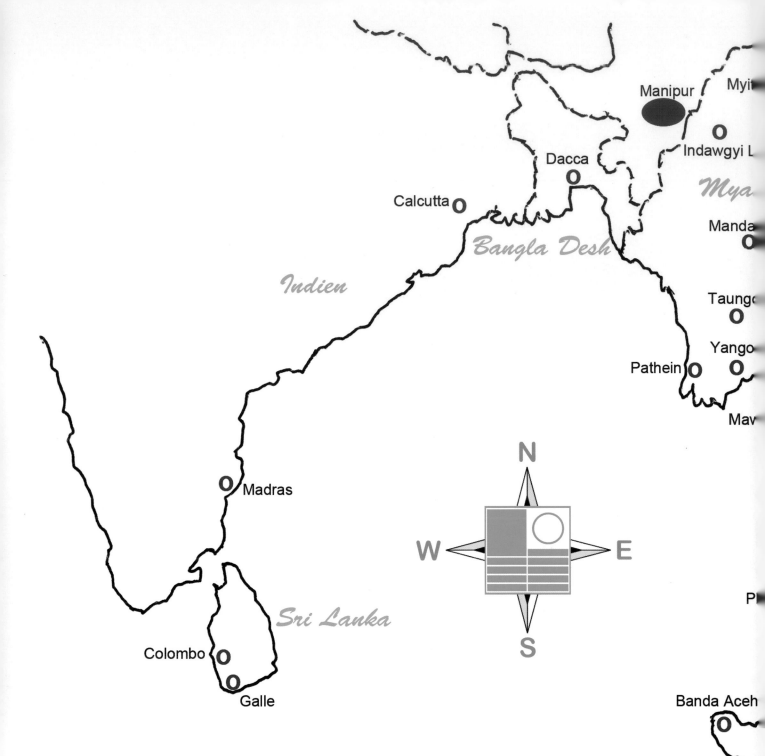

Manipur

Myi

Indawgyi L

Dacca

Mya

Calcutta

Manda

Bangla Desh

Indien

Taung

Yango

Pathein

Mav

N

Madras

W   E

Sri Lanka

S

Colombo

Banda Aceh

Galle

Singk

Sib

Map of the distribution of labyrinthfishes in the Indian Subcontinent, South-East Asia, and China

Taipei

Guangzhou
(Canton)

China

Hong Kong

Hanoi

Laos

Vientiane

iang Mai

Nong Khai

Thailand

Hue

Nakhon
Ratchasima

Da Nang

Bangkok

Siem Reap

Stung
Treng

Pak
Chong

Vietnam

Mahaohai

Cambodia

Trat

Phom Penh

Hua
Hin

Ho-Chi-Minh-City
(Saigon)

Ko Samui

Sandakan

Kota Kinabalu

Narathiwat

Brunei

Sabah

Sungai Kolok

Beaufort

Tawau

atun

East-Malaysia

Pinang

Terengganu

Pulau Natuna
Besar

Sibu

Malinau

Bukit Merah

West-Malaysia

Medan

Selangor

Kuantan

Kuching

Sarawak

Putussibau

Kuala
Lumpur

Kota Tinggi

Sanggau

Melak

Toba
Lake

Pontianak

Kuala Kurun

Buntok

Samarinda

Singapore
Pulau Bintan

Mandor

Pundu

Palangkataya

Padang

Langgam

Pulau
Bangka

Sukamara

Sampit

Banjamasin

Jambi

Palembang

Toboali

Pulau
Belitung

Kubu

Pulau Laut

Kalimantan

Sumatra

Jakarta

Java

Ambarawa

Yogyakarta

The Asian Labyrinthfishes

The genus *Anabas* CUVIER, 1816

The genus *Anabas* (Greek = one who walks, goes up, or climbs) contains several species and local forms. These are the climbing perches in the strict sense. Because they are only rarely maintained as aquarium fishes there is correspondingly little interest in these species. Their natural habitats lie in South-East Asia, but because *Anabas* are also used as food fishes it is now virtually impossible to establish their original, natural distribution. Nowadays they are found in the countries of Sri Lanka (Ceylon), Bangladesh, Myanmar (Burma), Thailand, Malaysia, Indonesia, Cambodia, Laos and from Vietnam to southern China.

These predators live in accumulations of water, where they lurk, well-camouflaged, among marginal vegetation or in the bottom mud, waiting for prey. The water parameters are highly variable and of subordinate importance for maintenance in the aquarium. The temperature fluctuates between 20 and 30 °C (68-86 °F) depending on the location and time of year.

The name "climbing perches" relates to their ability, when biotope conditions deteriorate, to leave their home waters and migrate to a more attractive abode, using their pectoral fins, opercula, and caudal peduncle to help them climb and/or move around. They clamber up the bank and wriggle in a somewhat ungainly fashion, travelling several metres to another body of water, doing this mainly during the damp or wet hours of the night or morning as well as when rain is falling. It is reported that these fishes can cover up to 100 metres (around 330 feet). Their unerring targeting of other bodies of water is particularly noteworthy. They also sometimes leave their watery home in order to hunt insects and other small creatures in the wet grass. Because these fishes are prized as food in their native lands, these nocturnal wanderings, which are often undertaken in groups, are regarded as particularly lucrative "fishing trips" by the local populations.

Two species from the overall distribution region are discussed below.

Anabas cobojius (HAMILTON, 1822)

Explanation of the species name:

The name *cobojius* is derived from a native name for this fish in Bangladesh and eastern India.

English name:

Gangetic Koi

Synonyms:

Cojus cobojius

Original description:

HAMILTON, F. 1822 : A account of the fish in the river Ganges and its branches. pp.120 and 372; London.

Systematics:

Family Anabantidae

Natural distribution:

At present it is not possible to give any definitive details of the distribution of the species. One major distribution region is the lowlands of Bangladesh, where in many places the species shares the same habitat with *Anabas testudineus*.

Biotope data:

The biotope in the natural habitat cannot be described specifically, as in Bangladesh these fishes are apparently to be found in every body of water. By preference they live in shallow, densely vegetated, overgrown, standing waters, frequently with a covering of water hyacinths. The water is usually soft to medium hard, and neutral to slightly alkaline.

Reproduction:

At present little is known about reproduction. Anyone who wishes to breed this species should provide a large aquarium, at least 150 cm (60 in) long and 50 cm (20 in) deep. The water level needs to be only around 20 cm (8 in) deep, and a dense covering of floating plants is recommended. These fishes spawn during an embrace in open water and do not construct any nest. The eggs float to the water's surface, where they then develop. The species is productive. After spawning the pair should be separated and removed from the breeding tank.

Total length: (size)

Size is supposedly similar to that of *Anabas testudineus*, but on the basis of my observations in Bangladesh I believe that this species remains somewhat smaller and is full-grown at around 15 to 18 cm (6 to 7 in) total length.

Anabas cobojius in the market at Dhaka

Habitat of numerous labyrinthfishes near Dhaka

Remarks: (differences from other species of the genus)

These fishes are a very attractive dark green colour, occasionally enhanced by a delicate coppery sheen. Unfortunately this attractive coloration has to date not been exhibited again in the aquarium. This species differs from *Anabas testudineus* in having a deeper body form, fewer scales in the lateral line, and a smaller number of hard rays and a larger number of soft rays in the dorsal fin. In addition *Anabas cobojius* has shorter pectoral fins and possesses a longer snout.

Bangladesh is still a country with simple methods of transport

Anabas testudineus (BLOCH, 1795)

Explanation of the species name:

testudineus (Latin) = made of tortoiseshell, tortoise-like, relating to the tortoise-like armour of hard scales.

English name:

Climbing Perch

Synonyms:

Amphiprion scansor
Amphiprion testudineus
Anabas elongatus
Anabas macrocephalus
Anabas microcephalus
Anabas scandens
Anabas spinosus
Anabas trifoliatus
Anabas variegatus
Anthias testudineus
Antias testudineus
Lutjanus scandens
Lutjanus testudo
Perca scandens
Sparus scandens
Sparus testudineus
Amphiprion scansor
Amphiprion testudineus

Original description

BLOCH, 1795: Naturgesch. Ausländ. Fische, VI, page 121, table CCCXXII

Systematics:

Family Anabantidae

Natural distribution:

The natural distribution includes many parts of South-East Asia. Because this species is also used as a food fish it is now virtually impossible to establish its original, natural distribution. Nowadays *Anabas testudineus* is found in the countries of Sri Lanka (Ceylon), Thailand, Malaysia, Indonesia, Cambodia, and from Vietnam to southern China.

Biotope data:

Anabas testudineus lives in South-East Asia, in almost all standing and slightly flowing waters, and hence water parameters play a subordinate role. The pH value usually lies between 6.5 and 8.0. Habitat preferences include dense marginal vegetation and dense clumps of plants in places in the water. Ideally the emerse areas of the bank will likewise be covered with a very dense growth of low-growing plants such as grasses or grass-like vegetation. We have also recorded fishes of this species in blackwater areas with very mineral-poor, very acid water, although at lower population densities.

Anabas testudineus

Anabas on their way across land in the morning

Anabas shortly after capture

The small, narrow irrigation channels in the rice fields are also often habitats for Anabas

Reproduction:

Distinguishing the sexes is difficult but in general males are darker and have a pointed anal fin. Reproduction is similar to that of *Anabas cobojius*.

Total length: (size)

The species grows up to 25 cm (10 in) long and is unsuitable for normal maintenance with small fishes.

Remarks: (differences from other species of the genus)

Because the species lives in very different geographical regions, its appearance is also highly variable and so there is little that can be said specifically about its coloration.

The following recommendations can be made as regards maintenance: dense planting is suggested, using hardy aquatic plants such as Giant Vallis (*Vallisneria americana* f. *gigantea*),

The harvested rice fields are often interesting study sites

The rice fields contain small watercourses that are home to numerous aquarium fishes

Anabas migrating across land

Because these are predatory fishes the menu should include suitably robust foods, which may include terrestrial insects such as flies plus soft maggots and beetles. Dry food is greatly enjoyed as a periodic food, and above all vegetable food should be provided as a supplement. With frequent, careful feeding the fishes will become hand-tame and take the food offered directly from their owner.

Anubias barteri var. *barteri*, and *Anubias barteri* var., as well as large-leaved plants such as *Echinodorus cordifolius* and ferns such as *Bolbitis heudelotii*. Depending on its geographical origin the species is likely to eat soft-leaved plants. Decoration with bogwood as well as rocky caves is also desirable. Java Fern (*Microsorum pteropus*) will perform good service as an additional decorative plant. Floating plants such as *Ceratopteris pteridoides* or *Pistia stratiotes* will give these fishes a better sense of security, as they are relatively timid.

A maintenance aquarium arranged in this way can be illuminated normally. Efficient filtration of the water is recommended. The water temperature can be allowed periodically to fluctuate between 22 and 30°C (71.5-86 °F) at long intervals, which will simulate the seasonal changes that occur in the natural habitat.

The water parameters are of subordinate importance, but obviously a water change (1/3 of the tank volume) every two weeks should be performed for this species as well. The tank must be tightly covered, as *Anabas testudineus* is not only prone to wandering but is also an accurate jumper. Tankmates should be only large and equally robust fishes, for example large gouramis of the genus *Trichogaster* (*Trichopodus*). Bushfishes such as *Ctenopoma kingsleyae*, *Ctenopoma multispinis*, and others are also suitable.

Area of residual water at the edge of a rice field

The genus *Belontia* MYERS, 1923

The genus *Belontia* currently contains two species. *Belontia hasselti* is the type of the genus, with a natural distribution in Malaysia and on the Indonesian islands of Kalimantan (Borneo), Sumatra, and perhaps Java. This, the so-called Malay Combtail, is the larger of the two species, and is in general a peaceful aquarium occupant. However, in tanks that are too small and unsuitable for the species, *Belontia signata* may exhibit aggressive behaviour among themselves and towards tankmates during brood care and territorial defence. Given optimal maintenance these fishes can be very attractive in terms of coloration and very interesting in their behaviour. They should be kept only in larger aquaria with a length of 120 cm (48 in) upwards. Dense planting and a partial dense covering of floating plants are a further prerequisite for successful maintenance and breeding. These fishes will otherwise be very timid and possibilities for observation limited.

Explanation of the species name:

hasselti = dedication, in honour of van HASSELT.

English name:

Malay Combtail

Synonyms:

Polyacanthus hasseltii
Polyacanthus kuhlii
Polyacanthus einthovenii
Polyacanthus helfrichii

Original description

CUVIER in CUVIER & VALENCIENNES, 1831: Hist. Nat. Poss. vol. 7, page 353, plate 195.

Systematics:

Family Belontiidae

Natural distribution:

Confirmed localities exist in Malaysia, the island of Borneo, and the island of Sumatra, and it may possibly also occur on the island of Java. I have recorded the species above all in Kalimantan Tengah, Borneo in Indonesia, and in the province of Jambi on Sumatra, again in Indonesia.

Biotope data:

On the basis of my observations in the field, the species lives by preference in blackwater areas, in very mineral-poor water with a low concentration of hydrogen ions, i.e. a pH value between 4.2 and 5.5. These fishes live in both smaller and larger watercourses, among dense plant aggregations and in gently flowing, clear water.

Reproduction:

Bubblenest breeder

The largest possible aquarium should be chosen for the

Belontia hasselti ♂

Belontia hasselti after capture in Danau Rasau

breeding of these interesting and beautiful fishes. The tank should be copiously planted, and a partial dense covering of floating plants at the water's surface will give the fishes a greater sense of security and reduce their shyness. They spawn during an embrace beneath the surface or sometimes in deeper areas of the water. The eggs rise to the water's surface and develop there. A bubblenest is only rarely constructed. The species is very productive.

Total length: (size)

Males can attain a total length of around 20 cm (8 in), while females remain somewhat smaller.

Remarks: (differences from other species of the genus)

The sexes can be distinguished only in adult-sized fishes. Males have longer and larger dorsal, caudal, and anal fins. The ventral fins in males exhibit a typical forked anterior ventral-fin ray. They also exhibit the species-typical honeycomb pattern on the membranes of the dorsal, anal, and caudal fins. This species is in general somewhat shy, and not aggressive towards tankmates. These fishes often give the impression of being in serious trouble as they lie on the bottom and sometimes lean against plants or rocks, but this is part of the normal behavioural repertoire of this species. Larger aquaria with a length of 120 cm (48 in) upwards should be used for the maintenance of this species, with dense planting

Belontia hasselti in the market at Kota Jambi in Sumatra

in places (but leaving sufficient open swimming space), and a partial dense covering of floating plants will cure the fishes of their shyness.

Belontia signata (GÜNTHER, 1861)

Explanation of the species name:
signata (Latin) = with (characteristic) markings.

English name:
Ceylon Combtail, Combtail

Synonyms:
Polyacanthus signatus

Original description
GÜNTHER, A. 1861: Catal.Fishes, vol. 3., page 379

Systematics:
Family Belontiidae

Natural distribution:
The species is endemic to the island of Sri Lanka (Ceylon).

Biotope data:
The natural habitat is gently flowing, narrow (but sometimes also larger), clear watercourses in the lowlands of the island. The species is found above all in the southern parts of the island, as well as in the Kottawa Forest area (see also *Malpulutta kretseri*). Here the fishes live among dead wood, leaf litter, and clumps of plants, as well as in or under overhanging bank vegetation.

Reproduction:
The species belongs to those bubblenest spawners where usually both parents guard their offspring as a family for several weeks. A sufficiently large aquarium (with a length of 130 cm (52 in) upwards) should be provided to ensure problem-free breeding. Prior to spawning the pair will defend a large area in the aquarium. The male constructs a bubblenest at the surface, and the fishes mate beneath it, during which the female is embraced by the male and turned onto her side or back. As the eggs are expelled they are immediately fertilised by the male and either float slowly upwards to the water's surface or are collected by the male in his mouth and placed in the bubblenest. The male alone guards the brood. The female guards the border of the breeding territory against predators, and later, after the fry are free-swimming, may also take part in the care of the offspring.

Total length: (size)
Up to 12 cm (4.75 in) long

Remarks: (differences from other species of the genus)
The species is sometimes described as aggressive. Nevertheless these fishes are considered shy and will exhibit their normal behaviour and their striking colours only if maintained in a suitable manner. The sexes are difficult to distinguish and often it is possible to recognise ripe females only by their plumper breast and belly region. Only larger aquaria with length of 120 cm (48 in) upwards should be used for maintenance.

Belontia signata ♂

Belontia signata, ♂ front & ♀ above

Rice fields on the island of Sri Lanka

Villages in Sri Lanka are often characterised by stupas

Elephant bathing place in Sri Lanka

The broad rivers that cross the island of Sri Lanka are often an Eldorado for aquarium plants

Newly-planted rice field

Women inserting the young plants

Group of *Cryptocoryne* in a stream at the edge of a rice field

The genus *Betta* BLEEKER, 1850

The genus *Betta* was erected in 1850 by van BLEEKER. The type species is *Betta trifasciata*, a synonym of *Betta picta*. Today *Betta* is the largest genus of labyrinthfishes with more than 80 scientifically described and undescribed species. It consists of two main groups, the first being the bubblenest builders and the second the mouthbrooding species. There are, however, further subdivisions based on differences among both the bubblenest spawners and the mouthbrooders, which accordingly are themselves split into additional groups. The genus also contains one group (the *Betta foerschi* group) which in evolutionary terms is intermediate between the bubblenest spawners and the mouthbrooders.

Phylogenetic arrangement of the *Betta* species
Scientifically described species only divided into groups.

Betta akarensis group (mouthbrooders)
- *Betta antoni* Tan & Ng, 2006: western Kalimantan (Kalimantan Barat), Kapuas area.
- *Betta akarensis* Regan, 1910: eastern Malaysia, northern and central Sarawak as well as Brunei, in the area around Belait, Tutong and Bandar in the Seri Begawan district.
- *Betta balunga* Herre, 1940: eastern Malaysia, southern Sabah, area around Tawau; Kalimantan Timur (Sebuku, Mahakam).
- *Betta chini* Ng, 1993: eastern Malaysia, Sabah, in the area around Beaufort.
- *Betta pinguis* Tan & Kottelat, 1998: western Kalimantan (Kalimantan Barat), Kapuas area.
- *Betta aurigans* Tan & Lim, 2004: Pulau Natuna Besar.
- *Betta ibanorum* Tan & Ng, 2004: eastern Malaysia, southern Sarawak
- *Betta obscura* Tan & Ng, 2005: central Kalimantan (Kalimantan Tengah) in the area of the Barito.
- *Betta midas* Tan 2009: western Kalimantan (Kalimantan Barath), Kapuas River area.

Betta albimarginata group (mouthbrooders)
- *Betta albimarginata* Kottelat & Ng, 1994: eastern Kalimantan (Kalimantan Timur) in the area around Sebuku.
- *Betta channoides* Kottelat & Ng, 1994: eastern Kalimantan (Kalimantan Timur)

Betta anabatoides group (mouthbrooders)
- *Betta anabatoides* Bleeker, 1851: southern and central Kalimantan (Kalimantan Selatan).

Betta bellica group (bubblenest spawners)
- *Betta bellica* Sauvage, 1884: western Malaysia, province of Selangor, Perak, Pahang, and Johor; island of Sumatra, northern Sumatra (?).
- *Betta simorum* Tan & Ng, 1996: island of Sumatra, southern Sumatra, area around Jambi and Riau.

Betta coccina group (bubblenest spawners)
- *Betta coccina* Vierke, 1979: island of Sumatra, area around Jambi and Riau; western Malaysia, province of Johor.
- *Betta hendra* Schindler & Linke, 2013: central Kalimantan (Kalimantan Tengah) Palangkaraya areal
- *Betta tussyae* Schaller, 1985: western Malaysia, province of Pahang.
- *Betta persephone* Schaller, 1986: western Malaysia, province of Johor.
- *Betta rutilans* Witte & Kottelat, 1991: western Kalimantan (Kalimantan Barat), Kapuas area.
- *Betta brownorum* Witte & Schmidt, 1992: eastern Malaysia, central and southern Sarawak; western Kalimantan (Kalimantan Barat)
- *Betta livida* Ng & Kottelat, 1992: western Malaysia, province of Selangor.
- *Betta miniopinna* Tan & Tan, 1994: island of Sumatra, area around Riau and the island of Bintan.
- *Betta burdigala* Kottelat & Ng, 1994: island of Sumatra and island of Bangka.
- *Betta uberis* Tan & Ng, 2006: central Kalimantan (Kalimantan Tengah) Pangkalanbun area.

Betta dimidiata group (mouthbrooders)
- *Betta dimidiata* Roberts, 1989: western Kalimantan (Kalimantan Barat), Kapuas area.
- *Betta krataios* Tan & Ng, 2006: western Kalimantan (Kalimantan Barat), Kapuas area.

Betta edithae group (mouthbrooders)
- *Betta edithae* Vierke, 1984: southern, central, and western Kalimantan (Kalimantan Selatan, Kalimantan Tengah, and Kalimantan Barat); island of Sumatra, island of the Riau group (Kepulauan Riau), island of Bintan, island of Banka, island of Biliton.

Betta foerschi group (mouthbrooders)
- *Betta dennisyongi* Tan, 2013, Aceh province, north-west Sumatra.

- *Betta foerschi* Vierke, 1979: central Kalimantan (Kalimantan Tengah), area of the Mentaya.

- *Betta strohi* Schaller & Kottelat, 1989: central Kalimantan (Kalimantan Tengah), area around Sukamara.

- *Betta rubra* Perugia, 1893: island of Sumatra, northern Sumatra, area around Aceh.

- *Betta mandor* Tan & Ng, 2006: western Kalimantan (Kalimantan Barat), Kapuas area.

Betta picta group (mouthbrooders)

- *Betta picta* (Valenciennes & Cuvier, 1846: island of Java, area around Bogor, Ambarawa, Bandung, and Jokjarkata.

- *Betta taeniata* Regan, 1910: eastern Malaysia, southern Sarawak; western Kalimantan (Kalimantan Barat), Kapuas area.

- *Betta simplex* Kottelat, 1994: Thailand, southern Thailand; area north of Krabi.

- *Betta prima* Kottelat, 1994: Thailand, area in south-eastern Thailand and in western Cambodia (?).

- *Betta falx* Tan & Kottelat, 1998: island of Sumatra, area around Jambi.

Betta pugnax group (mouthbrooders)

- *Betta pugnax* (Cantor, 1850): western Malaysia, island of Penang, Kedah, Terengganu, Pahang, Selangor, and Johor; island of Singapore; island of Sumatra, area around Riau and Jambi, Anambas.

- *Betta fusca* Regan, 1910: island of Sumatra, northern Sumatra.

- *Betta schalleri* Kottelat & Ng, 1994: island of Sumatra, southern Sumatra and the island of Bangka.

- *Betta enisae* Kottelat, 1995: western Kalimantan (Kalimantan Barat), Kapuas area.

- *Betta pulchra* Tan & Tan, 1996: western Malaysia, area around Johor.

- *Betta breviobesus* Tan & Kottelat, 1998: western Kalimantan (Kalimantan Barat), Kapuas area.

- *Betta pallida* Schindler & Schmidt, 2004: southern Thailand, area around Surat Thani, island of Samui, and area around Narathiwat.

- *Betta lehi* Tan & Ng, 2005: eastern Malaysia, southern Sarawak, and western Kalimantan (Kalimantan Barat), Kapuas area.

- *Betta raja* Tan & Ng, 2005: south-eastern Sumatra, area around Riau and Jambi.

- *Betta cracens* Tan & Ng, 2005: south-eastern Sumatra, area around Jambi.

- *Betta stigmosa* Tan & Ng, 2005: western Malaysia, area around Terengganu.

- *Betta ferox* Schindler & Schmidt, 2006: southern Thailand, Bori Pat, area around Rattaphum.

- *Betta apollon* Schildler & Schmidt, 2006: southern Thailand, area around Narathiwat / Yala

- *Betta kuehnei* Schindler & Schmidt, 2009: western Malaysia, Kota Bharu, south of Panjang.

Betta splendens group (bubblenest spawners)

- *Betta splendens* Regan, 1910: Thailand, central and southern Thailand.

- *Betta smaragdina* Ladiges, 1972: Thailand, eastern Thailand, area from Nong quay to Korat.

- *Betta imbellis* Ladiges, 1975: Thailand, south of the Iceland of Phuket to Suran Thani; Malaysia, Iceland of Penang, Selangor, Perak, Kedah, Terengganu, Johor; Iceland of Singapore; northern Sumatra, area around Medan.

- *Betta stiktos* Tan & Ng, 2005: north-eastern Cambodia, area around Stung Treng.

- *Betta mahachaiensis* Kowasupat et. al., 2012, Thailand, Provinz Samut Sakhon.

- *Betta siamorientalis*, Kowasupat et. al. 2012, Thailand, Ost-Thailand, Provinz Chachoengsao.

Betta unimaculata group (mouthbrooders)

- *Betta unimaculata* (Popta, 1905): north-eastern Kalimantan (Kalimantan Timur), area around Kayan and Howong.

- *Betta macrostoma* Regan, 1910: eastern Malaysia, northern Sarawak and Brunei, area around Belait.

- *Betta patoti* Weber & de Beaulort, 1922: eastern Kalimantan (Kalimantan Timur), area from Balikpapan to Samarinda.

- *Betta ocellata* de Beaufort, 1933: eastern Malaysia, Sabah, area around Sandakan, Kinabatangan, Lahad Datu, and Tawau, north-eastern Kalimantan (Kalimantan Timur), in the area around Sebuku.

- *Betta gladiator* Tan & Ng, 2005: eastern Malaysia, Sabah, area around Maliau.

- *Betta pallifina* Tan & Ng, 2005: central Kalimantan (Kalimantan Tengah), area of the upper Barito.

- *Betta compuncta* Tan & Ng, 2006: eastern Kalimantan (Kalimantan Timur), Mahakam area.

- *Betta ideii* Tan & Ng, 2006: south-eastern Kalimantan (Kalimantan Selatan), Batulicin area and island of Laut.

Betta waseri group (mouthbrooders)

- *Betta waseri* Krummenacher, 1986: western Malaysia, province of Pahang, Terengganu, Kuantan.

- *Betta tomi* Ng & Kottelat, 1994: western Malaysia, province of Johor.

- *Betta hipposideros* Ng & Kottelat, 1994: western Malaysia, province of Selangor; island of Sumatra, Riau area.

- *Betta spilotogena* Ng & Kottelat, 1994: island of Sumatra, area from Riau to Bintan and Singkep.

- *Betta chloropharynx* Kottelat & Ng, 1994: island of Sumatra and island of Bangka.

- *Betta renata* Tan, 1998: island of Sumatra, area around Jambi, province of Jambi.

- *Betta pi* Tan, 1998: Thailand, area around Sungai Kolok; north-eastern Malaysia, area around Kelantan.

- *Betta pardalotus* Tan, 2009: island of Sumatra, area around Palembang, province of Sumatra Selantan.

Eggs in the bubble nest

Differences in the reproductive behaviour of *Betta* groups

There are very striking differences to be seen in the sequence of reproductive behaviour in *Betta* species. Those of the bubblenest-spawning groups of fightingfishes construct their bubblenests predominantly immediately beneath the water's surface or among and beneath floating plants or other plants or

Males with eggs in mouth

Photo by Zhou Hang

Betta foerschi courting pair

photo by Horst Linke

foliage growing at the water's surface (*Betta splendens* group, *Betta bellica* group). However, bubblenest construction may also take place in deeper zones of the water among aquatic plants with large leaves, beneath overhangs (of rock or wood), and in smaller, cave-like openings (*Betta coccina* group and rarely also members of the *Betta splendens* group). The mating and hence the position of the body during the release of eggs are additional distinguishing characteristics.

The fishes perform an "embrace", in the course of which the male drapes himself around the female from above and in so doing turns her onto her back. The eggs expelled then mostly drop onto the anal fin of the male. The male then releases the female, and at this point the eggs fall away and sink towards the bottom. The female remains motionless for a few seconds more, in the so-called "spawning paralysis". During this time the male collects up in his mouth the eggs that have just been laid and fertilised, coats them with saliva, and carefully spits them into

the bubblenest. After the female "awakens" she too helps collect up the eggs. The bubblenest has an important function in the development of the brood. It seems that the more polluted the water, the larger the bubblenest. It is only the disinfectant effect of the saliva-coated air bubbles, creating a bacteria-hostile environment that permits the development of a large number of the eggs.

The mouthbrooders differ from the bubblenest spawners in their spawning procedure, and themselves fall into three groups. First of all there is the group which in evolutionary terms is intermediate to the bubblenest spawners and the mouthbrooders and exhibits striking characteristics of them both (*Betta foerschi* group.). The spawning and embracing of the partners almost always takes place in the upper part of the middle third of the water, in the shelter of a group of aquatic plants. In this group almost all of the preliminary activity is by the female. After several "dummy runs" the first actual spawning pass takes place. In this instance the female swims into the curve of the male's body, and he then embraces her tightly and turns her onto her back, a position that is otherwise almost exclusively reserved for bubblenest spawners. The embrace lasts for around 2-3 seconds, with both partners sinking to some degree during this time. The male then separates from the female, who for her part remains motionless for around 7 to 9 seconds before "awakening" again. After separation the female sinks to the bottom, where she almost always remains in a vertical position, head-down. The small number of eggs laid are taken into the male's mouth as they are released from the body of his partner. It has also, however, been observed that they are sometimes collected from her body (belly or anal region). There is no spitting of the eggs in front of the male as seen in the majority of mouthbrooding *Betta species*. On various occasions I have had the impression that the female was trying to deposit the eggs, subsequently picked up by the male, beneath a leaf in the spawning area, as if there were a bubblenest.

The spawning procedure of the second and largest group, the so-called "true" mouthbrooders, differs yet again (*Betta akarensis* group, *Betta albimarginata* group, *Betta anabatoides* group, *Betta dimidiata* group, *Betta edithae* group, *Betta picta* group, *Betta pugnax* group, and *Betta waseri* group). By preference, pairing takes place close to the bottom, and the pair almost always choose the same spot for each individual phase of the spawning. Again the actual spawning is preceded by several "dummy runs" in which the female swims around the slightly curved male and touches the centre of his body with her mouth at the level of the base of the dorsal fin. The male then wraps himself around the female from below in a brief embrace,

forming a U-shape around her body with his head and caudal fin pointing upwards. The female remains the right way up, in other words she is not turned onto her back. The embrace lasts for around 5 seconds, and then the pair separate. Eventually eggs are laid during one of these embraces. The female picks up the eggs that have just been laid and fertilised in her mouth, collecting them from where they lie on the anal fin of the male. Only after several minutes of vigorous "churning" of the eggs in the female's mouth (here too the saliva has a disinfecting effect) does she spit them, one at a time, immediately in front of the mouth of the male, and if the latter doesn't snap them up – as often happens in the beginning – she catches them up again herself. This often results in the pair playing a game of "who can catch the egg first". Usually there is no renewed spawning until all the eggs from a spawning pass (around 5 to 15) have been passed to the male.

In the final group of mouthbrooders (*Betta unimaculata* group) the spawning procedures are comparable with those of the species described above, but in this case it is not the female that collects up the eggs after the embrace, but the male. And in addition there is no subsequent transfer of the eggs from the female to the male. In this group too the female remains the right way up during the embrace. This group thus exhibits marked differences from those described above.

Betta akarensis REGAN, 1910

Explanation of the species name:
The species name relates to the provenance of the holotype, the Sungai Akar, a river in the south-east of Sarawak.

English name:
Akar Fightingfish, Akar *Betta*

Synonyms:
Betta climacura

Original description:
REGAN, C.T. 1910: The Asiatic Fishes of the Family Anabantidae. Proc. Zool. Soc. London. 1909, 4: 779, Pl. 77 (fig. 3). [Proc. Zool. Soc. London 1909 (pt 4)]

Systematics:
Betta akarensis group

Natural distribution:
The species was described on the basis of a fish from the Akar River in the south-east of Sarawak. *Betta akarensis* are found in large parts of central and eastern Sarawak as well as in the neighbouring Sultanate of Brunei to the east. The species is endemic. Only in the border region between Sarawak and the Sultanate of Brunei is the species replaced by *Betta macrostoma* in a small area. To what extent the species is also present in western Sabah is at present (2011) unknown.

Biotope data:
Betta akarensis lives a secretive existence, predominantly beneath dead leaves and among plants in calm bank zones, in gently flowing watercourses, large and small. The fishes live predominantly in mineral-poor, slightly acid water enriched with humic substances.

Reproduction:
Mouthbrooder.

Total length: (size)
Up to 10 cm (4 in).

Remarks: (differences from other species of the genus)
Soft, acid water and dense planting in places are advantageous for optimal maintenance in the aquarium. The addition of humic substances is important.

Betta akarensis ♂

Watercourse in Sarawak, habitat of *Betta akarensis*

Explanation of the species name:

Latin *albi* = white and *marginata* = bordered: hence "bordered with white stripes", referring to the colour of the fin edgings.

English name:

Whiteseam Fighter

Original description:

KOTTELAT, M. & NG, P.K.L. 1994: Diagnosis of five new species of fighting fishes from Bangka and Borneo (Teleostei: Belontiidae). Ichthyol. Explor. Freshwaters 5 (1), 65-78.

Systematics:

Betta albimarginata-channoides group

Natural distribution:

M. KOTTELAT collected *Betta albimarginata* in 1993 in the Sungai Sebuku basin, around 7 km outside Semunad on the road to Apas, in the drainage of the Sungai Sanul, a tributary of the Sungai (= river) Tikung in Kalimantan Timur, Borneo

Coordinates:

4° 04.08' N   117° 00.4' E (M. KOTTELAT & P. McKEE).

Biotope data:

This fish was also caught in May 1996 by DICKMANN, KNORR, and GRAMS in the region of Malinau, in a side-arm of the around 100 metres (110 yds) wide and approximately 2 metres (6.5 ft) deep, clear main river (the Sembuak), which has areas of rapids in places. The collecting site lay around 200 metres (220 yds) upstream. The stream was a roughly 2 metres (6.5 ft) wide, maximum 1.5 metres (5 ft) deep, whitish-murky watercourse with a slight current. The lurking places of the fishes were small accumulations of leaves on the water's surface or up to 15 cm (6 in) deep accumulations of dead leaves in the slower-flowing parts of the bank region. The banks were cloaked by the surrounding woodland. The water parameters were recorded as a pH value of 5.5 to 6, a total hardness of up to 3 °dGH, and a water temperature of 27 °C (80.5 °F). (Research in 1996 by DICKMANN, KNORR, & GRAMS)

Betta albimarginata ♀

Betta albimarginata ♂

Betta albimarginata, courting pair

Reproduction:

Mouthbrooder.

The species was bred shortly after its first importation. Fishes kindly passed on to the author were also bred successfully, when the familiar procedure for mouthbrooding fightingfishes was observed. A noteworthy feature was the prolonged transfers and/or intervals in the "spitting" of the eggs by the female. During these it was observed that the female twisted as if around an imaginary axis through the middle of her body and the male tried to follow the head of the female, thus swimming in circles around the twisting female in order to collect up in his mouth the eggs spat out by the female. Usually he was unsuccessful and the female snapped them up again herself. As a result spawning can last for 5 to 6 hours. Thereafter the male retires to brood. The larvae hatch after around 72 hours. The fry swim free about nine days after spawning. They are then barely 7 mm (1/4 in) long and dark in colour, and can immediately take freshly-hatched nauplii of *Artemia salina* as first food. Despite water quality and appropriate water parameters (pH between 5.5 and 6.0, conductivity 150 µS, and average water temperature of 29 °C (84 °F), the male usually carried only 20 % of the estimated 50 eggs to maturity.

If the species *Betta albimarginata* from Malinau is kept alone in a so-called species aquarium with dense groups of plants and areas of beech (*Fagus sylvatica*) leaves on the bottom, then young fishes will repeatedly appear, the product of unnoticed spawnings, and can grow on among the aquatic plants or leaf litter, unmolested by conspecifics. Never large numbers, but always sufficient to keep the species established without problems. All in all, an apparently easy-to-keep, small, very colourful, mouthbrooding fightingfish species. One negative point: these fishes are best maintained using only live food.

Total length: (size)

Males of *Betta albimarginata* from Malinau attain a total length of around 5.5 cm (2.125 in). The females are only slightly smaller.

Remarks: (differences from other species of the genus)

While males exhibit a striking orange-red and/or red-brown to black body colour and white fin-edgings, the females remain largely "peppered" brown-black. They lack the white fin-edgings. Typical features of male *Betta albimarginata* from Malinau are the dorsal fin, which is red at its base and otherwise black with a white margin, and a predominantly dark-coloured head. In comparison to the very similar species *Betta channoides*, this species has a more slender and pointed head in frontal view.

Betta albimarginata male with eggs in his mouth

Both species are a real bonus for the aquarium hobby. Once again aquarists are to be thanked for not sparing the expense and effort required to bring these fishes back alive for the aquarium from very remote regions in north-eastern Kalimantan.

Prior to its scientific description, the species was sometimes known as *Betta* sp. from Malinau.

References:

- GRAMS, F. & DICKMANN, P. 1997: Kleine Rote Maulbrüter – eine Labyrinthfischgruppe stellt sich vor. DATZ 50 (9): 562.

- DICKMANN, P.& van the VOORT, S. 2005: Die Gruppe der Weißsaum-Kampffische um *Betta albimarginata*. Aquarium Life 5: 36-43.

- DORN, A. 2007: Weißsaum *Betta* – Whiteseam-*Betta*. Betta News Journal, Journal European Anabantoid Club-AKL. 2007 (4): 26-28.

Head pattern of male *Betta albimarginata*

Explanation of the species name:

Similar to the genus *Anabas*.

English name:

Giant Betta

Original description:

BLEEKER, P. ,1850: Bijdrage tot de kennis der visschen met doolhofvormige kieuwen van den Soenda-Molukschen Archipel. Verhandenlingen Batavia Genootschap, 23:1-15

Systematics:

Betta anabatoides group

Natural distribution:

Indonesia: Kalimantan Selatan, Banjarmasin, watercourse around 55 km from Martapura on the Rantau-Martapura road.

Kalimantan Tengha, Mentaya area, Sungai Ramban, around 22 km west of Sampit on the road to Pembuanghulu and in the Palankaraya area.

As far as is known at present, the natural habitat is restricted to South Kalimantan, the southern, Indonesian part of the Island of Borneo. The species' occurrence on the nearby islands of Sumatra, Java, and Singapore has so far not been confirmed. On the other hand, the distribution region in southern Borneo is large, and the species is widespread there, along with *Betta edithae*.

Biotope data:

The watercourses are very clear and often heavily coloured dark brown. Sometimes they are genuine blackwater regions. The water is very soft and mineral-poor ; the conductivity was measured at between 5 and 30 µS/cm at water temperatures of 27 to 30 °C (80.5 to 86 °F). The pH value was in the strongly acid range. We measured values of 4.8 and lower. The species prefers shallow bank zones with heavy vegetation as habitat.

Betta anabatoide ♀

Betta anabatoide ♂

Blackwater biotope of *Betta anabatoides* in Kalimantan

Reproduction:
Mouthbrooder.

Total length: (size)
Around 10 cm (4 in).

References:
- TAN, H.H. and NG, P.K.L. 2005: The fighting fishes (Teleostei: Osphronemidae: Genus *Betta*) of Singapore, Malaysia and Brunei. The Raffles Bulletin of Zoology 2005, Supplement no. 13: 58 + 86.

- TAN, H.H., 2009: Redescription of *Betta anabantoides* BLEEKER, and a new species of *Betta* from West Kalimantan, Borneo (Teleostei: Osphronemidae). Zootaxa 2165: 59-68.

Courting pair of *Betta anabatoides*

Betta after capture

Betta species display their striking colours after capture

Blackwater biotope in Kalimantan

Explanation of the species name:
Dedication in honour of Irwan Anton from Pontianak.

English name:
Sanggau Betta

Original description:
TAN, H.H. and NG, P.K.L. 2006: Six new species of fighting fish (Teleostei: Osphronemidae: *Betta*) from Borneo. Ichthyol. Explor. Freshwaters,. 17 (2): 98-102.

Systematics:
Betta akarensis group

Natural distribution:
The specimen used for the description was exported via Patric YAP in Singapore, from the Sanggau area in the Kapuas region, on the road from Pontianak to Sanggau.

BAER, LINKE and NEUGEBAUER recorded the species in 1990 west of the town of Sanggau, south of the Pontianak-Sosok-Sanggau-Sintang road, around 7 km from Sanggau, in a hilly bush and woodland landscape, behind the memorial to the patriots.

The distribution of *Betta antoni* is recorded as extending from the area of the town of Sanggau to the Nangapinoh region on the Sungai Malwai, south-east of Sintang.

A small tributary of the Sungai Pinoh, around 20 km south of Nanga Pinoh on the road to Kota Baru

Coordinates: 0° 28' 4" S   114° 75' 2" E (H. KISHI, 3/2001).

The memorial at Sanggau, Kalimantan Barat

Betta antoni ♂

The swamps near Sanggau are also habitats of *Betta antoni*

The foodfish ponds near Sanggau represent a threat to the indigenous fish fauna of the area

Biotope data:
The species lives in densely vegetated, narrow watercourses with very soft, very acid water, which is predominantly coloured slightly to strongly brown.

Reproduction:
Mouthbrooder.

Total length: (size)
8 cm (3.125 in).

Remarks: (differences from other species of the genus)
The species is sensitive to high germ counts. Soft, acid water should be used f or maintenance. The addition of humic substances is very important.

Prior to its scientific description, the species was sometimes known as *Betta* sp. from Sanggau.

References:
- SCHÄFER, F. 1997: All Labyrinths – *Betta*, Gouramis, Snakeheads, Nandids. Aqualog, Frankfurt/M. Germany : 144

Explanation of the species name:

From Apollo (Greek), the Greek and Roman god of the sun and the arts, used by extension to denote any young, handsome man.

English name:

Apollo Fightingfish

Synonyms:

Betta pugnax

Original description:

SCHINDLER, I. & SCHMIDT, J., 2006: Review of the mouthbrooding *Betta* (Teleostei; Osphronemidae) from Thailand, with descriptions of two new species. Zeitschrift für Fischkunde 8 (1/2): 47-69.

Systematics:

Betta pugnax group

Natural distribution:

The specimen used for the description came from an area around 20 km west of the town of Narathiwat on the road to Marubo in southern Thailand.

Coordinates: 06° 23' N   101° 38' E (SCHINDLER, I.)

Additional confirmed distribution regions lie around 10 km east of Ruso, on the road from Ruso to Marubo, Yi Ngo, and Narathiwat in southern Thailand.

Biotope data:

In 2002 Norbert NEUGEBAUER visited the biotope at Sungai Kolok in the province of Narathiwat. He was able (pers. comm.) to record this up to 10 cm (4 in) long mouthbrooding fightingfish in a roughly 6-metres (20-ft) wide watercourse, overgrown and shaded in places with emersed scrub and trees. The water was flowing, colourless, and clear. The water depth in February, during the dry season, was still a metre (40 in) on average. The bottom consisted of fine gravel and larger stones. The fishes were collected mainly among bank vegetation trailing

Betta apollon ♀

Betta apollon ♂

Betta apollon pair – ♀ in front, ♂ behind

The fishes were collected mainly among bank vegetation trailing in the water. Accumulations of dead leaves and twigs on the bottom in areas of low current were also preferred habitats. The biotope lies in the Hala Bala nature reserve. Apparently only one road runs through this region. The watercourse investigated crosses the road beneath a bridge a few kilometres after the entrance to the nature reserve.

The following water parameters were recorded:

pH   6.9

Conductivity: 30 μS/cm

Betta apollon with longitudinal banding

Betta apollon with interesting underhead pattern

Water temperature: 26.1 °C (79 °F)
(Research in 2002: NEUGEBAUER, N.)

Reproduction:
Mouthbrooder.

Total length: (size)
In this species males attain a total length of around 10 cm (4 in); females remain slightly smaller.

Remarks: (differences from other species of the genus)
Differentiation from the other species of the *Betta pugnax* group is not easy. There are many parallels in external appearance to *Betta ferox* in particular. However, good distinguishing characteristics do exist, namely comparison of the under-head markings and the visible "arrowhead" of the prolonged branchiostegial ray at the upper posterior margin of the operculum in *Betta apollon*. At the same time the green coloration of the operculum is only weakly expressed in comparison with other members of the *Betta pugnax* group.

Biotope of *Betta apollon* in the Sungai Kolok Balah Halah area – photo by Ingrid Neugebauer

Biotope of *Betta apollon* in the Sungai Kolok Balah Halah area – photo by Ingrid Neugebauer

Betta aurigans TAN & LIM, 2004

Explanation of the species name:
The Latin *aurigans* relates to the metallic gold scales on the lower body as well as the green-gold operculum.

English name:
Natuna Fightingfish

Original description:
TAN, H. H. & LIM, K. K. P. 2004: Island fishes from the Anabas Natuna Islands, South China Sea, with description of a new species of *Betta*. Raffles Bull. Zool. Suppl. No. 11: 107-115.

Systematics:
Betta akarensis group

Natural distribution:
The natural habitat of *Betta aurigans* lies in the blackwater swamp regions of the island of Natuna Besar, north-west of the island of Borneo in the South China Sea.

Biotope data:
Betta aurigans lives in blackwater regions in the lowlands of the island, in very soft, very acid waters with a high component of humic substances.

Reproduction:
Mouthbrooder.

Total length: (size)
Probably up to 9 cm (3.5 in).

Remarks: (differences from other species of the genus)
The species has apparently not yet been imported alive.

References:
- VOORT, S.v.d., 2004: *Betta aurigans* TAN & LIM, 2004 – eine neu beschriebene Art der maulbrütenden Kampffische. Der Makropode 9/10: 198-199.
- SCHINDLER, I., 2004: Neue Kampffische beschrieben, Der Makropode 9/10: 199.

Betta aurigans - photo by Tan Heok Hui

Explanation of the species name:

Referring to the Balung River in Sabah, eastern Malaysia, the area from which the species was described.

English name:

Balunga Fightingfish, Balunga Betta

Original description:

HERRE, A.W.C.T., 1940: Additions to the fish fauna of Malaysia and notes on rare or little known Malayan and Bornean fishes. Bulletin of the Raffles Museum, Singapore, 16: 27-61

Systematics:

Betta akarensis group

Natural distribution:

Borneo: Sabah (eastern Malaysia) Balung River, east of the town of Tawau, in the south of Sabah, in northern Borneo, close to the border with north-eastern Kalimantan.

Borneo: Kalimantan Timur, Sebuku area, near Pembeliangan, watercourse by road from a "base camp" to Semunad, emptying into the Sungai Tikung

Coordinates:
04° 00' 48" N   117° 02' 54" E (M. KOTTELAT & P. McKEE, 1993).

Mahakam area, small clearwater river, around 10 km from Muara Badak

Coordinates:
00° 19.62' S   117° 20.93' E (H.H. TAN & D. WOWOR, 1999)

Biotope data:

The species lives by preference in dense areas of vegetation in shallow clearwater rivers with a slight current.

Betta balunga ♂

Betta balunga ♂

Biotope of *Betta balunga* in the Tawau/Sabah area

Reproduction:
Mouthbrooder.

Total length: (size)
About 11 cm (4.25 in).

Remarks: (differences from other species of the genus)
To date this species has only very rarely been maintained in the aquarium.

References:

• MÜCK, H. 1993: *Betta balunga* - ein Problemfisch. Der Makropode 9/10 : 106.

• DONOSO-BÜCHNER, R. 1994: Mein vorläufiger Kenntnisstand über den Kampffisch *Betta balunga* HERRE, 1940. Der Makropode 1/2 : 3-6.

• SCHMIDT, J. 2005: Die *Betta pugnax*-Gruppe, Teil 5: *Betta balunga* HERRE, 1940 , , Der Makropode 7/8: 116-119.

• TAN, H.H. & NG, P.K.L. 2005: The fighting fishes (Teleostei: Osphronenmidae: Genus *Betta*) of Singapore, Malaysia and Brunei. The Raffles Bulletin of Zoology, 2005, Supplement no.13: 43-99.

Betta bellica SAUVAGE, 1884

Explanation of the species name:

Latin adjective *bellicus* (feminine *bellica*) = "bellicose", "aggressive", "quarrelsome".

English name:

Slender Betta, Slender Fightingfish

Synonyms:

Betta fasciata
Betta bleekeri

Original description:

SAUVAGE, H.-E.,1884: Note sur une collection de poissons recueillie à Perak, presquile de Malacca. Bulletin de la Societie Zoologie de France, 9: 216-220.

Systematics:

Betta bellica group

Natural distribution:

In the south of western Malaysia (peninsular Malaysia), by preference in smaller blackwater rivers, as well as in northern Sumatra, near Medan (though this may relate to *Betta simorum*).

Biotope data:

Betta bellica is predominantly an inhabitant of blackwater biotopes, but on rare occasions also found in clear-water rivers. It prefers soft, acid water.

VIERKE recorded the species near Ayer Hitam, in a roadside ditch. The water was very soft and strongly acid. The pH value measured 4.6 and the electrical conductivity was recorded as 33 µS/cm at a water temperature of 28 °C (82 °F).

Reproduction:

Bubblenest spawner.

Total length: (size)

Up to 11 cm (4.25 in).

Remarks: (differences from other species of the genus)

The lectotype of the synonym *Betta fasciata* came from Deli (near Medan) in northern Sumatra. By contrast, the species *Betta simorum* is found in the Jambi area in southern Sumatra.

The sexes can be distinguished in *Betta bellica* in half-grown specimens. Male fishes are more boldly coloured and have somewhat longer fins. The slightly "frayed" caudal and anal fins are particularly characteristic. While females normally retain a rounded shape to the caudal fin, in males the central part of the tail is longer and becomes lance-shaped.

Betta bellica is very often confused with *Betta simorum*. *Betta bellica* differs from *Betta simorum* not only by having fewer fin-

Betta bellica ♂

Betta bellica ♂

rays and scales, a shorter distance between the pectoral-fin insertion and the first anal-fin rays, and longer ventral fins, but above all by a more strongly inclined head profile, ie the front part of the head slopes more steeply.

For optimal maintenance these fishes should be maintained in very soft, very acid water with lots of hiding-places and a good growth of plants. The addition of large amounts of humic substances is highly beneficial.

References:

- LINKE, H.1985: Neu- und wiederimportierte Zierfische. Asiatische Raritäten. Aquarium Heute, 3 (1). 5-6

- NIEUWENHUIZEN, A.v.d., 1987: *Betta bellica.* Aquarium wereld, 40 (5): 98-102.

- NG, P.K.L. & KOTTELLAT, M. 1992: *Betta livida*, a new fighting fish (Teleostei: Belontiidae) from blackwater swamps in Peninsular Malaysia. Ichthyological Exploration of Freshwaters 3 (2), 177-182.

- TAN, H.H. & NG, P.K.L., 1996: Redescription of *Betta bellica* SAUVAGE, 1884 (Teleostei: Belontiidae), with decription of a new allied species from Sumatra. The Raffles Bulletin of Zoology 44 (1): 143-155.

- TAN, H.H. & NG, P.K.L. 2005: The fighting fishes (Teleostei: Osphronemidae: Genus *Betta*) of Singapore, Malaysia and Brunei. The Raffles Bulletin of Zoology 2005, Supplement no.13: 58-60.

Explanation of the species name:

Derived from the Latin *brevis* = "short" and obesus = "fat", "obese", referring to the thick, "dumpy" appearance of the body.

English name:

Tayan Betta, named after the type locality in Kalimantan Barat on Borneo.

Synonyms:

Betta enisae

Original description:

TAN, H.H. & KOTTELAT, M. 1998: Two species of *Betta* (Teleostei: Osphronemidae) from the Kapuas Basin, Kalimantan Barat, Borneo. The Raffles Bulletin of Zoology, 1998, 46 (1): 46-50.

Systematics:

Betta pugnax group

Natural distribution:

The specimen on which the description was based came from a small watercourse, around 1 km from Sungai Tayan, near to the village of Tayan, around 87 km east of Pontianak, in the Kapuas basin in northern West Kalimantan (Kalimantan Barat), Indonesia.

Tayan is situated to the south of Batangtarang and the village of Sosok, which lies on the road from Pontianak to Sanggau. Tayan lies only a few kilometres north of the Sungai Kapuas. However, the distribution region is apparently very large, as there are also sites known for the species elsewhere in the vicinity as well as from the Putussibau area, around 150 km to the east.

Biotope data:

Betta breviobesus has been collected chiefly in wooded watercourses with slightly acid water (pH value 6-7). *Betta dimidiata* also lives in the same habitat.

Reproduction:

Mouthbrooder.

Total length: (size)

Up to 7 cm (2.75 in).

Remarks: (differences from other species of the genus)

Betta breviobesus is a peaceful fightingfish, and can be kept in company without problems. According to SCHINDLER & van der VOORT (2010) it is very similar, in terms of appearance, to the species *Betta enisae*, but readily distinguished by a higher number of scales in a lateral series and by the lack of the rows of spots on the membranes of the caudal fin that create a vertical striping in *Betta enisae*.

References:

- TAN, H.H. & KOTTELAT, M. 1998: Two species of *Betta* (Teleostei: Osphronemidae) from the Kapuas Basin, Kalimantan Barat, Borneo. The Raffles Bulletin of Zoology, 1998, 46 (1): 41-51.

- LINKE, H. 2007: The genus Betta. Betta News Journal (special edition) EAC/AKL : 5-30.

- SCHÄFER, F. 2009: Ein wenig bekannter Aquarium-fisch *Betta breviobesus*. Betta News Journal EAC/AKL, 2009 (4): 7-8, plate 1.

- SCHINDLER, I. & VOORT, S.v.d. 2010: *Betta breviobesus* in the hobby: its real identity revealed. Betta News 2010 (2) 25-26

Betta breviobesus ♂

Explanation of the species name:
Dedication in honour of Barbara and Allen Brown

English name:
Brown's Betta

Original description:
WITTE, K.-E. & SCHMIDT, J. 1992: *Betta brownorum*, a new species of anabantoids (Teleostei: Belontiidae) from northwestern Borneo, with a key to the genus. Ichthyological Exploration of Freshwaters, 2: 305-330

Systematics:
Betta coccina group

Natural distribution:
Eastern Malaysia: Sarawak, ca. 200 m into peat swamp forest on left side of road (01° 12' 08.7 N, 110° 39' 52.2 E), ca. 11 km towards Gedong after turnoff from Serian Sri Aman road (after 78 km to Kuching from Gedong mark) (research by (H.H. TAN & S.H. TAN, January 1996).

A swamp area by the road to Mantang, 3.8 km beyond the road bridge over the Sungai Sarawak in Kuching (A. & B. BROWN, 1986).

Biotope data:
The species lives exclusively in very soft, very acid blackwaters, with a pH between 4.2 and 4.9 and an electrical conductivity between 8 and 20 µS/cm.

Reproduction:
Bubblenest spawner.

However, the males of this species also on occasion take the brood into their mouths for prolonged periods during brood care, and hence are sometimes termed mouthbrooders. This supposed mouthbrooding may, however, serve to disinfect the brood using the oral mucus, if there is a high germ count in the breeding-tank water.

Both parents, sometimes together with young from previous broods, co-operate in the brood care of the young.

Betta brownorum ♂

Betta brownorum ♂

Total length: (size)
Around 4.5 cm (1.75 in).

Remarks: (differences from other species of the genus)

Both sexes usually exhibit an intense dark red body coloration. Compared to *Betta coccina*, these fishes remain smaller, and the species-typical, striking green lateral spot on the body is larger in males. Females also exhibit a lateral spot on the body, but it remains a lot smaller.

For optimal maintenance these fishes should, if possible, be maintained only in very soft, very acid water with plenty of humic substances added. The introduction of dead leaves on the bottom of the maintenance aquarium, along with a wealth of vegetation, is very important. If there are insufficient hiding-places the males may become very aggressive among themselves and towards the females. For this reason the maintenance aquarium should not be too small. Like the other small red fightingfishes this species is at risk from Velvet Disease (*Oodinium*). By and large, only the best water quality can prevent this, and hence regular partial water changes are obligatory.

Betta brownorum ♂

References:

- BROWN, A. & B., 1987: A survey of freshwater fishes of the family Belontiidae in Sarawak. The Sarawak Museum Journal : 155-174

- BROWN, A. & B., 1987: A summary of the anabantoids found in Sarawak in July 1986. AAGB Labyrinth, 7 (4), No. 34: 2-6

Explanation of the species name:

burdigala, meaning "Burgundy red", referring to the red body colour.

English name:

Red-Brown Dwarf Fightingfish, Burgundy Betta.

Original description:

KOTTELAT, M. & NG, P. 1994: Diagnoses of five new species of fighting fishes from Bangka and Borneo (Telostei: Belontiidae) Ichthyol. Explor. Freshwaters, 5 (1) :70

Systematics:

Betta coccina group.

Natural distribution:

Indonesia: island of Bangka (Banka), 4 km north of Bikang village on road from Koba to Toboali.

Biotope data:

Betta burdigala lives in small blackwater rivers and sometimes also in the large inundation zones in the remaining rain-forest areas in the south of the island of Bangka. These are predominantly very soft and very acid, clear and clean waters, with a slight to moderate current.

Klaus GERSTNER, Mike LINKE and the author recorded the species in 1993 in the south of Bangka and imported it alive for the first time. The fishes were living in the company of several other labyrinthfish species such as *Parosphromenus deissneri*, *Betta schalleri*, and *Sphaerichthys osphromenoides*. Our study site lay around 60 km south of Koba on the road to Toboali between the settlements of Djeridja and Bikang, around 4 km from Bikang. The site was a large area of woodland, swamp, and inundation zone in which a stream crossed the road. Because of previous heavy rainfall (rainy season – end June) the water level was very high and flowing moderately in most places. At this time it had a temperature of 24 to 25 °C (75 to 79 °F). KH and GH were less than 1°, the pH value was 5.0, and the electrical conductivity measured 18 µS/cm. The water was clear

Betta burdigala ♀

Betta burdigala ♂

Betta burdigala ♂

Betta burdigala beneath the bubblenest

Betta burdigala tending eggs

Betta burdigala country in the south of the island of Bangka

and coloured dark red-brown. The water level was on average easily a metre (39.5 in). The substrate consisted of light loamy sand. As well as a dense marginal growth of normally emersed vegetation, under water at this time of year, there were also numerous long-stemmed and long-leaved *Cryptocoryne* in the biotope.

In 2008 I recorded the species in almost the same area, a woodland swamp with a watercourse, around 1 km from Desa Bikang, 115 km south of Pangkalpinang, on the road from Pangkalpinang to Toboali.

Coordinates: 02° 53' 54 S   106° 27' 30 E

The following biotope parameters were recorded in 2008:

pH value: 4.28

Blackwater biotope of *Betta burdigala*

Group of plants in a *Betta burdigala* biotope

The watercourses are usually very shallow with little current. Habital of *Betta burdigala*

Water temperature: 27.2 °C (81 °F)

Air temperature: 33.8 °C (93 °F)

Conductivity: 0.8 μS/cm.

Blackwater, colour dark red-brown, visibility up to 60 cm (24 in), current slight to moderate. The researches took place during the rainy season after heavy rainfall.

Reproduction:
Bubblenest spawner.

Total length: (size)
Up to 5 cm (2 in).

Remarks: (differences from other species of the genus)

The species needs to be kept in a tank that is not too small, well planted, and with plenty of hiding-places. In addition, very soft, very acid, water, enriched with humic substances, is important for optimal maintenance. The introduction of dead leaves as a bottom covering is of great benefit in the maintenance of these fishes. Like the other small red fightingfishes this species is at risk from Velvet Disease (*Oodinium*). Generally speaking, only the best water quality can prevent this, and hence regular partial water changes are obligatory.

References:
• DEMANDT, K.-H., DONOSO-BÜCHER, R., & SCHMIDT, J., 2001: Rotweinrote Kämpfer - *Betta burdigala*. Der Makropode, 5/6: 73-79.

Explanation of the species name:

Latinised Greek, *Channa*-like, similar to the snakeheads (genus *Channa*) in mouth and head form.

English name:

Snakehead Betta, Snakehead Fightingfish

Original description:

KOTTELAT, M. & NG, P.K.L. 1994: Diagnosis of five new species of fighting fishes from Bangka and Borneo (Teleostei: Belontiidae). Ichthyol. Explor. Freshwaters, 5 (1), 65-78.

Systematics:

Betta albimarginata – channoides group

Natural distribution:

Type locality: Blackwater river, emptying into the Sungai Mahakam on the left-hand side near Mujub; Mahakam river area; Kalimantan Timur; Borneo.

Biotope data:

Dark-coloured stream near the settlement of Pampang; pH 6, general hardness less than 1 °dGH, carbonate hardness less than 2 °dKH, water temperature 26 °C (79 °F), with a Kahmhaut aus Eisenhydroxid (research data: GRAMS, KNORR, and DICKMANN, 4-5. 1996).

KOTTELAT (1994) reports woodland watercourses with brown, acid water as the habitat.

Reproduction:

Mouthbrooder

Total length: (size)

Around 5.0 to 5.5 cm (2 to 2.125 in).

Remarks: (differences from other species of the genus)

The addition of iron in the form of plant fertiliser is apparently an important prerequisite for the health and well-being of these fishes (DORN, 2005).

Prior to its scientific description, the species was sometimes known as *Betta* sp.from Pampang.

References:

- GRAMS, F. & DICKMANN, P. 1997: Kleine Rote Maulbrüter – eine Labyrinthfischgruppe stellt sich vor. DATZ 50 (9): 562.
- DORN, A. 2005: *Betta* (cf.) *channoides* von Pambang – ein kleiner roter Maulbrüter. Aquarium Life 5: 44-49.
- DORN, A. 2007: Weißsaum *Betta* – Whiteseam-*Betta*. Betta News Journal, Journal European Anabantoid Club-AKL, 2007/4: 26-28.

Betta channoides ♂

Mating sequence of *Betta channoides*

Betta chini NG, 1993

Explanation of the species name:
Dedication in honour of Professor Chin Phui Kong, Director of the Fisheries department of Sabah..

English name:
Chin's Betta

Synonyms:
Betta pugnax
Betta akarensis
Betta sp. *Kinabalu*

Original description:
NG, P.K.L., 1993: On a new species of *Betta* (Teleostei: Belonttiidae) from peat swamps in Sabah, Malaysia, Borneo, Ichthyol. Explor. Freshwaters, 4 (4): 290, Figs. 1-3.

Systematics:
Betta akarensis group

Natural distribution:
Eastern Malaysia: Sabah, Borneo: peat swamps, ca. 12 km from Beaufort.

The fishes were recorded in the area of the town of Beaufort in the south-west of Sabah, south of the capital, Kota Kinabalu.

Coordinates:
5° 33' 06 N   115° 0' 23 E (NG, P. K. L. & STUEBING, R. B., 1992)

Biotope data:
The species lives predominantly in swamp regions and blackwater rivers. The water here is predominantly mineral-poor with a very acid reaction – pH value between 4.5 and 6.0.

Reproduction:
Mouthbrooder.

Total length: (size)
Around 9 cm (3.5 in).

Remarks: (differences from other species of the genus)
The species has to date been only very rarely maintained and studied in the aquarium.

Betta chini ♂ showing underhead pattern

Betta chini ♂

Explanation of the species name:

Latin *chloropharynx* = green throat, referring to the variably visible, small green spots on the underside of the head.

English name:

Green-Throat Betta, Green-Throat Mouthbrooder

Original description:

KOTTELAT, M. & NG, P.K.L. 1994: Diagnoses of five new species of fighting fishes from Bangka and Borneo (Teleostei: Belontiidae), (Ichthyol. Explor. Freshwaters, 5 (1): 65-78.

Systematics:

The species was assigned to the *Betta akarensis* group by KOTTELAT & NG (1994), but according to TAN & NG 2005 belongs to the *Betta waseri* group.

Natural distribution:

The holotype was caught on the Indonesian island of Bangka, 99.4 km south of Pangkalpinang on the road to Toboali (41.4 km south of Koba).

Biotope data:

As far as is known at present, the species is endemic to the south of the Indonesian island of Bangka. Clear, clean, usually flowing blackwater streams and inundation zones. The water in this area is very mineral-poor and acid (pH value between 4.8 and in places 5.5). The species was first imported alive in 1994, by K. FRANK and N. NEUGEBAUER.

Reproduction:

Mouthbrooder.

Total length: (size)

Up to 9 cm (3.5 in).

Remarks: (differences from other species of the genus)

Species-typical features are the dark-coloured upper and lower lips, the pattern beneath the mouth, the contrast-rich, dark

Betta chloropharynx ♀

Betta chloropharynx ♂

markings on the posterior operculum, and the red iris of the eye. The unpatterned parts of the underside of the head sometimes exhibit a green coloration. The species does not exhibit any pattern of dots on the underside of the head and the opercula.

Typical underhead pattern of *Betta chloropharynx*

Betta biotope on the island of Bangka

Betta chloropharynx, ♀ in front, ♂ behind

The distinctive underhead pattern of *Betta chloropharynx*

Male *Betta chloropharynx* with metallic scales

Betta chloropharynx with stripe pattern

Juvenile *Betta chloropharynx*

Betta biotope on the island of Bangka

Betta coccina VIERKE, 1979

Explanation of the species name:
The species name is from the Latin coccinus = "scarlet", referring to the body coloration.

English name:
Wine-Red Betta

Original description:
VIERKE, J. 1979: *Betta coccina* nov. spec., ein neuer Kampffisch von Sumatra. Das Aquarium 121 (7): 288-289.

Systematics:
Betta coccina group

Natural distribution:
Type locality: Jambi, central Sumatra. The first specimens were known from the aquarium trade. Subsequent enquiries via the exporter, Vivaria Indonesia in Jakarta, provided information about the natural distribution region.

More recent data are:
Sumatra: Province of Riau, Pulau Padang, Sungai Ponder, 12 km east of Karau, on the road to Meribur (KOTTELAT, 1991).

Sumatra: Province of Jambi, Danau Rasau (KOTTELAT & TAN, 1994).

Western Malaysia: Province of Johor, Segamat-Muar area (NG, 1992).

A possible variant occurs in the Muar area in western Malaysia. It is as yet unclear whether this is the same species.

We too were able to record *Betta coccina* in Sumatra, in the Province of Jambi in the Danau Rasau (Lake Rasau), a lake-like widening with a link to the large Batang Hari near Rantanpanjang, around 76 km north-east of Kota Jambi to Tanjung and from there to the Batang Hari around 15 km upstream to the Danau Rasau near Rantanpanjang.

Betta coccina ♀ with no lateral spot

Betta coccina ♂ with lateral spot

Adult male *Betta coccina*

Biotope data:

In 1986 NAGY reported on localities in the south of the Malayan peninsula (translated from German): "We found *Betta coccina* in both blackwater and clearwater. The water parameters in the source region were a pH value between 4.1 and 4.6 and a conductivity between 27 and 35 µS/cm, during both the rainy season (high water) and dry periods when only the main streams still carried any water. Only a few hundred metres downstream the pH value had dropped to 3.8, while by contrast the conductivity was higher at 50 to 75 µS/cm. The temperature was between 25 and 27 °C (77 and 80.5 °F)."

The Danau Rasau is an extreme blackwater swampy lake, which is in places densely overgrown with trees and brush; the water's surface is predominantly covered with floating plants. The water had a pH value of 4.1 and a conductivity of 30 µS/cm at a water temperature of 29.3 °C (85 °F). The water was coloured strong dark brown and had a slight current everywhere. (Research by T. SIM, N. CHIANG, M. LINKE, & H. LINKE, May 2007). The species lives here in the company of a *Parosphromenus gunawani*, plus *Trichogaster leeri*, *Trichogaster trichopterus* (only a very low population density), *Belontia hasselti*, and *Sphaerichthys osphromenoides*.

Reproduction:

Bubblenest spawner.

Total length: (size)

Around 5.5 to 6.0 cm (2.125 to 2.375 in).

Remarks: (differences from other species of the genus)

This very slender species is characterised by a shiny, emerald-green lateral spot on a dark wine-red body. There are, however, also variants in which the green extends over large areas of the flanks. The iris is brilliant emerald green.

Betta coccina are typical blackwater fishes, and for this reason the species should be kept only in very soft, very acid water with peat filtration. In addition there should be not only large, dense groups of aquatic plants but also dead leaves (eg those of beech and oak) as a bottom covering. The latter, along with the peat filtration, should ensure the addition of important humic substances and hence optimal water chemistry (pH value 5.0 to 5.5 and conductivity 30 to 50µS/cm). For successful breeding, however, lower values are advisable (around pH 4.5 and conductivity 30 µS/cm).

Like the other small red fightingfishes this species is at risk from Velvet Disease (*Oodinium*). By and large, only the best water quality can prevent this, and hence regular partial water changes are obligatory.

References:

- VIERKE, J. 1979: *Betta coccina* nov. spec., ein neuer Kampffisch von Sumatra. Das Aquarium 121 (7): 288-289.
- TAN, H.H. and NG, P.K.L. 2005: The Labyrinthfishes (Teleostei: Anabantoidei, Channoidei) of Sumatra, Indonesia. The Raffles Bulletin of Zoology, 2005, Supplement no. 13: 115-138.

Travelling along the Danau Rasau in Jambi Province, Sumatra

Black water from the Danau Rasau

The Rasau blackwater swamp lake is a habitat of *Betta coccina*

Betta compuncta TAN & NG, 2006

Explanation of the species name:

Latin *compunctus* = "tattooed". The name was chosen in reference to both the often heavily tattooed indigenous tribes living in the distribution region of these fishes, and to the dark to black scale margins, especially on the posterior flank in females of this species.

English name:

Tattoo Betta

Original description:

TAN, H.H. and NG, P.K.L. 2006: Six new species of fighting fish (Teleostei: Osphronemidae: *Betta*) from Borneo. Ichthyol. Explor. Freshwaters, 17 (2): 108-111.

Systematics:

Betta unimaculata group

Natural distribution:

According to TAN & NG (2005), the specimen on which the description was based came from the eastern part of the province of Long Iram (Longiram), from a tributary of the Sungai Hajuq, around 800 m east of the NE Lampunut camp. The paratypes came from the same or nearby areas. On the basis of these data, however, the distribution includes not only the Mahakam river basin in eastern Kalimantan (Kalimantan Timur) but also the northern Barito river basin in central Kalimantan (Kalimantan Tengah), as the watershed between the Mahakam and the Barito lies between the collecting sites in the form of the 1728 m high Sepat Hawung mountain range.

All locations to date lie in the wider area to the east of Longiram (00° 03.92' S, 114° 55.34 E (TAN, H.H. 3/2000) – 00° 00.05' S, 114° 55.23 E (TAN, H.H. 3/2000) – 00° 05.29' S, 114° 52.01' E (TAN, H.H. et al. 4/2000)).

Biotope data:

The natural habitats of *Betta compuncta* are small, slow-flowing streams and rivers in the local wooded swamp regions. The pH values there between 3.7 and 4.3. The fishes were, however, also found in regions with pH values between 4.8 and 5.3 (TAN & NG, 2006). The species is numerous in its native habitat.

Reproduction:

Mouthbrooder.

Total length: (size)

Up to 10 cm (4 in).

Remarks: (differences from other species of the genus)

The species is distinguished from the other species of the *Betta unimaculata* group by its body pattern (pattern of streaks on the flanks), which is particularly striking in females.

Betta compuncta ♀ - photo by Tan Heok Hui

68 LABYRINTH FISH WORLD | Horst Linke

Betta cracens TAN & NG, 2005

Explanation of the species name:

Latin *cracens* = "slender", "elegant", referring to the slender body.

English name:

Slender Sumatra Betta

Original description:

TAN, H.H. and NG, P.K.L. 2005: The Labyrinthfishes (Teleostei: Anabantoidei, Channoidei) of Sumatra, Indonesia. The Raffles Bulletin of Zoology, 2005 Supplement no. 13: 115-138.

Systematics:

Betta pugnax group

Natural distribution:

The holotype came from the Jambi area, island of Sumatra, Indonesia. Jambi, Sungai Berliung Bata, Bertam, ca. 1 km into turnoff to Permata Biru Indah, 10 km from Jambi towards Palembang after main bus terminus (TAN & NG, 2005).

Biotope data:

The specimen on which the description was based was collected in a clearwater woodland swamp, part of which was being cultivated as a rubber plantation. Aquatic plants such as *Barclaya motleyi* were growing in large numbers over wide areas in some watercourse regions. The water depth varied here between 5 and 80cm (2 and 31.5 in). The pH value was 5.8. (Research by TAN, H.H. & NG, P.K.L., 1997.)

The fishes also live in small, up to 3 cm (1.25 in) wide and up to 60 cm (24 in) deep watercourses in swampy lowlands among the hills in the area of the settlement of Sungaibertam, as well as in clear, shallow, gently-flowing rainforest streams in the still fairly natural woodland regions, surrounded by dense brush and stands of trees (i.e. heavily shaded), around 20 km south-west of Kota Jambi, around 10 km from the turn-off from the road from Kota Jambi (the town of Jambi in the province of Jambi), to Kota Palembang (the town of Palembang in the southern province of Sumatra Selatan), near Sungaitiga. The water here is very clear and tinged

Betta cracens after capture

Betta cracens ♂

Biotope of *Betta cracens* in Jambi Province, Sumatra

Betta cracens country

Head close-up of *Betta cracens*

with brown, and has a slight current. The water temperature in January measured 25.8 °C (78.5 °F) (air temperature 29 °C (84 °F)), pH value 5.01 and conductivity 2 μS/cm.

Coordinates: 01° 42.31 S   103° 32.60 E.
(Research by H. LINKE, N. CHIANG, & T. SIM, May 2007 + Jan.2008.)

Reproduction:
Mouthbrooder.

Total length: (size)
Around 10 cm (4 in).

Remarks: (differences from other species of the genus)
When adult, both sexes have a very slender, elongate body with a dark to dark blue margin to the anal fin and the lower part of the caudal fin. The margins of the body scales are coloured green. Females develop a slightly prolonged (hinting at lance-shaped) caudal-fin form, while males exhibit a pointed extension in the lower part of the upper half of the caudal fin.

Explanation of the species name:

Dedication in honour of Dennis Yong Ghong Chong, a naturalist and labyrinthfish enthusiast.

English name:

Pseudo-Rubra Betta

Synonym:

Betta rubra

Original description:

TAN, H. H., 2013: The Identity of *Betta rubra* (Teleostei: Osphronemidae) Revisited, with Description of a New Species from Sumatra; Indonesia, The Raffles Bulletin of Zoology 2013 61(1): 323-330

Systematics:

Betta foerschi group

Natural distribution:

Indonesia: north-west Sumatra: Aceh, in the Kabupaten (governmental district) Naga Raya: Lamie, Alue Rayeuk, watercourse along the Meulaboh to Blangpidie road; Kabupaten Aceh Barat Daya: Alue Laby, watercourse running through the oil-palm plantation by the Melaboh to Blangpidie road. (SIM, T. et al., 2009)

The southern part of the distribution of *Betta dennisyongi* borders the northern natural habitat of *Betta rubra*, which has been recorded in the Singkil area, in blackwater peat swamps with a pH of around 5.5. (SIM, T. et al., 2009)

Biotope data:

Betta dennisyongi has been recorded in clear, flowing watercourses with a pH of around 7, but elsewhere in slightly brown-coloured water with a pH of around 6. Preferred habitats were among accumulations of dead leaves, often over sandy bottoms or among marginal vegetation.

Female *Betta dennisyongi* with lanceolate caudal fin

Male *Betta dennisyongi*

Head close-up of *Betta dennisyongi*

Reproduction:
Mouthbrooder

The reproductive procedure is comparable to that of *Betta rubra*.

Mating usually takes place in a cave or beneath an overhang at the bottom of the tank. The male embraces the female and turns her onto her back (comparable with the mating procedure in bubblenest spawners). The eggs are picked up by the male in his mouth directly from the area of the female's genital opening. The female remains in the spawning paralysis for longer than the male (often almost five times as long). The mouthbrooding is performed by the male. The female remains near the male even after spawning. Females have a smaller and more pointed head. During spawning they become darker in colour and exhibit a light longitudinal band on the back. When extended, the ovipositor looks like a large white dot, the same size as one of the eggs it expels.

These fishes are mouthbrooders of the transitional type between bubblenest spawners and mouthbrooders (*Betta foerschi* group).

Total length: (size)
Around 5.5 to 6.0 cm (2.125-2.375 in)

Remarks: (differences from other species of the genus)
The species differs very little from *Betta rubra*, and mainly in the head pattern as well as a number of small morphological differences. *Betta dennisyongi* exhibits a narrow dark bar from the posterior edge of the eye to the posterior margin of the operculum, and a broad band on the underside of the head, extending to the lower margin of the eye. This bar is absent in *Betta rubra* and replaced by two large, dark spots with a space between. *Betta dennisyongi* are very slender fishes with a delicate red colour on the flanks. They attain a somewhat greater total length in captivity. The unpaired fins have a narrow, light blue margin. Depending on mood, these fishes exhibit five to seven broad dark vertical bars on the sides of the body. The red coloration appears to be variably intense and is thought to be highly mood-dependent. The caudal fin is often lance-shaped in males and has prolonged central rays. These fishes look paler in normal coloration.

Explanation of the species name:

Latin *dimidiata* = divided in half, relating to the small size of these fishes.

English name:

Small Longfinned Fightingfish

Original description:

ROBERTS, T., 1989: The freshwater fishes of western Borneo (Kalimantan Barat, Indonesia). Calif. Acad. Sci. San Francisco: 210.

Systematics:

Betta dimidiata group

Natural distribution:

Betta dimidiata is known from the Sungai Seriang region, 37 km west of the village of Putussibau in West Kalimantan (Kalimantan Barat), very often in blackwater or clearwater regions in the upper Kapuas River area.

We recorded *Betta dimidiata* in a swamp region along the road from Sosok to Tajan, a branch of the Pontianak-Sanggau road, between Bael and Tabang, around 36 km from the village of Tajan. (Research in 1990: BAER, I., LINKE, H., & NEUGEBAUER, N.)

Biotope data:

The region investigated, between Bael and Tabang, was a swamp area ringed by hills and surrounded by secondary woodland. The water was still, brownish to brown in colour, and clear. The pH value was 5.24. The water temperature measured 31.8 °C (89 °F) in areas exposed to the sun, and 25 °C (77 °F) in wooded areas. Hardness and carbonate hardness were both less than 1° dH and the conductivity was 08 µS/cm. *Betta dimidiata* is syntopic there with *Betta krataios*, inter alia.

Reproduction:

Mouthbrooder.

Total length: (size)

Males up to 7 cm (2.75 in) body length, total length up to 9 cm (3.5 in) by virtue of the prolonged finnage; females around 6 cm (2.5 in) body length.

Remarks: (differences from other species of the genus)

This comparatively small mouthbrooder is particularly striking by virtue of its prolonged, pointed finnage, especially noteworthy in males of the species.

Soft, acid water is beneficial for maintenance, and above all for breeding. These fishes are good jumpers. The species is peaceful and rather shy, and if possible should not be kept with other, larger mouthbrooders.

Betta dimidiata Male

Displaying males of *Betta dimidiata*

References:
• CHIANG, N., 2010: *Betta dimidiata*. Betta News, Journal of the European Anabantoid Club mit AK Labyrinthfische im VDA, 2010 (2): 22

Female *Betta dimidiata*

Betta edithae VIERKE, 1984

Explanation of the species name:
Dedication in honour of Mrs Edith Korthaus.

English name:
Edith's Betta

Original description:
VIERKE, J., 1984: *Betta taeniata* REGAN 1910 und *Betta edithae* sp. n., zwei Kampffische aus Südborneo. Das Aquarium, 18 (176): 58-63.

Systematics:
Betta edithae group

Natural distribution:
Betta edithae was described from South Kalimantan (Kalimantan Selantan), from the vicinity of the town of Banjamasin and from the Sungai Barito delta.

The species has subsequently also been recorded in Central and West Kalimantan (Kalimantan Tengah and Kalimantan Barat) as well as on the islands of Sumatra, Bintan, Bangka, Biliton, and the Riau Islands group.

Biotope data:
In the natural habitat *Betta edithae* lives in a wide variety of biotopes with very different water parameters. The species is thus found in neutral as well as in very acid water types.

Reproduction:
Mouthbrooder.

Total length: (size)
Around 6 cm (2.5 in).

Remarks: (differences from other species of the genus)
Despite its adaptation to a wide range of water parameters, by preference this species should be maintained in soft, slightly acid water with added humic substances. These fishes are problem-free in their maintenance and will readily proceed to breed. A varied diet is important. They will take foods as well as live foods, without problem.

Prior to scientific description, the species was sometimes known as *Betta picta*, *Betta* sp.from Borneo or *Betta* sp. *aff. taeniata*.

References:
* VIERKE, J., 1979: Zur Systematik maulbrütender Kampffisch-Arten *Betta* sp. *affin. taeniata* (Mentaya) von Borneo. Das Aquarium 119 (5): 210-212.

Betta edithae ♂

Betta edithae, ♂ in front, ♀ behind

Mouthbrooding male *Betta edithae* with eggs in his mouth

Betta edithae after capture

Blackwater stream along a sandy track through bush landscape

High water at Palangkaraya, Kalimantan

Flooded bushland with trackway, habitat of *Betta edithae*

Blackwater trench next to the road in the Plantation area

- VIERKE, J., 1979: Ein maulbrütender Kampffisch (spec.*affin.*) aus Borneo. Das Aquarium 119 (5): 206-209.

- VIERKE, J., 1984: *Betta taeniata* REGAN 1910 und *Betta edithae* sp. n., zwei Kampffische aus Südborneo, Das Aquarium, 18 (176): 58-63.

Betta enisae KOTTELAT, 1995

Explanation of the species name:
Dedication in honour of Enis Widjanarti.

English name:
Enis's Betta

Original description:
KOTTELAT, M. 1995: Four new species of fishes from the middle Kapus basin, Indonesian Borneo (Osteichthyes: Cyprinidae and Belontiidae), Raffles Bull. Zool. 43 (1): 51-64.

Systematics:
Betta pugnax group

Natural distribution:
The fishes were collected in the area of the Sungai Kapuas in the wider vicinity of Semitau. M. KOTTELAT gives the location for the specimen used for the description as the Sungai Santik, a tributary of the Sungai Tawang, near to the western "Danau Sentarum Field Centrum", West Kalimantan (Kalimantan Barat), Indonesia. (Research by KOTTELAT & WIDJANARTI, 1993.)

Coordinates: 00° 50' 21 N   112° 03' 50 E (KOTTELAT, M.)

N. NEUGEBAUER and I. BAER recorded these fishes as long ago as 1993, in Batangtarang, south of Sosok, at a branch off the road from Pontianak to Sanggau via Mandor in the east of West Kalimantan (Kalimantan Barat). The water parameters recorded were pH 5.58 and conductivity less than 1 µS/cm at a water temperature of 27 °C (80.5 °F).

Biotope data:
The species lives in soft, acid water with a large component of humic substances.

Reproduction:
Mouthbrooder.

Total length: (size)
Up to 9 cm (3.5 in).

Remarks: (differences from other species of the genus)
Typical features of males of *Betta enisae* are the striking green operculum, the usually light-blue- to black-margined anal and caudal fins with a narrow white outer edging. *Betta enisae* is very similar to *Betta breviobesus* in external appearance and very easily confused with it. The coloration is almost the same, only *Betta breviobesus* has a dark pattern of streaks and/or dots on the fin membranes of the caudal fin, creating a pattern of vertical stripes that is absent in *Betta enisae*.

References:
- SCHINDLER, I. & VOORT, S.v.d. 2010: *Betta breviobesus* in the hobby: its real identity revealed. Betta News, Journal of the European Anabantoid Club /AKL, 2010 (2): 25-26

Betta enisae ♂

Start of the mating embrace

Betta enisae ♀

Egg transfer, with eggs being spat out

Mating procedure

Start of the mating embrace

Betta enisae, ♂ in front, ♀ behind

Egg transfer, with eggs being spat out

Explanation of the species name:

Latin *falx* = "sickle", referring to the striking colour demarcation in the anal and caudal fins in male fishes of this species.

English name:

Pijoan Betta, Sickle Betta

Original description:

TAN, H.H. & KOTTELAT, M., 1998: Redescription of *Betta picta* (Teleostei: Osphronemidae) and description of *Betta falx* sp. n. from central Sumatra. Revue Suisse de zoologie, 105 (3): 557-568.

Systematics:

Betta picta group

Natural distribution:

The holotype was collected in the region of the Sungai Alai, 19.5 km from Muarabungo to Muaratebo, at the bridge over the Alai river, Sumatra, Indonesia.

Coordinates:

01° 28' 42.6" S,   102° 18' 31.7" E (KOTTELAT & TAN, 1995).

The species has to date also been recorded in numerous regions in the province of Jambi, west of the town of Jambi in Sumatra, Indonesia.

We recorded *Betta falx* in the area of the Pijoan River and of the Soak Putat, on the road from Kota Jambi to Muarabulian (T. SIM, N. CHIANG, & H. LINKE: 2007).

Biotope data:

The species is apparently a typical inhabitant of almost standing waters in swampy wooded regions. The preferred habitat is densely vegetated zones along the banks. Here the pH value lies between 4.7 and 6.8. (Research in 1998 by M. KOTTELAT & H.H. TAN.)

The habitats in the area of the Pijoan river were likewise almost standing bodies of water, mainly inundation zones and residual

Betta falx ♀

Betta falx ♂

waters at the edge of the river bank. We recorded the pH value at 5.0 and the water temperature at 27 °C (80.5 °F) (T. SIM, N. CHIANG, & H. LINKE, May 2007).

Reproduction:
Mouthbrooder

Total length: (size)
6.5 to 7 cm (2.5 to 2.75 cm).

Remarks: (differences from other species of the genus)

Betta falx exhibits very many parallels with *Betta picta* and was for a long time regarded as a population of *Betta picta* from Sumatra. The two species differ particularly in the pattern and coloration of male fishes. In *Betta falx* the lower part of the caudal fin and the anal fin exhibit a red to red-brown margin (by contrast blue-green in *Betta picta*, as in *Betta simplex* and *Betta taeniata*). *Betta falx* has a rounded caudal-fin form, *Betta picta* likewise, but sometimes also slightly lance-shaped. The distance between the tip of the snout and the anterior edge of the eye is somewhat longer in *Betta falx* than in *Betta picta*.

The species has sometimes been confused with *Betta picta* and *Betta edithae*.

Mouthbrooding male *Betta falx*

Underhead pattern of *Betta falx*

Betta falx, ♀ in front, ♂ right

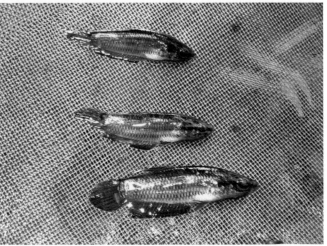
Betta falx after capture

Biotope of *Betta falx* at Pijoan, Sumatra

Bank zone with mixed water

Shallow bank zone with dead leaves – habitat of *Betta falx*

Explanation of the species name:
Latin ferox = "wild", "fierce", "spirited".

English name:
Bori Pat Betta

Synonyms:
Betta pugnax
Betta sp. from Boriphat

Original description:
SCHINDLER, I. & SCHMIDT, J. 2006: Review of the mouthbrooding *Betta* (Teleostei; Osphronemidae) from Thailand, with descriptions of two new species. Zeitschrift für Fischkunde, 8 (1/2): 47-69.

Systematics:
Betta pugnax group

Natural distribution:
Betta ferox is probably endemic to the area of the Boriphat watercourse, which crosses National Highway 406 around 40 km south-west of Rattaphum in the direction of Satun in southern Thailand. The Boriphat watercourse flows through the Boriphat National Park, a large wooded area with a waterfall and a leisure park with the Boriphat waterfall as its centre point. A place for numerous groups of people to congregate, who like to have family picnics here and whose children enjoy bathing in the dammed pool of water by the waterfall. However, huge, water-worn natural rocks limit the area over which visitors have freedom, and as a result the damage to nature is apparently kept within bounds. The leisure area is completed by a large paved area with souvenir stalls and a large car park.

Coordinates: 07° 00.05 N   100° 08.55 E. (H. LINKE)

Biotope data:
Only a few hundred metres from the waterfall the Boriphat again flows totally undisturbed, with a width of on average 5 to 6 metres (16 to 20 ft) and a water depth of around 30 to 60 cm (12 to 24 in). The water is clear and coloured slightly brownish. The banks on either side are cloaked in dense greenery. The substrate consists of fine light brown gravel, coarse in places. In deeper zones of water, where there is less current, the river bottom is often covered in a centimetres-thick layer of dead leaves from the surrounding trees. These zones, along with the overgrown bank regions, are the preferred habitat of a wide variety of fish species. Here we were also able to record *Betta ferox* during our researches in the Boriphat. The population density was not very large, but nevertheless within an hour we had around 10 fishes of this species in the net. These were exclusively half-grown specimens, which were to be found predominantly among dead leaves or beneath the dense bank vegetation. Our researches took place in the month of February,

Betta ferox ♂

Head close-up of *Betta ferox* ♂

Underhead coloration of *Betta ferox* ♂

Boriphat Waterfall

Habitat of *Betta ferox* in the Boriphat

The Boriphat Waterfall in southern Thailand

ie at the beginning of the dry season. The water had already dropped to a very low level compared to in the rainy season and the months thereafter. In the rainy season the water depth here can often measure up to 2 metres (6.5 ft) deep for long periods. Our water measurements, taken in the early afternoon in mid February, gave the following values:

pH 6.3

Conductivity: 51 µS/cm
Water temperature: 25.5°C (78 °F)
Carbonate hardness: 1 °dKH
General hardness: 2 °dGH
(Research in February 2005; LINKE, H. & LINKE, M.).

The area around the Boriphat Waterfall is also a tourist centre

The Boriphat, habitat of *Betta ferox*

Bank zone of the Boriphat

Reproduction:
Mouthbrooder.

Total length: (size)
Males attain a total length of up to 10 cm (4 in), females remain slightly smaller.

Remarks: (differences from other species of the genus)
The species was discovered by Professor Rolf Geisler in the Boriphat in southern Thailand as long ago as 1975.

The names *Betta pugnax* or *Betta* sp. from Boriphat have also been used for this species.

References:
- GEISLER, R. 1981: Ein Maulbrütender Kampffisch von der malaiischen Halbinsel: *Betta pugnax*. Aquarien Magazin, 2: 78-82, Kosmos Verlag Stuttgart.
- LINKE, H. 2006: Der "Bori Pat" - Ein Bach mit Vergangenheit. Aquarium Life, 2006/6 : 12-15.

Betta foerschi VIERKE, 1979

Explanation of the species name:
Dedication in honour of Dr. Walter Foersch.

English name:
Chameleon Betta, Foersch's Betta

Original description:
VIERKE, J. 1979: *Betta anabatoides* und *Betta foerschi* spec. nov., zwei Kampffische aus Borneo. Das Aquarium, 123 (9): 386-390.

Systematics:
Betta foerschi group

Natural distribution:
Type locality: Mentaya river system. 250 km north-west of Banjarmasin, southern Borneo.

Unfortunately these locality data from 1979 were superficial for commercial reasons regarding the local exporters. Not until many years later did details of the actual collecting site become known. According to these, the natural habitat of these fishes lies around 75 km north-west of Sampit in the area of Palangan, and consists of very soft, acid waters (pH 5.2) with temperatures around 25 °C (77 °F).

In 1988 Jürgen KNÜPPEL and the author recorded these fishes around 100 km further east in the area north of the town of Palangkaraya, in a stream around 3 km north of Tangkiling in the direction of Kasungan, and were able to import them alive.

In 2009 the author recorded *Betta foerschi* as also very numerous in many blackwater streams in the area around Kota Palangkaraya.

Biotope data:
According to the original description (1979), the specimen on which the description was based was collected in the swampy bank region of a small, rather fast-flowing stream.

In 1990 we also recorded these fishes in a blackwater stream flowing slowly through a swamp surrounded by brush on either side of the road from Tangkiling to Kasungan. The water was up to 50 cm (20 in) deep in places and heavily overgrown with emerse, grass-like plants. The following water parameters were measured: dGH and dKH less than 1, pH 4.2, and conductivity 17 µS/cm at a water temperature of 21.9 °C (71.5 °F), and 24.9 °C (77 °F) in very shallow water among plants.

Betta foerschi ♂

Male *Betta foerschi*

The "deer stream" in Palangan, Kalimantan Tengah, biotope of *Betta foerschi*

Reproduction:

Mouthbrooder.

The species spawns in the middle layers of the water and its reproductive behaviour represents a transitional stage between bubblenest spawner and mouthbrooder. The fry are usually released from the mouth of the male after 7 to 8 days at a water temperature of around 26 °C (79 °F) and are thereafter independent. They will immediately take freshly-hatched *Artemia* nauplii as first food. For successful breeding this species requires very soft (dGH less than 3) and very acid (pH value below 5.0) water as well as the copious addition of humic substances.

Blackwater stream north of Palangkaraya, Kalimantan Tengah, habitat of *Betta foerschi*

Total length: (size)

Individuals of this species grow to around 6.5 cm (2.5 in) long.

Remarks: (differences from other species of the genus)

Betta foerschi are often aggressive among themselves, but these fishes frequently prove to also be very shy. The sexes are relatively simple to distinguish. When the fishes are excited the oblique bands on the opercular margins become gleaming gold, more rarely red, in males, and blood red in females. Males are usually more intensely coloured and have somewhat longer fins.

References:

• FOERSCH, W. 1979: Erfahrung bei der Aquarienpflege von *Betta anabatoides* und *Betta foerschi* spec. nov. Das Aquarium, 124 (10): 447-449.

Explanation of the species name:

Latin *fuscus* (feminine fusca) = dark, referring to the fins in the specimen used for the description.

English name:

Dusky Betta

Synonyms:

Betta pugnax.

Original description:

REGAN, C.T. 1910: The Asiatic Fishes of the Family Anabantidae. Proc. Zool. Soc. London, 1909 (4): 780, Pl. 78 (fig. 2).

Systematics:

Betta pugnax group

Natural distribution:

The holotype was caught in Sumatra (REGAN 1910).

Betta fusca was recorded by V. ETZEL in 1971 in the Pajakumbu area in western Sumatra.

A further distribution region lies in the Pakanbaru area in the province of Riau in central Sumatra (Dr. YUWONO, Aquaria Indonesia pers. comm., 1980).

Biotope data:

The second-named site was a watercourse around 20 to 30 cm (8 to 12 in) wide and up to 30 cm (12 in) deep. The bottom was predominantly covered in large stones that provided *Betta fusca* with shelter and hiding-places. The water temperature was 25 °C (77 °F). (Research in 1971: V. ETZEL.)

Reproduction:

Mouthbrooder.

Total length: (size)

Around 9 cm (3.5 in).

Remarks: (differences from other species of the genus)

Betta fusca exhibits many parallels with *Betta pugnax*. As in *B. pugnax*, males have bright green scales on the operculum. Dr. J. SCHMIDT (pers. comm.; translated) has provided additional details for differentiation: "The body form of *Betta fusca* remains more slender than in other large, pointed-finned, mouthbrooding fightingfishes. The body of *Betta fusca* is reddish brown, the spots on the scales in males create a species-typical pattern of lines. The fins are prolonged, but not as extremely as in other species and forms of the *Betta pugnax* group. The mouth is terminal, slightly upward-pointing, as in *Betta pugnax*."

Betta fusca ♂

Betta fusca, ♀ left, ♂ right

Betta fusca ♂

Betta fusca ♂ displaying

Explanation of the species name:

Latin *gladiator* = fighter. The species is aggressive and territorial in its behaviour.

English name:

Gladiator Betta

Original description:

TAN, H.H. & NG, P.K.L., 2005: Fighting fishes (Teleostei: Osphronemidae: Genus *Betta*) of Singapore, Malaysia and Brunei. The Raffles Bulletin of Zoology, 13: 75-76.

Systematics:

Betta unimaculata group

Natural distribution:

Borneo: Sabah, Maliau Basin, north-east of base camp 1996.

Coordinates:

05° 14' N   116° 53' E (TAN, S.H. & TAN, T.H.T.: 1996)

Stream at right trail ca. 2 km into Jalan Babi towards Maliau falls and small streams at right trail ca. 3-4 km into Jalan Babi towards Maliau falls.

The species may be endemic to the southern Sabah area, in the Maliau Basin Conservation Area between Kota Kinabalu and Tawau.

Bioope data:

In the natural habitat *Betta gladiator* lives in soft and very acid water (pH between 4.2 and 4.8). (Research in 1996: TAN, S.H. & TAN, T.H.T.)

Reproduction:

Mouthbrooder.

Total length: (size)

Around 10 cm (4 in).

Remarks: (differences from other species of the genus)

The species is aggressive and territorial in its behaviour and should be maintained only in suitably large, heavily planted aquaria. The water should be very soft and very acid. The addition of humic substances is also very important, as this will make for better healing in the event of injuries.

The species has sometimes been confused with *Betta unimaculata* or *Betta ocellata*.

Betta gladiator - photo by Tan Heck Hui

Betta gladiator ♂

Explanation of the species name:

Dedication in honour of Hendra Tommy, who discovered the species and first made it available to the aquarium hobby.

English name:

Palangka Betta

Original description:

SCHINDLER, I. & LINKE, H., 2013: *Betta hendra* – a new species of fightingfish (Teleostei: Osphronemidae) from Kalimantan Tengah (Borneo, Indonesia). Vertebrate Zoology 63(1): 35- 40

Systematics:

Betta coccina group

Natural distribution:

This species was first exported in 2009 by the company KURNIA AQUARIUM in Palangkaraya, and is very similar to the species *Betta foerschi* in appearance. These fishes live in the blackwater swamp areas along the road from Palangkaraya to Bandjamasin, around 3 km south-east of Palangkaraya, between Palangkaraya and Berengbenkel and in the adjacent areas to the south.

Coordinates: 02°16.58 S   113°56.65 E (LINKE, H. 2011)

Biotope data:

Betta hendra is a blackwater fish that lives in very soft, very acid water.

Water parameters:

pH value: 3.95
Conductivity: 6 µS/cm
Water temperature: 28.5 °C (83.3 °F)
(Research in May 2011: LINKE, H.)

Reproduction:

Bubblenest spawner.

This species can be maintained very well in mineral-rich water, but on the basis of experience to date needs to be kept in very soft and very acid water for breeding.

Betta hendra female

Betta hendra Male

Displaying *Betta hendra* male

Total length: (size)
Around 5.5 cm (2.25 in).

Remarks: (differences from other species of the genus)

Betta hendra was first imported by members of the European Anabantoid Club/AKL in 2009.

It is possible that *Betta* sp. Sengalang (also a bubblenest spawner),

Biotope of *Betta hendra* southeast of Palangkaraya

which was exported at the same time, is identical with *Betta hendra*.

References:

- LINKE, H. 2009: Neue (?) *Betta* –Arten aus Kalimantan. Betta News, Journal of the European Anabantoid Club/AKL, 2009 (3): 22.

- LINKE, H. 2010: Die neuen *Betta* aus Kalimantan/The new *Bettas* from Kalimantan. Betta News Journal of the European Anabantoid Club/AKL, 2010 (1): 26.

- WEIBLEIN, T.2011:Bemerkungen zu *Betta* sp Palangka. Betta News, Journal of the European Anabantoid Club/AKL, 2011 (3): 20-22.

Black water biotope of *Betta hendra*

Black water biotope of *Betta hendra*

Black water biotope of *Betta hendra*

Betta hendra female

Black water from the habitat of *Betta hendra* compared to drinking water

Betta hipposideros NG & KOTTELAT, 1994

Exzlanation of the species name:

Greek, referring to the stripe pattern on the lower part of the head.

English name:

Hippo Betta

Original description:

NG, P.K.L. & KOTTELAT, M., 1994: Revision of the *Betta waseri* group (Teleostei: Belontiidae). The Raffles Bulletin of Zoology. 42: 593-611.

Systematics:

Betta waseri group

Natural distribution:

The holotype was collected in 1991 by P. K. L. NG, in the Selangor area in western Malaysia:

Malaysia: Selangor, north Selangor peat swamp forest, 39 km stone, on road from Sungai Besar to Tanjung Malim.

Indonesia: island of Sumatra, Riau Province, blackwater swamp regions draining into the Sungai Bengkwan, a tributary of the Sungai Indragiri.

Biotope data:

The species lives in blackwaters, and hence in very soft, acid water.

Reproduction:

Mouthbrooder.

Total length: (size)

Around 11 cm (4.25 in).

Remarks: (differences from other species of the genus)

These fishes should be maintained only in very soft, acid water. Species-typical features are the markings on the underside of the head and the dot-like patterning of the various longitudinal lines on the lower body.

Prior to its scientific description, the species was sometimes known as *Betta* sp. *aff. waseri*

References:

- TAN, H.H. and NG, P.K.L. 2005: The fighting fishes (Teleostei: Osphronemidae: Genus *Betta*) of Singapore, Malaysia and Brunei. The Raffles Bulletin of Zoology 2005, Supplement 13: 84-85.

Betta hipposideros

Displaying male *Betta hipposideros*

Typical underhead pattern of *Betta hipposideros*

The underhead pattern of *Betta hipposideros* is very striking

Betta ibanorum TAN & NG, 2004

Explanation of the species name:

Latinised Malay, meaning "of the Iban", the largest group of people in Sarawak. The Iban are known as head-hunters.

English name:

Iban Betta, Iban Fightingfish

Synonyms:

Betta anabatoides
Betta akarensis
Betta sp. aff. taeniata
Betta climacura

Original description:

TAN, H.H. & NG, P.K.L. 2004: Two new species of freshwater fish (Teleostei: Balitoridae, Osphronemidae) from southern Sarawak. In YONG, H.S., NG, F.S.P., & YEN, E.E.L. (eds.) Sarawak Bau Limestone Biodiversity. Sarawak Museum Journal, Vol. LIX, No. 80 (New Series); Special Issue No. 8: 267-284.

Systematics:

Betta akarensis group

Natural distribution:

Eastern Malaysia: Borneo, Sarawak, west of the capital Kuching, Bako National Park, Bukit Gondol (N. SIVASOTHI, 1994).

Betta ibanorum has been found in the Bako National Park and in the neighbouring regions around Kuching; its distribution is restricted to south-western Sarawak.

Its occurrence has also been confirmed in the regions east of Bau. The distribution of *Betta ibanonorum* apparently merges with that of *Betta lehi* to the west of Bau.

Biotope data:

The species lives in watercourses large and small, in mineral-poor, slightly acid water.

Reproduction:

Mouthbrooder.

Betta ibanorum ♀

Betta ibanorum ♂

Betta ibanorum ♂

Typical stripe pattern of *Betta ibanorum*

Betta ibanorum exhibits a striking underhead pattern

Total length: (size)

Around 10 cm (4 in).

Remarks: (differences from other species of the genus)

When adult, male *Betta ibanorum* have a lance-shaped caudal fin (rounded in females), dots forming rows on the membranes of the dorsal and caudal fins, less prolonged ventral fins, yellow eyes, sometimes with an element of red, and a horizontal band across the underside of the head beneath the mouth, with no additional markings.

The species has sometimes been confused with *Betta akarensis*, *Betta anabatoides*, and *Betta* sp. *aff. taeniata*.

References:

- TAN, H.H. & NG, P.K.L. 2004: Two new species of freshwater fish (Teleostei: Balitoridae, Osphronemidae) from southern Sarawak. In YONG, H.S., NG, F.S.P., & YEN, E.E.L. (eds.) Sarawak Bau Limestone Biodiversity. Sarawak Museum Journal, Vol. LIX, No. 80 (New Series); Special Issue No. 8: 267-284.

- VAN DER VOORT, S. 2005: Die *Betta akarensis* Gruppe. Der Macropode 2005, 1/2:

Narrow, shallow watercourses are preferred habitats of *Betta ibanorum*

Often the biotopes are very narrow, and full of plants and dead leaves

Betta ideii TAN & NG, 2006

Explanation of the species name:
Dedication in honour of Takashige Idei.

English name:
Pulau Laut Betta

Synonyms:
Betta sp. aff. patodi
Betta unimaculata
Betta cf. patodi
Betta sp. Pulau Laut

Original description:
TAN, H.H. and NG, P.K.L. 2006: Six new species of fighting fish (Teleostei: Osphronemidae: *Betta*) from Borneo. Ichthyol. Explor. Freshwaters, 17 (2): 111-113.

Systematics:
Betta unimaculata group

Natural distribution:
The specimen on which the description was based came from the region north of Baturicin (Batulicin), a tiny settlement on the coast of the Laut Straits at Pulau Laut in the south-east of Kalimantan (Kalimantan Selatan), Borneo.

The species was also recorded on the island of Laut by BAER, I., LINKE, H. and NEUGEBAUER, N. in 1990. Its natural habitat lies in the sometimes fast-flowing mountain streams of the Gunung Sepatung (the Sepatung Mountains), south of Kota Baru, the largest settlement on Pulau Laut. We found these fishes in the channels and reservoirs of the old Dutch water-supply system. These watercourses are very narrow and shallow in places, with large and small rocks. The surrounding montane woodlands shade this biotope.

Biotope data:
The flowing water in these mountain streams was very clear and without colour. Field measurements provided the following data: general hardness 2 °dGH, carbonate hardness 4 °dKH, conductivity 82 µS/cm at a water temperature of 24.5°C (76 °F), and a pH value of 7.5. *Betta ideii* lives predominantly among large rocks, dead leaves and aquatic plants, and emerse bank vegetation trailing in the water. (Research in July 1990; H. LINKE.)

Reproduction:
Mouthbrooder.

Total length: (size)
10 to 12 cm (4 to 4.75 in).

Betta ideii ♂

Female *Betta ideii*

The light spot on the tip of the snout is typical of the species

A male *Betta ideii* with the typical light spot on the lower lip

Remarks: (differences from other species of the genus)

It appears that a light spot is on the lower lip is species-typical in both sexes.

Prior to its scientific description, the species was sometimes known as *Betta* sp. *aff. patoti*, *Betta unimaculata*, *Betta* cf. *patoti*, or *Betta* sp. from Pulau Laut.

Biotope of *Betta ideii* at Pulau Laut, south-east Kalimantan

Watercourse at Pulau Laut, habitat of *Betta ideii*

Biotope with running water on the hill at Pulau Laut

Betta ideii lives among large rocks in the running water

Explanation of the species name:

Latin *imbellis* = peaceful, not warlike.

English name:

Peaceful Betta, Crescent Betta

Original description:

LADIGES, 1975: *Betta imbellis* nov. spec., der Friedliche Kampffisch. Aquar.Terrar. 28 (8): 262-264.

Systematics:

Betta splendens group

Natural distribution:

Terra typica: Kuala Lumpur, Malaysia. The natural distribution was given in the original description as (translated) "swamp regions in the wider vicinity of Kuala Lumpur" (the capital of Malaysia). Since then many more locations have been recorded. The species is distributed throughout the entire western Malaysia region. The species occurs here predominantly in waters with a neutral to slightly alkaline pH, but also – surprisingly – in slightly to strongly acid blackwater areas, albeit only at a low population density.

The species has also been recorded in a region south-west of Medan on the island of Sumatra, around 60 km south-west of Medan (Research in 1980, 2005: LINKE, H.).

There is a further location on the island of Singapore (LIM, K.K.P 1994), stream at Lorong Banir.

Astonishingly the species has also been recorded on Borneo, in the Tawau area in Sabah (TAN, H.H., 1996): Jalan Sin On Jaya, swampy area.

Coordinates:
4° 16' 07.6" N 117° 54' 25.6" E (TAN, H.H., 1996)

Hitherto the northern limit of the distribution region was regarded as being a line from Ko Samui (in the east) to Phuket (in the west) in southern Thailand. To the north of this line there are – or were formerly – no *Betta imbellis*. Hence all the more astonishing the discovery of *Betta imbellis* in south-

Betta imbellis ♀

Displaying male *Betta imbellis*

Betta imbellis ♂ from Phuket, Thailand

eastern Cambodia (K. KUBOTA, pers. comm.), along with a *Betta imbellis*-like species (*Betta* cf. *imbellis*) in western Cambodia (research by HERMANN, L. & LINKE, H.: 2007) and southern Vietnam (research by CHIANG, N. & LINKE, H.: 2009).

Biotope data:

Observations on the island of Phuket reveal that the habitat of *Betta imbellis* there lies in the sometimes water-covered rice fields and associated permanently water-filled ditches, or in pool-like accumulations of water.

Betta imbellis lives predominantly among groups of plants or in vegetated bank zones. *Betta imbellis* is generally known as an inhabitant of rice fields, associated ditches, swamp regions, and flooded meadows. In the region of lakes and ponds at Tham Sra Kaew behind the village of Ban Nai Sa on the National Highway 4033, north of Krabi on the west coast of southern Thailand, *Betta imbellis* lives in a gently flowing,

slightly clouded watercourse around 6 metres (20 ft) wide, near to the turquoise-coloured main lake. The water is mainly only around 50 cm (20 in) deep and in the dam area the water's surface is densely covered with floating plants. In places there are also dense stands of *Cryptocoryne cordata*. The fishes live predominantly in the densely vegetated bank zones and among the thickets of *Cryptocoryne* and *Eichhornia*. *Betta imbellis* was very numerous in this biotope.

The following water parameters were recorded in this watercourse near the turquoise-coloured main lake in the region of lakes and ponds at Tham Sra Kaew (research in February 2005; LINKE, H. & LINKE, M.):

pH value: 7.8
Conductivity: 438 µS/cm
Water temperature: 26.5 °C (79.7 °F)
Carbonate hardness: 7 °dKH
General hardness: 11 °dGH

Coordinates: 08° 10.15 N   098° 40.45 E (LINKE, H.)

On the island of Pinang these fishes are again to be found in small ditches near the rice fields as well as in small, sometimes only fist-sized, water-filled depressions in the ground, and in swamps and inundation zones in the south-west of the island.

This species is also found on the east (South China Sea) coast of western Malaysia. Here the natural habitat includes both "normal water" and blackwater, that is, water with a neutral pH and also very acid.

The locality to the south-west of the town of Medan on the Indonesian island of Sumatra consists of irrigation channels, predominantly densely overgrown with emerse vegetation, and swamp regions associated with rice fields and palm plantations.

Betta imbellis is also sold for competitions in the Betta market at the Jalan Mentawai in Medan.

Reproduction:

Bubblenest spawner.

The sexes are very easy to distinguish. Males develop longer fins and are dark blue to black in body coloration. The operculum, the margins of the scales on the body, and the fins have very attractive antique green to bold light blue zones. The caudal fin has a red edge with a black margin and the posterior tip of the anal fin is again bright red in colour. The females, by contrast, have short fins and a brown body coloration with faintly red-blue fins. When displaying they exhibit a yellow-brown oblique pattern on the body and a broad, light, longitudinal band on the back.

Breeding this species is not difficult. The male builds a bubblenest beneath floating plant leaves at the water's surface, sometimes also among plants or dead wood in lower layers of the water. The nest usually has a diameter of around 5 cm (2 in). A female that is ready to spawn, recognisable by her yellow-brown body pattern, will follow the male beneath the bubblenest after a short period of courtship. The pair then spawn during an embrace. The entire spawning act lasts for around two hours and is usually very harmonious. There is no injury to the female such as is known in, the Siamese Fightingfish.

The male embraces his partner and thus turns her over onto her back. The eggs are then expelled only a few millimetres out of the oviduct of the female and are usually caught beneath the edges of her folded pectoral fins or on her folded anal fin. They are usually released from these areas only when the male breaks free from the embrace and in so doing turns the female onto her side. The eggs then drop away and are collected by the male in his mouth. Before the female arouses herself from the immobility that occurs during the embrace, the so-called spawning paralysis, almost all the eggs from a spawning episode, around 5-25, will have been collected up by the male and placed in the nest. During the days that follow the bubblenest may expand to up to 10 cm (4 in) across. The male

Mating embrace in *Betta imbellis*

Scenes of *Betta imbellis* mating

Scenes of
Betta imbellis mating

tends and guards the brood very attentively. After around 75 hours the development is complete and the fry begin to swim free. At this point in time the parent fishes should be removed, although they won't generally harm the fry.

The freshly-hatched nauplii of a small *Artemia* species will be taken without problem as first food.

Total length: (size)
Around 5.5 cm (2.25 in) long.

Remarks: (differences from other species of the genus)
Species-typical features are the red edging to the caudal fin and the red coloration on the posterior tip of the anal fin, but above all the broad, turquoise blue, vertical stripes on the gill-plates. At the same time the scales on the dark blue to black body exhibit more or less continuous turquoise blue margins.

The maintenance of this species is unproblematical, but it should be kept only with other small and peaceful species.

Multiple pairs of *Betta imbellis* can be maintained together without problem only in very large aquaria, as in this species too the males can be aggressive among themselves. Males are particularly prone to lose their peaceful nature beneath the bubblenest, and this can lead to major battles. If a group of *Betta imbellis* are grown up together then it will be less difficult to keep them together later as well. Only when adult "stranger" individuals are added are there likely to be problems and serious injuries result.

Betta imbellis is used in competitive battles in various parts of Malaysia and Indonesia.

For optimal maintenance the aquarium should be well planted. Small caves and bogwood, as well as dead leaves on the bottom (to provide humic substances as well as hiding-places), should all form part of the décor. Floating plants are also recommended.

The water temperature can be between 25 and 27 °C (77 and 80.5 °F), but if you want to see these fishes in the full splendour of their coloration, the temperature should be raised to 29 °C (84 °F).

The breeder of *Betta imbellis* in Phuket Thailand

Betta imbellis from Ko Samui SouthEast Thailand

Displaying male *Betta imbellis* from the island of Samui

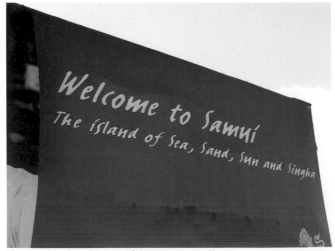

There are also lots of labyrinthfish species on Ko Samui

Arriving on the Thai island of Samui

Swampy area on Ko Samui

There are lots of stretches of open water in the swamp

Water flows into the swamp via small ditches

The swamp is the habitat of *Betta imbellis* and many other fish species

Catching fishes is only possible with a basket in this dense tangle of plants

Temple precinct on Ko Samui

The visitor will find numerous statues of Buddha on Ko Samui

Displaying *Betta imbellis* male from Ko Samui

Working elephant on Ko Samui

Strangers are watched closely on Ko Samui

Betta imbellis from Krabi SouthWest Thailand

Displaying *Betta imbellis* male from Krabi

Betta imbellis biotope in the Tham Sra Kaew area, north of Krabi

Water-filled ditch in the Tham Sra Kaew area, north of Krabi

The limestone karst scenery north of Krabi

This swamp filled with aquatic plants is also a habitat of *Betta imbellis*

Displaying *Betta imbellis* male of the area of the area Krabi

Betta imbellis from Medan NorthEast Sumatra

Betta imbellis from Sungai Kolok South Thailand

Betta imbellis ♂ of Sungai Kolok

Field *Betta imbellis* from Sungai Kolok

Norbert Neugebauer and helper catching *Betta imbellis*

Photo by Ingrid Neugebauer

Swampy area north of the area Sungai Kolok, habitat of *Betta imbellis*

Photo by Ingrid Neugebauer

The species has sometimes been confused with *Betta splendens* or *Betta rubra.*

References:

• LADIGES, 1975: *Betta imbellis* nov.spec., der Friedliche Kampffisch. Aquar.Terrar. 28 (8):262-264.
• ROLOFF, E. 1975: Haltung und Zucht von *Betta imbellis* LADIGES, 1975. Aquar.Terrar. 28 (8): 265-266.

Betta imbellis SouthEast Cambodia

Betta imbellis from Cambodia, ♂ in front, ♀ behind

Betta imbellis ♀

Embrace beneath the nest

Spawning scenes in *Betta imbellis* from Cambodia

Spawning scenes in *Betta imbellis* from Cambodia

Spawning scenes in *Betta imbellis* from Cambodia

BETTA IMBELLIS LADIGES, 1975   117

Spawning scenes in *Betta imbellis* from Cambodia

Male collecting the eggs; the female is still in spawning paralysis after the embrace

Retrieving eggs falling from the nest during spawning

Male *Betta imbellis* from Cambodia

Betta krataios TAN & NG, 2006

Explanation of the species name:

Greek *krataios* = strong, powerful, in reference to the robust, powerful, and sturdy body from of these fishes.

Original description:

TAN, H. H. & NG, P. K. L. 2006: Six new species of fighting fish (Teleostei: Osphronemidae: *Betta*) from Borneo. Ichthyol. Explor. Freshwaters, 17 (2): 98-102.

Systematics:

Betta dimidiata group

Natural distribution:

Sites for *Betta krataios* are at present known only from the Sanggau and Mandor regions in the lower Kapuas basin in West Kalimantan (Kalimantan Barat), Borneo. We found the species with *Betta dimidiata* in a swamp region along the road from Sosok to Tajan, a branch of the Pontianak-Sanggau road, between Bael and Tabang, around 36 km from the village of Tajan. (Research by BAER, I., LINKE, H. & NEUGEBAUER, N.: 1990)

Biotope data:

The region investigated, between Bael and Tabang, was a swamp area ringed by hills and surrounded by secondary woodland. The water was still, brownish to brown in colour, and clear. The pH value was 5.24. The water temperature measured 31.8 °C (89.25 °F) in areas exposed to the sun, and 25 °C (77 °F) in wooded areas. Hardness and carbonate hardness were both less than 1° dH and the electrical conductivity was 8 µS/cm. *Betta krataios* was syntopic there with *Betta dimidiata*.

Reproduction:

Mouthbrooder.

The species spawns by embracing in a sheltered spot, usually near the bottom. The pairing can last for up to 12 hours. In my case the successful development of the brood took place in soft, strongly acid water (average pH value 4.6 and a conductivity of 80 µS/cm at a water temperature of around 27 °C (80.5 °F)). The fry were released from the paternal mouth after 11 days and

Betta krataios ♀

Betta krataios ♂

Displaying *Betta krataios* male

Head coloration in *Betta krataios*

were already around 6 mm (0.125 in) long at that time. Freshly-hatched *Artemia* nauplii were taken as first food without problem.

Total length: (size)

Males around 7.5 cm (3 in), females around 6.5 cm (2.5 in).

Remarks: (differences from other species of the genus)

In comparison to *Betta dimidiata*, *Betta krataios* does not exhibit elongated or prolonged fins. In *Betta krataios* females remain somewhat smaller than males.

Betta krataios differs from *B. dimidiata* (to date the only other member of the *B. dimidiata* group) by the presence of the following combination of characters (in life): iridescent turquoise operculum versus gold in *B. dimidiata*; yellowish brown body (versus reddish); yellowish fins (versus reddish); rounded caudal fin (versus lancet-shaped in *Betta dimidiata*); more lateral scales (30-32 versus 29-30). (Data after TAN & NG, 2006.)

In addition to the characters enumerated by TAN & NG (2005a: 58) in the group description, the group is defined by the following additional characters: small adult size; broader blue margin to the anal fin; lateral scales 29-32.

Explanation of the species name:

Dedication in honour of Jens Kühne.

English name:

Bluethroated Fightingfish, Kühne's Betta

Original description:

SCHINDLER, I. & SCHMIDT, J. 2009: *Betta kuehnei*, a new species of fighting fish (Teleostei, Osphronemidae) from the Malay Peninsula, Bull. of Fish Biology, 10 (1/2): 39-46.

Systematics:

Betta pugnax group

Natural distribution:

The holotype came from a small watercourse, around 35 km south of the village of Panjang (perhaps Rantau Panjang, next to the border with Thailand) in the province of Kelantan, in the north-east of western Malaysia. According to the original description, additional fishes of this species were caught around 40 km south of Sungai Kolok (Sungai Kolok lies around 10 km north of Rantau Panjang, but in Thailand). This suggests the same collecting region.

We found the species around 2 km south of the village of Jedok, on National Highway 4 from Tarah Merah to Jeli, shortly after the Sungai Jelok, left-hand fork in the village of Jedok coming from the direction of Tana Merah, then around 2 km after the school in Gual Jelok. Jelok lies around 70 km from Kota Bharu in the direction of Jeli, District of Tana Merah. The species is not uncommon in the north-east of western Malaysia and is distributed in various areas.

Biotope data:

The natural habitat was a small watercourse with clear, colourless water. The stream was well shaded by woodland and scrub. Water parameters cited were a pH value of 7.0, a water temperature of 25 °C (77 °F), and a general hardness of 3 °dGH. (Research in August 2008: KÜHNE, J.)

Betta kuehnei ♀

Betta kuehnei ♂

Head coloration of a *Betta kuehnei* ♂

sited. We were unable to find any *Betta kuehnei* in the flowing water of the stream itself.

Coordinates: 05°48′41 N   101°57′22 E (LINKE, H.)

The water was clear and slightly brownish in colour. The pH measured 5.32 and the conductivity 4 µS/cm at a water temperature of 28.7 °C (83.7 °F). (Research in October 2010: LIM, T. Y. and LINKE, H.)

Reproduction:
Mouthbrooder.

Total length: (size)
Around 8 cm (3 in).

Remarks: (differences from other species of the genus)

Betta kuehnei has the following characters: a rounded caudal-fin form in both sexes, operculum with a strong antique green coloration in males, no under-head stripes, no longitudinal band extending onto the caudal fin (it is interrupted on the tail and then ends in the form of dots on the caudal peduncle, as in *Betta pallida*). *Betta kuehnei* exhibits no pattern of dots or spots on the membranes of the anal and caudal fins as in the similarly coloured *Betta* sp. "Waeng".

The collecting site lay at the edge of a rubber-tree plantation. Here a narrow stream, mainly 80 cm (31 in) wide and on average 5-10 cm (2-4 in) deep, with a moderate current, flowed through the plantation. The *Betta kuehnei* were, however, found chiefly in the mainly up to 20 cm (8 in) deep residual pools, which contained a strikingly dense growth of *Cryptocoryne* and were half shaded by the dense scrub among which they were

Betta kuehnei biotope near Jedok

Betta kuehnei prefer to live among dead leaves on the bottom in very shallow water

Betta kuehnei biotope with dense vegetation near Jedok

Stream with gently flowing water near Jedok

Betta kuehnei has been compared with an as yet scientifically undetermined species, *Betta* sp. "Satun", but it is very doubtful whether they are identical. *Betta* sp. "Satun" is to date known only from Satun Province in south-western Thailand, around 300 km away. The species was recorded there for the first time in 2004, by K. KUBOTA.

References:

• VOORT, S.v.d., 2009: *Betta kuehnei*, eine neue Kampffischart von der Maliischen Halbinsel. Betta News – Journal des EAC/AKL, 2009 (4): 19-20.

Explanation of the species name:

Dedication in honour of C. U. Leh, Curator of Zoology at the Sarawak Museum, Kuching, Sarawak.

English name:

Leh's Betta

Original description:

TAN, H.H. & NG, P.K.L., 2005: Fighting fishes (Teleostei: Osphronemidae: Genus *Betta*) of Singapore, Malaysia and Brunei. The Raffles Bulletin of Zoology, supplement 13: 43-87.

Systematics:

Betta pugnax group

Natural distribution:

The species lives in the western part of Sarawak (eastern Malaysia). The location for the specimen used for the description lies between the villages of Bau and Lundu in the Sungai Stuum Muda, 21.1 km before the bridge over the Batang Kayan (coming from Bau, the second bridge over the Batang Kayan).

Coordinates:

01° 28' 51.3 N 109° 58' 18.1 E (LINKE, H.: 2008).

We recorded *Betta lehi* in a forest watercourse in the rainforest area of the Kampong Stoum, around 5 km from the main road from Bau to Lundu, heading south along a narrow sandy path to its end (after turning right once), near to a farm; 50 m (165 ft) to the west of the footpath, around 15 minutes in the direction of the Kalimantan Barat border.

Coordinates:

01° 27' 50 N 109° 58' 52 E (LINKE, H.: 2008).

The species probably also lives in the adjacent (to the south) swamp and lowland area of Kalimantan Barat, Indonesia.

Biotope data:

In the forest watercourse of the Kampong Stoum we recorded the following water parameters: pH 5.57 and conductivity 6 µS/cm at a water temperature of 24.4 °C (76 °F). The water was clear and coloured slightly brownish, with a weak to moderately strong current. (Research in August 2008: LO, M., CHIANG, N., & LINKE, H.)

Reproduction:

Mouthbrooder.

Total length: (size)

Up to 9 cm (3.5 in).

Betta lehi ♂

photo by Lothar Hermann

Habitat of *Betta lehi* in the rainforest near Kampong Stoum, Sarawak

Watercourse in the rainforest

Betta lehi after capture near Kampong Stoum

Remarks: (differences from other species of the genus)

Like the members of the *Betta pugnax* group, the species exhibits a lancet-shaped caudal fin, but unlike them has not only an antique green operculum but also gold spots thereon.

The anal fin exhibits a dark, blue-green margin and the fin membranes are a delicate green in colour, as are the majority of the scale margins on the body. The iris of the eye is red.

Heavily shaded watercourse in the rainforest near Kampong Stoum

Betta livida NG & KOTTELAT, 1992

Explanation of the species name:

Latin *lividus* (feminine *livida*) = bluish, referring to the green-blue eyes and the green-blue flank spot.

English name:

Emerald-Spot Betta

Original description:

NG, P.K.L. & KOTTELLAT, M. 1992: *Betta livida*, a new fighting fish (Teleostei Belontiidae) from blackwater swamps in Peninsular Malaysia. Ichthyological Exploration of Freshwaters, 3 (2): 177-182.

Systematics:

Betta coccina group

Natural distribution:

Western Malaysia: Selangor: The fishes were discovered in the large blackwater swamp region south of the Sungai Bernam, around 120 km north of Kuala Lumpur.

Biotope data:

These are typical inhabitants of blackwaters, living in very soft, very acid water (pH between 4 and 5). The species lives in still water and also in fast-flowing watercourses. It is usually to be found among overhanging bank vegetation and aquatic plants. The species is syntopic with *Betta hipposideros* and *Parosphromenus harveyi*, as well as many other species.

Reproduction:

Bubblenest spawner.

Total length: (size)

Males around 5 cm (2 in), while females remain somewhat smaller.

Remarks: (differences from other species of the genus)

The females of *Betta livida* have somewhat smaller fins and exhibit a light dorsal band in courtship coloration. The species is often confused with *Betta coccina* from Sumatra, which also occurs in the south-east of the Malaysian peninsula. Depending on mood *Betta livida* also exhibits an emerald green spot on wine-red flanks, prolonged unpaired fins, iridescent green tips to the ventral fins, and two light-coloured to golden vertical bars on the gill-plate. In normal coloration males rarely exhibit a flank spot, and then only faintly. This flank spot is completely absent in females.

Soft, acid (pH value around 5.0) water is necessary and important for successful maintenance. The addition of humic substances via peat filtration or the introduction of dead leaves (leaf litter) is very important. The water temperature should lie between 23 and 25 °C (73.5 and 77 °F). Short periods at higher temperatures are tolerated without problem when used to simulate seasonal fluctuations such as occur in the wild. Like the other small red fightingfishes this species is at risk from Velvet Disease (*Oodinium*). By and large, only the best water quality can prevent this, and hence regular partial water changes are obligatory.

References:

- NG, P.K.L. 1993: Schwarzwasserfische aus Nordselangor (Malaiische Halbinsel). DATZ 46 (2): 112-117.
- SCHMIDT, J. 2006: *Betta livida* – der Dunkelrote. Der Makropode 9/10: 154-161.
- SCHMIDT, J. 2006: Der Dunkelrote - *Betta livida*. Aquarium life, 3: 12-19.

Betta livida ♂

Explanation of the species name:

Latinised Greek *macrostoma* = large mouth, referring to the large mouth and mouth opening.

English name:

Spotfin Betta

Original description:

REGAN, C.T. 1910: The Asiatic Fishes of the Family Anabantidae. Proc. Zool. Soc. London, 1909: 778, Pl. 78 (fig. 3).

Systematics:

Betta unimaculata group

Natural distribution:

The holotype was described from Sarawak in eastern Malaysia. In the very recent past the species has been recorded very frequently in clear, flowing mountain rivers in the border region between north-eastern Sarawak and the Sultanate of Brunei, and in the Sultanate of Brunei itself.

Biotope data:

Thomas SCHULZ (pers. comm.) was able to collect the species in the upper course of the Mendaram River (Sungai Mendaram) near the town of Rampayoh in the Sultanate of Brunei. According to his data, the water there was very soft and acid with a pH value of 4.3.

Reproduction:

Mouthbrooder.

Reproduction is comparable with that of *Betta unimaculata*. As is also known from other mouthbrooding *Betta* groups, after the fertilisation of the eggs the female *Betta macrostoma* collects them in her mouth from the anal fin of the male, but she does not then spit them out singly or in small numbers in front of the mouth of the male for him to pick up, as in other species, but passes them to the male in small clumps, directly from mouth to mouth. The development of the offspring takes around 20 days. When they leave their father's mouth they are already between seven and nine millimetres long and can immediately take freshly-hatched *Artemia* nauplii. With a varied diet and frequent

Betta macrostoma ♀

Betta macrostoma ♂

water changes they can themselves attain sexual maturity after around six months.

Total length: (size)

Up to 10 cm (4 in).

Remarks: (differences from other species of the genus)

The striking body coloration and pattern readily distinguish this species from the other large mouthbrooding fightingfishes. T. SCHULZ was able to bring the species back alive for the aquarium in 1984 and study it in detail. C. Tate REGAN had only preserved material back in his day, but nowadays we know the splendid coloration of these fishes. Notable features are the blue mouth, the sometimes yellow gill-plates, and the orange-brown body. The supposedly species-typical, dark, light-ringed spot on the lower posterior part of the dorsal fin in males is not present in all males. The sexes can be clearly distinguished only in larger individuals. While the female frequently still exhibits the double longitudinal stripe pattern of juveniles, males appear more colourful, without longitudinal stripes and sometimes with colourfully patterned fins.

These fishes are peaceful towards other species of similar size. They occasionally indulge in minor "squabbles" among themselves, perhaps ending in vigorous mouth-fighting.

For maintenance an aquarium of adequate size is required, densely planted in places and containing good quality, mineral-poor, acid water. It appears that water that is extremely low in bacteria, clean, and rich in humic substances is absolutely essential for successful maintenance. The species is very susceptible to infections. At the same time a varied diet, including above all live insects and their larvae, is very important. Unfortunately the species is still not without problems in its maintenance.

References:

- HERMANN, L. 2009: *Betta macrostoma* - ein Traum vieler Labyrinthfischfreunde / *Betta macrostoma* - a dream of many labyrinthfish fans. Betta News, Journal of the European Anabantoid Club/ Arbeitskreis Labyrinthfische, 2009 4/6: 25-26.

Betta macrostoma courting, ♂ in front   photo by Lothar Hermann

Betta macrostoma courting   photo by Lothar Hermann

Head close-up of *Betta macrostoma* ♂

Betta macrostomaa transferring an egg packet

photo by Lothar Hermann

Female collecting eggs from the anal fin of the male   photo by Lothar Hermann

Pair embracing. The eggs have already been expelled and are lying on the anal fin of the male

photo by Lothar Hermann

Betta mahachaiensis KOWASUPAT et al., 2012

Explanation of the unofficial name:

The name means "of Maha Chai" and relates to the distribution region in the province of Samut Sakhon, Thailand.

English name:

Mahachai Fightingfish

Original description:

KOWASUPAT, C., PANIJPAN, B., RUENWONGSA, P.& SRIWATTANAROTHAI, N. 2012: *Betta mahachaiensis*, a new species of bubble-nesting fighting fish (Teleostei: Osphronemidae) from Samut Sakhon Province, Thailand, Zootaxa 3522:49-60

Systematics:

Betta splendens group

Natural distribution:

The natural habitat of *Betta mahachaiensis* lies in a comparatively small area at Mahachai, between Samut Sakhon and Samut Songkhram, as well as around Ban Phaen, south-west of Bangkok. This is a heavily built-up area with lots of large industrial businesses. Artificial sea-water evaporation pans for harvesting salt are the main feature of the landscape close to the coast. The few remaining natural green areas have only a slim chance of survival.

In the natural habitat *Betta mahachaiensis* lives not in natural swamps and rice fields, but predominantly in half-dark palm swamps sheltered from the sun.

Biotope data:

The special features of Mahachai include small wooded swamps with Nippa Palms (ton chark in the Thai language) growing in the water. This palm species does not develop a tall trunk but instead produces fronds around 5 to 6 metres (17-20 feet) high. The stout thornless stems sprout from the base and thus create a dense tangle of thick, robust fronds around the heart of the plant near its base. In August, in the first third of the rainy season, the water depth in the area is around 60 to 80 cm (24 to 32 in). The breeding sites of *Betta mahachaiensis* are often to be found in the spaces between the stems of the fronds. Here the males construct their bubblenests and hence they could be located without problem. The swamp water was clear and brownish in colour. The pH value was 6.5 to 6.8. (Research in August 2003: LINKE, H. & PUMCHOOSRI, A.)

Reproduction:

Bubblenest spawner.

Betta mahachaiensis ♂

Mating and brood-care sequence of *Betta mahachaiensis*

Bubblenest with brood beneath the water surface

Bubblenest with larvae

Releasing the embrace after spawning beneath the bubblenest

Male embracing female during spawning

Courtship prior to spawning in *Betta mahachaiensis*

Total length: (size)
Around 5.5 cm (2.5 in).

Remarks: (differences from other species of the genus)
Betta mahachaiensis exhibits many parallels with *Betta smaragdina* in its appearance, but also noteworthy differences. Thus *Betta mahachaiensis* has no or only very little red in the fins by comparison. The operculum is not green all over, but merely exhibits two broad green stripes. Adult males often have a slightly lancet-shaped caudal fin, but this caudal-fin form is not species-typical; in other words, individuals of this species with a round caudal fin are also known.

Betta mahachaiensis are very colourful fightingfishes and well worth keeping.

Nippa Palm stumps in the palm woodland provide hiding-places and brooding sites for *Betta mahachaiensis*

Betta mahachaiensis biotope in scrubland

Collecting in the open scrubland at Mahachai

Sand track in the *Betta mahachaiensis* biotope

There is very little light in the Nippa Palm woodland

Nippa Palm woodland, habitat of *Betta mahachaiensis*

There is very little light in the Nippa Palm woodland

Sometimes fishes can be caught out of the water by hand

Nippa Palm woodland with lake-like clearing

Nippa Palm woodland usually grows in water

Betta mahachaiensis ♂ after capture

Betta mahachaiensis is sometimes very common in the biotope

Betta mandor TAN & NG, 2006

Explanation of the species name:

Referring to the distribution region, around Mandor (the village and river of that name) in the Kapuas basin in north-western Kalimantan (Kalimantan Barat), Borneo.

English name:

Mandor Betta

Original description:

TAN, H. H. & NG, P. K. L. 2006: Six new species of fighting fish (Teleostei: Osphronemidae: *Betta*) from Borneo. Ichthyol. Explor. Freshwaters, 17 (2): 104-105.

Systematics:

Betta foerschi group

Natural distribution:

The specimen used for the description was collected in the area of the village of Mandor on the road from Pontianak to Sanggau. Additional collecting sites lie on the Pontianak to Sanggau road near Anjungan in the direction of Mandor.

Biotope data:

These fishes live in blackwaters containing clear, very soft, very acid water with a slight current and a strong red-brown colour. General and carbonate hardness both less than 1°dH, pH value between 4.5 and 5.3, and water temperature between 27.6 and 28.2 °C (81.7 and 82.8 °F). (Research in August 1990; LINKE, H.)

Reproduction:

Mouthbrooder.

Total length: (size)

Up to 7 cm (2.75 in).

Remarks: (differences from other species of the genus)

Prior to its scientific description, the species was sometimes known as *Betta* sp. from Mandor or *Betta* sp. Mandor.

Betta mandor ♀

Betta mandor ♂

Betta mandor ♂

Courting female *Betta mandor*

Bridge over the blackwater river at Mandor

Blackwater biotope in Mandor area

Explanation of the species name:

After Midas, a mythical King of Phrygia whose touch changed everything into gold; referring to the golden operculum and lower body scales.

English name:

Midas Betta, Midas Fightingfish

Original description:

TAN, H.H. 2009: Redescription of *Betta anabatoides* BLEEKER, and a new species of *Betta* from West Kalimantan, Borneo (Teleostei: Osphronemoidae) Zootaxa 2165: 59-68.

Systematics:

Betta akarensis group

Natural distribution:

The natural distribution is restricted to the north-western regions of the Kapuas River in West Kalimantan (Kalimantan Barat), northwards to the region of the town of Lundu in western Sarawak (eastern Malaysia). The holotype was collected in a blackwater region on the road from Pontianak to Anjungan, around 7 km from the village of Anjungan, around 58 km east of Pontianak.

Coordinates: 00° 18.84 N   109° 08.09 E

We were also able to record *Betta midas* in southern Central Kalimantan (Kalimantan Tengah) in a blackwater river by the road from Pundu to Sampit (Research in June 2008: LINKE, M., LINKE, H., & YAP, P.).

Biotope data:

Betta midas lives by preference in blackwater rivers and swamp regions with very soft, very acid water and a large component of humic substances.

Reproduction:

Mouthbrooder.

Total length: (size)

Around 9 cm (3.5 in).

Remarks: (differences from other species of the genus)

The aquarium maintenance of the species is without problem as long as the natural water parameters are taken into consideration. These fishes are somewhat shy and not aggressive. For maintenance they should be provided with a richly planted aquarium. The species exhibits no under-head striping.

Betta midas ♂

References:

- VOORT, S.v.d. 2009: *Betta midas* – A new species from West-Kalimantan;Borneo, and redescription of *Betta anabatoides* (BLEEKER)
- Betta News, Journal of the EAC/ AKL, 2009 (4): 21-22.

Head close-up of *Betta midas*

Betta midas biotope

Small blackwater streams are the preferred habitat of *Betta midas*

Betta midas after capture

Betta midas after capture

Betta midas also lives in the bank of larger zones

Access to the biotope is often very difficult

The black water often flows over a very light sandy substrate

Rainforest area in East Kalimantan

Betta midas after capture

Betta miniopinna TAN & NG, 1994

Explanation of the species name:
Latin *minium* = vermillion red and *pinna* = fin, relating to the red fins.

English name:
Bintan Betta

Original description:
TAN, H.H. & TAN, S.W. 1994: *Betta miniopinna*, a new species of fighting fish from Pulau Bintan, Riau Archipelago, Indonesia (Teleostei: Belontiidae) Ichthyol. Explor. Freshwaters, 5 (1): 41-44, 3 figs.

Systematics:
Betta coccina group

Natural distribution:
Swamp forest at Tanjong Bintan and near Pasir Segiling, Pulau Bintan, Riau Islands, Indonesia.

Coordinates:
01° 10' N   104° 30' E (NG, P.K.L. et al. 1993)

Northern Pulau Bintan: stream at south-east of rubber plantation near road linking Tanjung Pinang and Tanjung Uban.

Biotope data:
These fishes have been recorded in the natural habitat in inundation zones and swamp regions with blackwater conditions (pH value between 4.9 and 5.5).

Reproduction:
Bubblenest spawner.

Total length: (size)
Around 3.5 to 4 cm (1.3 and 1.5 in).

Remarks: (differences from other species of the genus)
Betta miniopinna is a small, slender fish, possibly related to

Betta miniopinna ♀

Betta miniopinna ♂

Betta miniopinna ♂

Betta persephone. Betta miniopinna exhibits a predominantly reddish to dark brown-red body coloration. There are usually large iridescent light green spots on the dorsal and anal fins. Hitherto the red ventral fins were regarded as being an important, species-typical character of *Betta miniopinna*, but it is now known (K. KUBOTA, pers. comm.) that *Betta persephone* from the area of Yong Peng in the south-west of western Malaysia exhibits not only black but also red ventral fins.

For optimal maintenance *Betta miniopinna* should be kept only in very soft, very acid, water containing humic substances. The maintenance aquarium should contain plenty of hiding-places and a dense growth of plants in places. Like the other small red fightingfishes this species is at risk from Velvet Disease (*Oodinium*). By and large, only the best water quality can prevent this, and hence regular partial water changes are obligatory.

References:

• BARWITZ, V. 2011: *Betta miniopinna* – Ein Schmuckstück im Aquarium, Betta News, Journal des European Anabantoid Club/ Arbeitskreis Labyrinthfische 4: 10-12.

Betta miniopinna ♀

Explanation of the species name:

Latin *obscurus* (feminine *obscura*) = dark in colour.

Original description:

TAN, H.H. & NG, P.K.L., 2005: Fighting fishes (Teleostei: Osphronemidae: Genus *Betta*) of Singapore, Malaysia and Brunei. The Raffles Bulletin of Zoology, 13: 90-91.

Systematics:

Betta akarensis group

Natural distribution:

The holotype was collected in the area of the Sungai Barito basin, near Desa Kerendan, in the Sungai Lahai and tributaries, East Kalimantan (Kalimantan Tengah), Borneo.

The paratypes originate from the Sungai Bunes, an affluent of the Sungai Kerendan and from a tributary of the Sungai Lahai, around 1 km from Desa Kerendan.

Sungai Muara Laung and at the market in Puruk Cahu. Sungai Mongkumuh and Sungai Bahadeng, a tributary of the Sungai Barito, East Kalimantan (Kalimantan Tengah).

Biotope data:

Probably resident in clear- and blackwater habitats.

Reproduction:

Mouthbrooder.

Total length: (size)

Probably around 9 cm (3.5 in).

Remarks: (differences from other species of the genus)

The species has sometimes been confused with *Betta edithae*.

Head close-up of *Betta obscura*

Betta obscura ♀ in striped coloration

Betta obscura ♂

Betta ocellata de BEAUFORT, 1933

Explanation of the species name:

Latin *ocellatus* (feminine *ocellata*) = with an eye-spot or ocellus, referring to the spot on the caudal peduncle.

English name:

Eyespot Betta, Ocellated Betta

Original description:

De BEAUFORT, L. F. 1933: On some New or Rare species of Ostariophysi from the Malay Peninsula and a new Species of *Betta* from Borneo. Bull. Raffles Mus. 8: 35-36.

Systematics:

Betta unimaculata group

Natural distribution:

Holotype: Borneo, Sabah (eastern Malaysia); near Bettotan, Sandakan, north-eastern Sabah, eastern Malaysia.

Other locations: SAFODA, forest reserve, near Kampong Batu Puteh, drains into Sungai Kinabaltargan, Sabah, eastern Malaysia (Research in 1994: TAN, H.H.).

Biotope data:

Probably lives in clear and blackwater habitats, with the natural waters predominantly exhibiting a slight brown coloration and having soft, weakly acid, water parameters.

Reproduction:

Mouthbrooder.

Total length: (size)

Around 10 cm (4 in).

Remarks: (differences from other species of the genus)

The species exhibits very numerous parallels with *Betta unimaculata* in its splendid coloration. Species-typical of *Betta ocellata* are the round spot on the caudal peduncle and a coloured scale, intensely coloured like the operculum, posterior to and at the level of the eye.

Betta ocellata probably has a very much larger distribution region than hitherto assumed. The species probably also lives in large parts of north-eastern Kalimantan (Kalimantan Timur).

The species has sometimes been confused with *Betta patoti* or *Betta unimaculata*. Many locations ascribed to *Betta unimaculata* in the past should correctly be ascribed to *Betta ocellata*.

References:

- TAN, H.H. & NG, P.K.L., 2005: Fighting fishes (Teleostei: Osphronemidae: Genus *Betta*) of Singapore, Malaysia and Brunei, The Raffles Bulletin of Zoology, supplement 13: 43-87.

Head close-up of *Betta ocellata*

Betta ocellata ♂

Betta pallida SCHINDLER & SCHMIDT, 2004

Explanation of the species name:

Latin *pallidus* (feminine *pallida*) = pale.

Synonyms:

Betta prima

Original description:

SCHINDLER, I. & SCHMIDT, J., 2004: *Betta pallida* spec. nov., a new fighting fish from southern Thailand (Teleostei: Belontiidae). Zeitschrift für Fischkunde, 7 (1): 1-4.

Systematics:

Betta pugnax group

Natural distribution:

Thailand, Narathiwat Province, about 30 km west of Narathiwat on the road to Ruso.

Coordinates: 06° 21' N   101° 38' E (SCHINDLER, I.)

Betta pallida has also been recorded in clearwater rivers on Ko (= island) Samui. The site was a watercourse in the eastern part of Ko Samui, via the turn-off from National Highway 4169 near Lamai at the temple complex by the Wat Lamai at the Ban Lamai Cultural Hall, around 2 km inland, where the asphalt road changes to a sandy track. Here a small stream flows across the path.

Betta pallida has also been recorded in a larger watercourse in the north of the island of Samui. The collecting site was in this case to be found via Street Maenam 3, a side-road off National Highway 4169 in the village of Mae Nam. The watercourse lies in a deep gulley surrounded by dense brush and trees, barely 2 km after the turn-off, around 100 m (330 ft) parallel to the concrete road (Street Maenam 3).

Biotope data:

The habitats of *Betta pallida* on Ko Samui that have been investigated were small to middle-sized flowing watercourses up to 80 cm (31.5 in) deep, with accumulations of dead wood and dead leaves. The biotopes were well-shaded by growths of brush and trees.

Betta pallida ♀

Betta pallida ♂

Betta pallida ♀

Head close-up of *Betta pallida*

Water parameters:

pH 6.3
Conductivity: 82 µS/cm
General hardness: 2 °dGH
Carbonate hardness: 1 °dKH
Water temperature: 25.5 °C (78 °F).
The population density of the fishes was not very high. They were found almost exclusively among the accumulations of dead leaves. (Research in Jan/Feb 2005: LINKE, H. & LINKE, M.)

Coordinates: 09° 33.69 N   099° 59.83 E (LINKE, H.)

Reproduction:

Mouthbrooder.

Total length: (size)

Around 7 cm (2.75 in).

Remarks: (differences from other species of the genus)

The species is problem-free in its maintenance as long as the natural water parameters are borne in mind.

Betta pallida exhibits numerous parallels with *Betta prima*. A striking character of *Betta pallida* is the longitudinal band, which ends with a gap before a spot on the caudal peduncle. In *Betta prima* the longitudinal band ends without interruption on the caudal peduncle, that is, without any caudal-peduncle spot.

The species has sometimes been confused with *Betta prima* or *Betta taeniata*.

References:

• SMITS, H.M., 1945: The freshwater fishes of Siam, or Thailand. Bull. US. Nat Mus., 188: 455.

Betta pallida biotope

Habitat of *Betta pallida* on Ko Samui

Watercourse on Ko Samui, habitat of *Betta pallida*

Dead leaves and twigs in the water are the preferred habitat of *Betta pallida*

Explanation of the species name:

Adjectival combination of Latin *pallidus* = pale and *finis* = boundary. The species name refers to the strikingly coloured border to the anal fin and the lower part of the caudal fin in females.

Original description:

TAN, H.H. & NG, P.K.L., 2005: Fighting fishes (Teleostei: Osphronemidae: Genus *Betta*) of Singapore, Malaysia and Brunei. The Raffles Bulletin of Zoology, supplement 13: 93-95.

Systematics:

Betta unimaculata group

Natural distribution:

The holotype was collected in the area of Muara Teweh, Kec. Laung Tuhup, Desa Maruai, Rawa/Sungai Laung in the region of the upper Barito River, Central Kalimantan (Kalimantan Tengah), island of Borneo, Indonesia.

The species is found in the area north of the town of Muara Teweh in the Sungai Kerendan region on the road from Kerendan to Longiram and in the regions north of Muara Teweh to Muara Laung on the Sungai Berito.

Biotope data:

Betta pallifina is found mainly in blackwater habitats, where the hydrogen-ion concentration (pH) is usually less than 5.

Reproduction:

Mouthbrooder.

Total length: (size)

Around 9 cm (3.5 in).

Remarks: (differences from other species of the genus)

If possible, only very soft, very acid, water enriched with humic substances should be used for the optimal maintenance of *Betta pallifina*. These fishes are very colourful and somewhat shy in captivity. The species should be kept only with peaceful *Betta species*. The aquarium should be heavily planted and provided with numerous hiding-places.

References:

• VOORT, S. v. d., 2009: *Betta pallifina* from Muara Teweh; Kalimantan Tengah. Betta News, Journal of the European Anabantoid Club/AKL, 2009 (1): 25-26.

Betta pallifina ♀

Betta pallifina ♂

Betta pardalotus TAN, 2009

Explanation of the species name:

Latin *pardalotus* = leopard-like (derived from the scientific name of the leopard *Panthera pardalis*), referring to the spotted operculum.

English name:

Leopard Betta

Original description:

TAN, H.H. 2009: *Betta pardalotus*, a new species of fighting fish (Teleostei: Osphronemidae) from Sumatra, Indonesia. Raffles Bulletin of Zoology 57 (2): 501-504.

Systematics:

Betta waseri group

Natural distribution:

The natural habitat of *Betta pardalotus* lies in the basin of the Musi River, Laut Kenten, Sungai Gelam, in the area of Kota Palembang, the provincial capital of Sumatra Selantan, and in the province of Jambi in the south-east of the island of Sumatra, Indonesia.

Biotope data:

The species lives in small streams, beneath trailing bank vegetation and in dense stands of aquatic plants, as well as among dead leaves on the bottom.

Reproduction:

Mouthbrooder.

Total length: (size)

Around 9 cm (3.5 in).

Remarks: (differences from other species of the genus)

The species exhibits the underhead markings typical for the *Betta waseri* group, only in this case with a very bold pattern of spots on the operculum.

Betta pardalotus ♀

Betta pardalotus ♂

Betta pardalotus ♂

References:

• VOORT,S.v.d. 2009:) *Betta pardalotos* Zucht und Pflege, Betta News, Journal of the EAC/AKL, 2009 (4): 27-28.

Head close-up of *Betta pardalotus* ♂

Explanation of the species name:
Dedication in honour of W. J. Tissot van PATOT.

English name:
Banded Fighter, Patot's Betta

Original description:
WEBER, M. & de BEAUFORT, L. F., 1922: The fishes of the Indo-Australian Archipelago, vol. 4. Brill, Leiden, Netherlands. xiii + 359 pp.

Systematics:
Betta unimaculata group

Natural habitat:
The holotype originated from the area of the Mangar River, a stream 25 km east of Balikpapan Bay; River Bluu, East Kalimantan (Kalimantan Timur), Borneo, Indonesia.

Further locations are 50 km north-west of the town of Balikpapan: Kenagan Timber Concession, Selerong Camp: muddy seepage pool in dry streambed, Balikpapan Regency, Kalimantan Timur.

Biotope data:
The natural habitats of *Betta patoti* are usually small watercourses with medium-hard to soft water with a pH value between 6.0 and 7.0.

Reproduction:
Mouthbrooder.

Total length: (size)
Up to 12 cm (5 in).

Remarks: (differences from other species of the genus)
The 10-11 cross-bands on the body are species-typical. By comparison, *Betta* sp. from Tana Merah exhibits only 8 cross-bands.

These fishes should be maintained only in large tanks with a wealth of plants and numerous hiding-places. Good water quality is very important for optimal maintenance.

Betta patoti ♀

Betta patoti ♂

References:
- LINKE, H. 1998: *Betta* sp. von Tana Merah, Labyrinthfische-Farbe im Aquarium. Tetra Verlag GmbH, Münster, Germany: 75

Betta patoti ♂

Underhead pattern in *Betta patoti*

Spawning scenes of *Betta patoti*

After spawning the eggs lie on the anal fin of the male

Mouthbrooding male

Spawning near the bottom

After spawning the male collects the eggs from the bottom

Male with brood in his mouth

Betta persephone SCHALLER, 1986

Explanation of the species name:

After Persephone, the Greek goddess who was the wife of Pluto and Queen of the Underworld. In reference to the often secretive, "underworld" existence of this fish among and beneath dead leaves.

English name:

Dwarf Dark Green Betta

Original description:

SCHALLER, D., 1986: Laubschlupf. Eine Überlebensstrategie in einem besonderen Biotop und die Beschreibung einer neuen Kampffischart. DATZ - Die Aquarien- und Terrarien-Zeitschrift, 39: 297-300.

Systematics:

Betta coccina group

Natural distribution:

The natural distribution region lies on the Malayan peninsula. According to SCHALLER (1986), the site for the specimen used for the description was a water on Asian Highway No. 2, around three kilometres north of Ayer Hitam. Heading north, it was the second culvert (box-section bridge) in the second valley.

Biotope data:

In the wild *Betta persephone* leads a very secretive existence among roots and layers of leaf litter. According to SCHALLER's data, at times of frequent rainfall the fishes are usually to be found among the root tangles of the scrub element of the woodland vegetation, above the flooded ground and leaf litter. When the water level drops, by contrast, the fishes disappear into these layers of leaves and also sometimes inhabit any zones of water underneath.

Reproduction:

Bubblenest spawner.

Total length: (size)

Around 3.5 cm (2.5 in).

Betta persephone ♀

Betta persephone ♂

Betta persephone ♂

Displaying male *Betta persephone*

Remarks: (differences from other species of the genus)

Betta persephone is a dark grey-green to black *Betta* species with no markings. These fishes often exhibit a marked dark green iridescence in the dorsal, caudal, and anal fins. Males have a striking light border to the upper anterior part of the dorsal fin and white tips to the ventral fins. Females are a little rounder and have somewhat shorter fins. On the basis of observations to date these fishes are very peaceful and can be maintained without problem if their native water parameters are taken into account. Like the other small red fightingfishes this species is at risk from

Velvet Disease (*Oodinium*). By and large, only the best water quality can prevent this, and hence regular partial water changes are obligatory.

Betta persephone is sometimes regarded as identical to *Betta miniopinna*, but the latter species is more slender and elongate in its body form. Both species, along with *Betta brownorum*, are characterised by both parents participating in brood care.

Betta pi TAN, 1998

Explanation of the species name:

The name derives from the Greek letter pi (π), used to denote the mathematical constant representing the relationship between the circumference and the diameter of a circle. The name relates to the underhead pattern in this species.

English name:

Pi Betta, Pi Fightingfish

Original description:

TAN, H. H.: 1998: Two species of the *Betta waseri* group (Teleostei: Osphronemidae) from central Sumatra and southern Thailand. Ichthyol. Explor. Freshwaters v.8 (no.3) 281-287

Systematics:

Betta waseri group

Natural distribution:

The natural habitat of this species lies in the region north of the town of Sungai Kolok in the province of Narathiwat in southern Thailand. H. H. TAN writes of the collecting site: "holotype and paratype series from the same collection date were obtained from swamps near Nam Tod Deng, from an open area of apparently cleared peat swamps forest. The substrate was very soft and deep (up to 1.3 m) and there was very little free water area. The pH of the habitat was 6.0."

The holotype was caught near "Mae Nam Tod Deng, about 6 km north of Sungai Kolok, open area in swamps, ca. 100 – 200 m south before bridge". The types were collected by M. KOTTELAT and K. KUBOTA.

Biotope data:

During his travels through the province of Narathiwat, Norbert NEUGEBAUER (pers. comm.) found *Betta pi* in a blackwater swamp region near Sungai Pardi, some distance from the (left-hand) side of National Highway 4056, around 15 km north of the town of Sungai Kolok.

Head close-up of *Betta pi* showing underhead markings

Betta pi ♂

Blackwater biotope of *Betta pi* in the Sungai Kolok area
Photo by Norbert Neugebauer

Watercourse in the Sungai Kolok area, habitat of *Betta pi*
Photo by Norbert Neugebauer

Heinz-Jürgen BUSSE, who for several years worked for an ornamental fish export company in the province of Narathiwat, reports (pers. comm.) the occurrence of *Betta pi* in the Ban Buketa area, around 30 km south-west of Sungai Kolok in the direction of Waeng and from there to Ban Buketa. The habitat was a blackwater swamp region, fed by a mountain stream. The area was marshy and the water depth was usually shallow, up to 60 cm (24 in). Dense stands of *Cryptocoryne cordata* provided the fishes with numerous areas of shelter and hiding-places. They could be caught without problem in the accumulations of predominantly standing water among the aquatic plants. The following (average) water parameters were recorded by Heinz-Jürgen BUSSE:

pH value: 5.0
Conductivity: 10 to 20 µS/cm, after rainfall 60 µS/cm due to minerals washed in.
Water temperature: around 27 °C (80.5 °F)

Carbonate hardness: < 1 °dKH
General hardness: < 1 °dGH

Reproduction:
Mouthbrooder.

Total length: (size)
Around 11 cm (4.25 in).

Remarks: (differences from other species of the genus)
Betta pi is a slender fightingfish species. Species-typical are the underhead pattern and the longitudinal band, broadening downwards posteriorly, from the posterior operculum to the caudal peduncle. In addition, males exhibit a narrow blue marginal stripe in the anal fin, distinguishing them from the other members of the *Betta waseri* group.

Explanation of the species name:

Latin *pictus* (feminine *picta*) = coloured, painted.

English name:

Spotted Betta

Synonyms:

Panchax pictum
Betta rubra
Betta trifasciata

Original description:

VALENCIENNES, A.1846: Histoire naturelle des poissons, vol. 10, 8th edition. Levrault, Paris.18: xix+505 pp.520-553

Systematics:

Betta picta group

Natural distribution:

REGAN (1910) gives Ambarawa in central Java as a collecting site for *Betta picta*. The species is found in wide areas of the island of Java. Data regarding its occurrence on Sumatra have not been confirmed. So far the so-called *Betta picta* from Sumatra have always turned out to be the very similar *Betta falx*.

Biotope data:

A few notes on the subject from my collecting trip in January 1986. The route headed north from Yogjakarta in central Java. The road climbed continuously across the lowlands towards the mountains, heading east near the Borobudur region and past the villages of Magelang, Secang, and Jambu to Ambarawa. The highest elevations appeared to be between Magelang and Ambarawa. The road was in good condition and permitted good progress. In the town of Ambarawa a small road branches off to Muncul, and according to the maps leads to Salatiga in the south-east. The distance from Ambarawa to Muncul is 10 km. The road passes through numerous villages and is initially bordered by rice fields. In only a few places does it approach

Betta picta ♀

Betta picta ♂

Lake Rawapening. A few kilometres further on the lowland basin ends, and the road then runs along the edge of the hills by the lake. Numerous narrow, fast-flowing (particularly during the rainy season (January)) streams and small rivers cross the road. Every usable bit of land is used for cultivating rice. The vegetation is very dense and vigorous. Muncul lies close to the lake. A stream, with a swimming bath, runs through the village. Experimental collecting in the stream, which was very fast-flowing, proved unsuccessful. To the right of the stream, by the swimming-pool car park, there was a concrete water storage tank measuring around 10 x 35 m (33 x 115 ft). The water depth was around 70 cm (27.5 in). An inflow, plus water constantly welling up from the bottom (lava sand), ensured a continuous exchange and through-flow of water, creating water movement discernible as a "gentle" current. The water in this concrete tank was very clean and clear and without coloration. The temperature among the thick covering of floating plants measured 23.9 °C (75 °F), in 10 cm (4 in) of depth only 22.8 °C (73 °F), and in 50 cm (20 in) of depth 22.3 °C (72.1 °F). As well as large numbers of Guppies, numerous *Betta picta* were found among this covering of floating plants, as well as in areas with a loose growth of floating leaves and among banana leaves lying in the water. There were a number of adult males with broods in their mouths, some with eggs and others with almost fully developed larvae. The fishes had apparently entered this concrete tank via the inflow from the stream and were living there under apparently optimal conditions in the absence of any predatory fishes. It is conceivable that the *Betta* species was also to be found in the rice fields (constructed on hillside terraces) or their irrigation channels, and the rice farmers in the vicinity stated that this was the case. However, attempts at collecting there were unsuccessful.

There was a second collecting site around 50 kilometres to the west, again at an altitude of around 500 m (1650 ft). Here too there were extensive hilly regions that were used for rice cultivation. The collecting site lay few hundred metres outside the village of Pikatan. Here there were water-filled ditches around 50 cm (20 in) wide on either side of the road. Because of frequent rainfall the water flowing cross the sloping land was constantly in motion and in places also made its way across vegetated ground. Only occasional large (around 10 to 12 cm (4 to 5 in) long), fast-swimming barbs were to be seen

Betta picta male mouthbrooding

in the vegetation-free streams and ditches. It was possible to find numerous *Betta picta* in the often large areas of land inundated to a depth of up to 3 cm (1.25) by the water and densely vegetated with grass and similar plants. Here around 15 individuals of this species were collected in a short time, again including several brooding males.

The species concerned was the same as that collected previously, though there may have been differences in coloration. The specimens collected at this site exhibited golden gill-covers. And the water at the site was relatively cool, with a temperature of 23.2 °C (73.8 °F).

Reproduction:

Mouthbrooder.

Betta picta has no special requirements as regards water conditions, either for maintenance or breeding. The fishes will readily proceed to breed even in small aquaria, which should be well planted in places and if possible decorated with small caves. The spawning procedure is similar to that of the majority of other mouthbrooding *Betta* species. The fry are around 6 mm (0.25 in) long on release and will immediately take freshly-hatched *Artemia salina* nauplii.

Total length: (size)

Up to 6.5 cm (2.5 in).

The temple precinct of Borobudur in central Java

Still areas in fasf-flowing watercourses are also habitats of *Betta picta* near Ambarawa

The species exhibits clear sexual dimorphism at adult size. Males have a larger, more heavily developed head and exhibit bolder colours.

Various forms with different coloration and/or markings are known, apparently reflecting different areas in the natural distribution.

The species *Betta picta* differs from *Betta falx* in having a different coloration to the fin margins of the lower part of the caudal fin and the anal fin. In *Betta picta* these fin margins are turquoise green to light blue in colour, while in *Betta falx* they are red to red-brown. The caudal-fin form is predominantly round in both species, but *Betta picta* sometimes also exhibits a slightly lancet-shaped caudal-fin form.

References:

• TAN, H.H. & KOTTELAT,M.,1998: Redescription of *Betta picta* (Teleostei: Osphronemidae) and description of *Betta falx* sp. n. from central Sumatra. Revue Suisse de Zoologie 105 (3): 557-568.

Biotope of *Betta picta* on the island of Java

Biotope of *Betta picta* near Ambarawa

Rice fields in the highlands near Ambarawa, Java

Explanation of the species name:

Latin *pinguis* = "fat", in reference to the plump body in relation to total length.

English name:

Large Fightingfish

Original description:

TAN, H.H. & KOTTELAT, M. (1998): Two species of *Betta* (Teleostei: Osphronemidae) from the Kapuas Basin, Kalimantan Barat, Borneo. The Raffles Bulletin of Zoology, 46 (1): 43-46.

Systematics:

Betta akarensis group

Natural distribution:

The species has to date been recorded only in the Kapuas River region in the north-west of West Kalimantan (Kalimantan Barat) in Borneo. The specimen used for the description was discovered in 1995 by M. KOTTELAT *et al.* in the Sungai Letang, near to the village of Kandung Suli (Keeamatan Jongkong), in Kalimantan Barat.

Biotope data:

This species was caught in a small blackwater stream surrounded by open woodland landscape. This probably means very soft, very acid water. The fishes were sorted out from among numerous other fishes used by the fishermen as bait for catching *Chitala lopis*. As far as is known at present, this region is also home to *Betta enisae* KOTTELAT, 1995 and *Betta dimidiata* ROBERTS, 1989.

Reproduction:

Mouthbrooder.

Total length: (size)

These fishes grow relatively large and can attain a total length of 12 cm (5 in).

Betta pinguis Female

Betta pinguis Male

Spawning scenes of *Betta pinguis*

Spawning and brood-care scenes of *Betta pinguis*

After spawning the fertilised eggs lie on the anal fin of the male and the female then takes them into her mouth

Mouthbrooding ♂ ♀ *Betta pinguis*

Betta pinguis ♀

Explanation of the species name:

Latin *primus* (feminine *prima*) = "first", "original", "earliest".

English name:

Chantaburi Mouthbrooder

Original description:

KOTTELAT, M. 1994: Diagnoses of two new species of fighting fishes from Thailand and Cambodia (Teleostei: Belontiidae). Ichthyol. Explor. Freshwaters, 5 (4): 297-304, 6 figs., 2 tabs.

Systematics:

Betta pugnax group

Natural distribution:

Holotype: Thailand: Chantaburi Province: Creek at about km 1 on the road to Nam Tok Phliu, after leaving Chantaburi-Trat highway.

Betta prima was recorded as very numerous in a swamp and inundation region by National Highway 3 at km 329, around 90 km before the town of Trat, south of Chanthaburi in south-east Thailand. (Research in 1992: KUBOTA, K. & LINKE, H.)

Betta prima is distributed in the area from Chantaburi to Trat. It is not known how widely the species may also be distributed on the other side of the nearby national boundary, in Cambodia.

Biotope data:

The habitats we investigated were standing waters that were probably filled solely with rain water. The water temperature in this biotope was usually more than 30 °C (86 °F). The water was very acid with a pH value of 5.5. The electrical conductivity measured 68 µS/cm at the time of our research. The water was mainly clear and coloured slightly brownish. *Betta prima* was recorded predominantly in zones heavily overgrown with plants. (Research in 1992: KUBOTA, K. & LINKE, H.)

Reproduction:

Mouthbrooder.

Betta prima ♀

Betta prima ♂

Betta prima pair, ♂ below

Collecting *Betta prima* in marshland south of Chantaburi, south-east Thailand

Biotope of *Betta prima* between Chantaburi and Trat in south-east Thailand

Total length: (size)
Around 7 cm (2.75 in).

Remarks: (differences from other species of the genus)

Betta prima is problem-free in its maintenance. The species is peaceful and should not be housed with aggressive *Betta* species. Soft, neutral to acid water is beneficial for optimal maintenance. The maintenance aquarium should provide the fishes with hiding-places and be densely planted in places. The species breeds readily.

Betta prima was originally imported as *Betta* sp. from Nam Tok Krating. In his book The freshwater fishes of Siam, or Thailand, SMITHS writes of the existence of a so-called "*Betta taeniata*" that was caught as long ago as 1928 near Chanthaburi in the south-east of Thailand, but was apparently never imported alive. The first live specimens were imported for the aquarium by K. KUBOTA as recently as 1990.

References:

- SMITHS, H. M., 1929: Notes on some Siamese fishes. J. Siam Soc., Nat Hist. Suppl., 8: 11-14.
- SMITHS, H. M., 1945: The fresh-water fishes of Siam, or Thailand. Bull. U.S. National Mus., 188: 1-622.

Betta pugnax (CANTOR, 1849)

Explanation of the species name:
Latin *pugnax* = "pugnacious".

English name:
Pugnax Mouthbrooder, Penang Betta

Synonyms:
Macropodus pugnax
Betta fusca
Betta brederi
Betta macrophtalma

Original description:
CANTOR, T.E., 1850: Catalogue of Malayan fishes, J. Roy. Asiatic Soc. Bengal. 18: i-xii + 981-1143, pls. 1-14.

Systematics:
Betta pugnax group

Natural distribution:
The collecting site of the lectotype of *Betta pugnax* was given as Pulau Pinang, the island of Penang, in the north-west of western Malaysia. *Betta pugnax* is distributed throughout the entire Malayan peninsula (West Malaysia), including Singapore.

Biotope data:
These fishes usually live in the marginal zones of fast-flowing watercourses, among bank vegetation in the areas of weaker current found here and there. The distribution regions of this species are often in areas of higher land. The preferred natural habitat of this species is mountain watercourses, as on the island of Penang, with very clear, clean, and soft water, in places only 90 cm (36 in) wide and 50 cm (20 in) deep, with numerous stones on a bottom of fine sand, and large aggregations of aquatic plants.

The water temperature in the rainy season is around 22 °C (71.5 °F), and rises to only 26 °C (79 °F) during the warm dry season, despite strong irradiation by the sun.

Reproduction:
Mouthbrooder

For breeding the recommended water parameters are soft,

Juvenile *Betta pugnax*

Betta pugnax ♂

Female spitting out eggs in front of male

Mating in *Betta pugnax*

Male *Betta pugnax* mouthbrooding

slightly acid (pH around 6.0), and a temperature of 25 to 26 °C (77 to 79 °F). The water should be only 10 to 12 cm (4-5 in) deep and floating plants are essential.

Betta pugnax is one of the mouthbrooding species. It spawns near the bottom, without constructing a bubblenest. The partners embrace one another and the eggs are expelled, dropping onto the slightly curved anal fin of the male, from where the female takes them into her mouth after breaking free of the embrace. The spawning passes take place at long intervals, almost always at the same spot in the aquarium. Between the individual embraces the female spits the eggs out in front of the mouth of the male, usually singly, whereupon he snaps them up and keeps them in his mouth. Sometimes the male also spits the eggs back to the female, almost like a game of ball.

After the conclusion of the spawning act, which can last for up to five hours, the male retires to the shelter of an area of densely-packed plants. Around 10 days later the fully-developed fry leave the mouth of their father. They now require small foods such as freshly-hatched *Artemia salina* nauplii or fine pond and/ or prepared foods. In addition regular water changes (a third of the tank volume every 14 days) should be made, as healthy, clean water will encourage the growth of the young fishes.

Total length: (size)
Around 10 cm (4 in).

Remarks: (differences from other species of the genus)
Males exhibit rows of very attractive, brilliant antique green scales on a dark, grey-brown body. The opercula are the same green colour. Females have somewhat shorter fins and are not so boldly coloured.

The maintenance of this species presents no problems. The aquarium for these fishes should be richly planted and provide hiding-places in the form of rocky caves or bogwood. The water temperature can be between 23 and 27 °C (73.5 and 80.5 °F). The water should be as soft as possible and slightly acid.

Feeding should be heavy, rich, and varied. Flake food is also readily taken as a snack. Terrestrial insects are a very important food for optimal maintenance and successful reproduction.

Biotope of *Betta pugnax* in the mountainous area of the island of Penang, Malaysia

Biotope of *Betta pugnax* in the mountainous area of the island of Penang, Malaysia

Biotope of *Betta pugnax* in the mountainous area of the island of Penang, Malaysia

Biotope of *Betta pugnax* in the mountainous area of the island of Penang, Malaysia

Betta pugnax being checked over after capture

Betta pulchra TAN & TAN, 1996

Explanation of the species name:

Latin *pulcher* (feminine *pulchra*) = "beautiful", "attractive", referring to the coloration.

Original description:

TAN, H.H. & TAN, S. H. 1996: Redescription of the Malayan fighting fish *Betta pugnax* (Teleostei: Belontiidae), and description of *Betta pulchra*, new species from Peninsular Malaysia. Raffles Bull. Zool., 44 (2): 428, figs. 2a-c (423).

Systematics:

Betta pugnax group.

Natural distribution:

A specimen from the area of the Kampong Jasa Sepakat, Pontian, Johor, West Malaysia was designated holotype.

Around 4 km towards Pontian Kechil from Sri Bunian (Coordinates: 01° 27' 13.1 N 103° 24' 52 E) and Jasa Sepakat (01° 31' 30.5 N 103° 27' 47.7 E) and 5 km into side road leading into oil-palm estate, Pontian, Johor, West Malaysia.

The species is apparently endemic to the only natural distribution region known to date, the Pontian Kecil area, around 40 km west of the town of Johor Bahru, in the extreme south-west of the Malayan Peninsula.

Biotope data:

Betta pulchra is a blackwater species which lives in inundation zones and swamp regions, as well as in watercourses in plantations with very acid, clean, and slightly flowing water (pH between 3.9 and 4.2). Here the fishes live syntopic with *Belontia hasselti*, *Betta bellica*, *Betta imbellis*, *Sphaerichthys osphromenoides*, *Parosphromenus* sp., and *Trichopsis vittata*.

Reproduction:

Mouthbrooder.

Total length: (size)

Around 9 cm (3.5 in).

Remarks: (differences from other species of the genus)

These fishes are comparable to *Betta pugnax*, but exhibit a more slender body and a lancet-shaped caudal fin whose central prolongation is higher on the upper part of the fin by comparison. The brightly coloured, iridescent scales are not blue-green as in *Betta pugnax*, but green in colour. The same applies to the boldly coloured lower half of the head.

For optimal maintenance and successful breeding *Betta pulchra* should be kept in water that matches its native biotope – mineral-poor and acid, with a pH value between 4.5 and 6.0. Dense planting of the aquarium in places is advantageous. The addition of humic substances is very important.

Betta pulchra ♂

Explanation of the species name:

Derived from a Malayan and Indonesian term for a king or prince.

English name:

Royal Betta, Raja Betta

Original description:

NG, P.K.L. & TAN, H.H., 2005: The Labyrinthfishes (Teleostei: Anabantoidei, Channoidei) of Sumatra, Indonesia. The Raffles Bulletin of Zoology 2005, Supplement 13: 115-138.

Systematics:

Betta pugnax group

Natural distribution:

Indonesia: in the eastern part of central and southern Sumatra, in the provinces of Riau and Jambi. The holotype came from the Sungai Ayer Merah, a feeder stream to Danau Souak Padang, ca. 15 minutes upstream by boat (TAN, H.H. & TAN, S.H.: 1996).

The species has to date also been recorded west of the town of Jambi (Djambi, Telanaipura) in the direction of Muaratembesi and south towards Kota Palembang.

Biotope data:

Betta raja lives in the lowlands, usually in regions of wooded swamp, lakes, and small watercourses. The species is found predominantly in blackwater.

Reproduction:

Mouthbrooder.

Total length: (size)

10 to 11 cm (4 to 4.5 in).

Betta raja ♂ after capture

Betta raja ♂ mouthbrooding

Head close-up of *Betta raja*, ♀ left, ♂ right

Remarks: (differences from other species of the genus)

Betta raja exhibits the species-typical character of comparatively long ventral fins, which, when folded, extend to the posterior quarter of the anal fin. In addition there may be a single striking green scale on the posterior edge of the operculum, and a dark longitudinal band extending from the eye to the tail – a species-typical character in male fishes, which, when adult, have a lancet-shaped caudal fin, while in females the tail is rounded.

Underhead coloration in *Betta raja* ♂

Betta raja, ♂ above, ♀ below;
the long ventral fins are the typical feature of this species

Spawning scenes of *Betta raja*

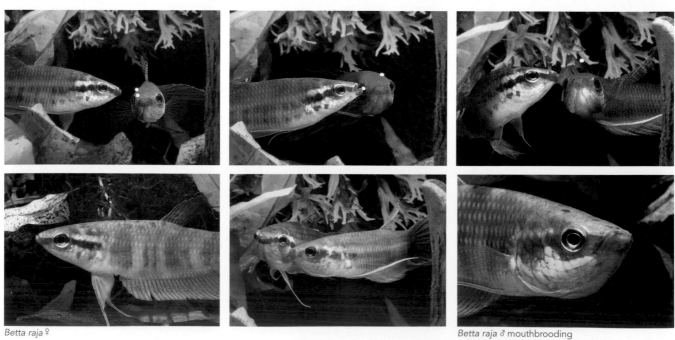

Betta raja ♀

Betta raja ♂ mouthbrooding

Explanation of the species name:

Made-up Latin *renatus* (feminine *renata*) = "having a kidney", referring to the kidney-shaped underhead marking.

Original description:

TAN, H.H. 1998: Description of two new species of the *Betta waseri* group (Teleostei: Osphronemidae), Ichthyol. Explor. Freshwaters 8 (3): 281-287, 5 figs.

Systematics:

Betta waseri group

Natural distribution:

The holotype came from the area of the province of Jambi in the swamp woodlands of Rantau Panjang, Sumatra, Indonesia. Further *Betta renata* have been found in a region of swamp woodland near Pematang Lumut in the north-east of the province of Jambi and in the Lahat region in the province of Sumatra Selatan, as well as in the Indragiri River region in the province of Riau.

Biotope data:

Betta renata lives predominantly in very soft, very acid water with pH values between 4.0 and 4.8. The species is a typical inhabitant of blackwaters.

Reproduction:

Mouthbrooder.

Total length: (size)

Up to 11 cm (4.5 in).

Remarks: (differences from other species of the genus)

Betta renata exhibits a distinctive underhead marking, as do the species *B. chloropharynx*, *B. hipposideros*, *B. spilotogena*, and *B. waseri*, and this is important for differentiating the species.

For optimal maintenance these fishes should be provided with a large tank containing soft, very acid water, with dense planting in places and adequate hiding-places.

Betta renata in normal coloration

Betta renata in fright coloration

Betta renata, ♂ above, ♀ below

Striking underhead coloration in *Betta renata* ♂

Betta renata ♀

Head close-up of *Betta renata*

Typical underhead pattern of *Betta renata*

Biotope of *Betta renata*

Explanation of the species name:

Latin *ruber* (feminine *rubra*) = "red", referring to the weak red coloration of the body.

English name:

Rubra Betta, Teba Betta

Synonyms:

Betta imbellis
Betta picta

Original description:

PERUGIA, A. 1893: Di alcuni pesci raccolti in Sumatra dal Dott. Elio Modigliani. Annali del Museo Civico di Storia Naturale Giacoma Doria, Genova, 13: 241-247.

Systematics:

Betta foerschi group

Natural distribution:

Indonesia: northern Sumatra: Toba Lake (?) and Sibolga (?) (data from 1886). There are additional collecting sites in Aceh Barat, Alur Sungai Lamueselatan (1982).

The most recent collections, for which precise locality details are not available, likewise suggest a distribution in southern Aceh, in the north-east of the island of Sumatra.

To date (2009) the natural distribution regions have not been made public for commercial reasons.

Biotope data:

We can at present (2010) only speculate about the habitat of *Betta rubra*. On the basis of maintenance and breeding experiences, the species probably lives in gently flowing, clear, mineral-poor, acid water (pH value around 4.8 to 5.3) with an electrical conductivity of around 50 µS/cm.

Reproduction:

Mouthbrooder.

The supposition that this mouthbrooder is a transitional form between bubblenest spawners and mouthbrooders (*Betta foerschi* group) has been confirmed by my own observations. The species embraces during mating in the manner normal for mouthbrooders of the genus *Betta*, specifically the male embraces his partner in a sheltered spot close to the bottom.

Betta rubra female

Betta rubra male

Head close-up of *Betta rubra*

But in contrast to the procedure in the other mouthbrooders, during the embrace the female doesn't remain the right way up (belly-down) but is turned onto her back as in the bubblenest spawners. Moreover the eggs expelled are not taken into the female's mouth after the embrace, but are collected up directly by the male. There is no spitting of the eggs by the female in order to transfer them to the male, such as is usual in the other mouthbrooding *Betta* species (except for the *unimaculata* group). This corresponds to the breeding behaviour of the *Betta* species in the *foerschi* group. The young are released from the mouth of the male after around 10 to 12 days and at this time have a total length of around 6 mm (0.25 in). My brood developed successfully at the following water parameters: conductivity around 45 µS at an average temperature of 28 °C (82.5 °F) and a pH value of 4.4.

Total length: (size)

Around 5.5 cm (2.25 in).

Remarks: (differences from other species of the genus)

H.H. TAN published the first photos of living *Betta rubra*. The members of this species are very slender fishes, coloured delicate red on the flanks. The unpaired fins have narrow light blue margins. The head as well as the underhead region exhibit a dark stripe pattern such as is usual for many mouthbrooders of the genus *Betta*. Depending on mood, these fishes exhibit six to seven broad dark vertical bars on the sides of the body. The red coloration is apparently variably bright and thought to be strongly mood-dependent. Perhaps the species occurs in different distributional areas and hence has local forms with variable colour patterns.

References:

- BAENSCH, H.A. & FISCHER, G.W., 1988: Aquarium Atlas Foto-Index 1-5. Mergus Verlag, Melle, Germany. 1211 pp.

- KOTTELAT, M., WHITTEN, A.J., KARTIKASARI, S.N., & WIRJOATMODJO, S., 1993: Freshwater Fishes of Western Indonesia and Sulawesi. Periplus Editions Limited (HK) Ltd, Hong Kong : 223-224

- DONOSO-BÜSCHNER, R. & SCHMIDT, J. 1999: Kampffische-Wildformen. Bede-Verlag, Ruhmannsfelden, Germany. 79 pp.

- TAN, H.H. & NG, P.K.L. 2005: The Labyrinthfishes (Teleostei: Anabantoidei, Channoidei) of Sumatra, Indonesia. The Raffles Bulletin of Zoology, Supplement 13: 115-138.

- LINKE, H., 2008: Der Phantomfisch – *Betta* rubra. Betta News, Journal of the European Anabantoid Club/Arbeitskreis Labyrinthfische im VDA, 2008 (2):

- DIEKE; H. , 2012: Eine unverhoffte Bekanntschaft. Aquaristik, Dähne Verlag, 19; (1), pp.14-18

Spawning scenes of *Betta rubra*

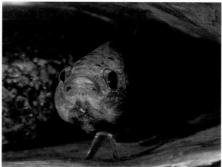

During spawning in the mouthbrooder *Betta rubra* the female is turned onto her back, as in the substrate spawners

Betta rubra male mouthbrooding

Betta rubra male mouthbrooding

Betta rubra male mouthbrooding

Explanation of the species name:
Latin rutilans = "glowing red".

English name:
Red Fightingfish

Original description:
WITTE, K.-E. & KOTTELAT, M: in KOTTELAT 1991: Notes of the taxonomy and distribution of some Western Indonesian freshwater fishes, with diagnoses of a new genus and six new species (Pisces: Cyprinidae, Belontiidae, and Chaudhuriidae). Ichthyol. Explor. Freshwaters, 2(3): 278-280.

Systematics:
Betta coccina group

Natural distribution:
The specimen used for the description came from the Sungai Kepayang, 7 km SE of Anjungan on the road to Pontianak, Kalimantan Barat, Borneo.

Coordinates: 00° 20' N   109° 08' E (KOTTELAT, M. 1990)

Betta rutilans is found north-east of Pontianak in the region between the villages of Sungai Penjuh and Anjungan. We were able to find this species 8 km south-west of Anjungan on the road from Pontianak to Mandor in West Kalimantan (Kalimantan Barat). (Research in 1990: BAER, I., LINKE, H., & NEUGEBAUER, N.)

Biotope data:
The natural habitat of *Betta rutilans* is a typical blackwater biotope with corresponding parameters. The following water parameters were obtained in the almost standing water (at low water): pH value 4.5; carbonate hardness < 1 °dKH; electrical conductivity 39 µS at a water temperature of 27.6 °C (81.5 °F). (Research in August 1990: LINKE, H.)

Reproduction:
Bubblenest spawner.

For breeding these fishes should be kept in tanks with a volume of at least 50 litres (11 gallons) and containing dense groups of aquatic plants. The water should be very soft and very acid (pH around 4.5 to 5.0). The addition of humic substances via peat filtration or the dead leaves of oak, beech, and/or the Asian Sea Almond tree is very important.

Total length: (size)
Up to 5 cm (2 in).

Betta rutilans ♂

Betta rutilans ♂

Remarks: (differences from other species of the genus)

The sexes are readily distinguished in adult individuals. The males have somewhat longer fins and are more intensely coloured.

Despite initial suppositions to the contrary these fishes belong to the bubblenest spawners and are comparable to *Betta brownorum*, *Betta uberis*, and *Betta tussyae*. The species was first imported alive by BAER, LINKE, and NEUGEBAUER.

Like the other small red fightingfishes this species is at risk from Velvet Disease (*Oodinium*). By and large, only the best water quality can prevent this, and hence regular partial water changes are obligatory.

Prior to its scientific description, the species was sometimes known as *Betta* sp. from Anjungan

Explanation of the species name:
Dedication in honour of Dietrich Schaller.

English name:
Schaller's Fightingfish

Original description:
KOTTELAT, M. & NG, P.K.L., 1994: Diagnoses of five new species of fighting fishes from Bintan and Bangka (Teleostei: Belontiidae), Ichthyol. Explor. Freshwaters, 5 (1): 65-78.

Systematics:
Betta pugnax group

Natural distribution:
The holotype was collected 5.5 km north of Payung on the road to Pangkalpinang, island of Bangka, Indonesia. The species is distributed all over the island of Bangka.

Biotope data:
Betta schalleri has been recorded in a wide variety of water types. It was first imported in 1993, by GERSTNER, K., LINKE, H., & LINKE, M.

These fishes live syntopic with several other labyrinthfish species such as *Parosphromenus deissneri*, *Betta burdigala*, and *Sphaerichthys osphromenoides*. Our study site lay around 60 km south of Koba on the road to Toboali between the villages of Djeridja and Bikang, around 4 km before the village of Bikang. It was a large region of woodland, swamp, and inundation zone where a stream crossed the road. Because of recent heavy rainfall (rainy season – end June) the water was very high and usually exhibited a moderate current. At the time in question it had a temperature of 25 °C (77 °F). Carbonate and general hardness were both less than one dH, the pH was 5.0, and the electrical conductivity 18 µS/cm. The water was dark red-brown in colour, and clear. The water depth was on average around a metre (40 in). The bottom consisted of light, loamy sand.

Reproduction:
Mouthbrooder.

Total length: (size)
Around 10 cm (4 in).

Remarks: (differences from other species of the genus)
The species has to date been imported only rarely. It is problem-free in its maintenance.

Betta schalleri usually has a light brown to light grey body coloration. In fright coloration two, occasionally three, dark narrow longitudinal bands run parallel along the body, with the central band extending from the eye to the caudal peduncle. The lower half of the head is coloured bright light green in males.

Betta schalleri ♂

Blackwater biotope in the south of the island of Bangka, habitat of *Betta schalleri*

The fins predominantly exhibit the body coloration and (apart from the dorsal fin and sometimes also the anal fin) only rarely have spot markings on the membranes. When folded, the ventral fins extend to the middle of the anal fin. The caudal fin is rounded to slightly lancet-shaped. The species has a spot on the caudal peduncle.

Rainforest biotope at the roadside in the south of Bangka

Explanation of the species name:

Latin adjective *siamorientalis*, from Siam, the name until 1949 of the country now called Thailand, and Latin orientalis meaning "eastern". Hence, "from eastern Siam", referring to the distribution of the species

English name:

Black Imbellis

Original description:

KOWASUPAT, C., PANIJPAN, B., RUENWONGSA, P. & JEENTHONG, R., 2012: *Betta siamorientalis*, a new species of bubble-nesting fighting fish (Teleostei: Osphronemidae) from eastern Thailand, Vertebrate Zoology 62(3) 2012, 387-397

Systematics:

Betta splendens group

Natural distribution:

The natural distribution of this fish has so far been confirmed as extending from eastern Thailand across Cambodia to Vietnam. Ichthyologists from the Betta Project at Mahidol University in Bangkok have recorded the species in the provinces of Sa Kaeo, Prachin Buri, and Chachoengsao in eastern Thailand.

We found this fish in a swampy area at the edge of the village of Priay Khmang, around 12 km before the town of Siem Reap

Displaying male *Betta siamorientalis*

Betta siamorientalis were very numerous in a swampy meadow used as pasture for cattle and/or water buffalo. In copious hoof prints and other small, shallow accumulations of water. Also in the marginal zones of wide moat in Vietnam

Betta siamorientalis after capture in Vietnam

Betta siamorientalis after capture in Vietnam

on the northern side of the road from Poipet to Siem Reap in western Cambodia (HERMANN, L. & LINKE, H. 2007), and in a swamp by a canal in an area of scrub and grassland near Duc Hoa, to the north-east of Ho Chi Minh City in Vietnam, around 45 km east of the border with Cambodia. (CHIANG, N. & LINKE, H. 2008)

Biotope data:

Cambodia:

We caught these fishes in a densely vegetated area of swamp/swampy lake. The lake increases in size during the rainy season and then has an area of several square kilometres. These fishes were very numerous there and could be found without difficulty among the floating plants. The temperature at the water's surface in the afternoon was 34.6 °C (94.3 °F), and around 30 °C (86 °F) at a depth of about 30 cm (12 in). (Research in February 2007: HERMANN, L. & LINKE, H.)

Vietnam:

These fishes were very numerous in a swampy meadow used as pasture for cattle and/or water buffalo. The copious hoof prints and other small, shallow accumulations of water, along with water-filled ditches, provided plenty of swimming space for the *Betta siamorientalis*.

Coordinates: 10°54′01 N   106°29′20 E

We measured the following water parameters: pH 3.52 and a conductivity of 592 μS/cm at a water temperature of 29 °C (84 °F). The water was mainly stagnant, yellowish, and murky, with a slight current only in the adjacent canal. It may be that the areas of land adjacent to this canal were seasonally flooded by the canal overflowing during the rainy season. (Research in February 2008: CHIANG, N. & LINKE, H.)

Reproduction:

Bubblenest spawner

Total length: (size)

Around 5.5 cm (2.125 in).

Remarks: (differences from other species of the genus)

The species exhibits many parallels to *Betta imbellis* in its appearance. These include the red margin to the caudal fin as well as the red tip to the anal. But it cannot be unequivocally identified as *Betta imbellis* because of the absence of the bright light blue stripes on the gill cover that are species-typical for *Betta imbellis*. On the other hand, the fin coloration is atypical for *Betta splendens*.

Ichthyologists from the Betta Project at Mahidol University in Bangkok have been able to establish, using "DNA barcodes" and other techniques, that the "Black Imbellis" is significantly different from *B. imbellis* in genetic terms.

References:

- CHANDRA, E. 2007: *Betta splendens* von Siam Rep-Kambodscha, Der Makropode.4/07: 115-116.
- KÜHNE, J. 2008: Der *Betta splendens* von Seam Reap – Kambodscha. Aquarium live 12(6): 44-49.
- LINKE, H. 2007: Drei Kampffische aus Kambodscha, Aquaristik Fachmagazin 197(5): 34-37.
- PANIJPAN, B. & KOWASUPAT, C. 2012: The Black Imbellis / Der Schwarze Imbellis, Betta News 2012 (1): 11.
- SCHÄFER, F. 2009: *Betta* cf. *imbellis* "Vietnam Black" - ein neuer Kampffisch aus Vietnam, AqualogNews 88: 14.

We caught *Betta siamorientalis* in a densely vegetated area of swamp / swampy lake in Cambodia

Betta simorum TAN & NG, 1996

Explanation of the species name:

Latin genitive plural, = "of the Sims". Dedication in honour of Thomas G. K. SIM and his wife Farah, of Sindo Aquarium in Jambi, Sumatra.

Synonyms:

Betta fasciata
Betta bleekeri

Original description:

TAN, H.H. & NG, P.K.L., 1996: Redescription of *Betta bellica* SAUVAGE, 1884 (Teleostei: Belontiidae), with description of a new allied species from Sumatra. The Raffles Bulletin of Zoology 44 (1): 143-155.

Systematics:

Betta bellica group

Natural distribution:

Indonesia: island of Sumatra, east coast of southern Sumatra, in the area of the province of Jambi and in the Indragiri river basin in the province of Riau, as well as in the province of Sumatra Selantan.

The holotype was collected in a swamp region at Pematang Lumut, Jambi Province (KOTTELAT, M.: 1994). The paratypes

likewise originated from regions of woodland swamp close to Pematang Lumut in the province of Jambi (NG, P.K.L.: 1995).

We were also able to find *Betta simorum* in Danau (Lake) Rasau in the province of Jambi and in the north of the province of Sumatra Selatan. (Research in 2007 and 2008: SIM, T., CHIANG, N., & LINKE, H.)

Biotope data:

Betta simorum is a typical inhabitant of blackwaters, often found in very soft and very acid water (pH value between 3.7 and 4.2).

Reproduction:

Bubblenest spawner.

Betta simorum ♀

Betta simorum ♂

Larva of *Betta simorum*

Betta simorum being checked over after capture

Blackwater swamp in the north-east of the province of Jambi, Sumatra, habitat of *Betta simorum*

Biotope of *Betta simorum* near Sentang, Sumatra

Total length: (size)
Around 9 cm (3.5 in).

Remarks: (differences from other species of the genus)
The sexes can be distinguished in both sub-adult and adult *Betta simorum* by the longer fins and more intense coloration in males.

Watercourse in Sumatra

Betta simplex KOTTELAT, 1994

Explanation of the species name:

Latin *simplex* = "single", "simple".

English name:

Krabi Fightingfish

Original description:

KOTTELAT 1994: Diagnoses of two new species of fighting fishes from Thailand and Cambodia (Teleostei: Belontiidae). Ichthyol. Explor. Freshwaters, 5 (4): 301-304, Figs. 3-4.

Systematics:

Betta picta group

Natural distribution:

Thailand/Krabi Province: Northwest of Krabi, spring (small lake) of Tham Sra Kaew and Nine Ponds, behind Ban Nai Sra village, 2.2 km from National Highway 4034, 1800 m behind Public Health Center.

The natural habitat of these fishes lies north-west of the town of Krabi in southern Thailand. The site is a spring with associated small lakes, known as Tham Sra Kaew (Nine Ponds), and lies behind the village of Baan Nai Sra. The main lake, the turquoise-green spring-fed Tham Sra Kaew, lies close to National Highway 4033, around 2.2 km after the junction with National Highway 4034 and behind the village of Ban Nai Sa (also written as Baan Nai Sra). A small road branches off in the village, leading to the Tham Sra Kaew. It is also the route to the camp for "Elephant Trekking" and the Sra Kaew Cave. The surrounding area is very hilly, a picturesque karst landscape characterised by white limestone rocks, sometimes covered in plants, and surrounded by the dense greenery of tropical woodlands. The main lake lies at the beginning of the chain of ponds. The water here exhibits a slightly clouded turquoise coloration. Only during heavy rainfall does water running into the lake from the shoreline colour the pool orange-red by virtue of the surrounding laterite soil. But only a short while afterwards the turquoise-coloured water from the strong upwelling on the lake-bed dispels the orange-coloured water, and the lake regains its original turquoise-green coloration.

Betta simplex lives predominantly in the sometimes densely-vegetated, shallow bank zones and is easy to catch there. Large areas of the bottom of the lake, and particularly the shallow watercourses linking the chain of ponds downstream, are covered by dense carpets of *Cryptocoryne*. It appears that recently this lake is being used ever increasingly by the local population as a washing and bathing place, unfortunately with noticeable detrimental consequences for the lake.

Because of the local geology, the water in this area is comparatively mineral-rich. Hence we are not dealing here with so-called soft-water fishes, such as are to be found in the majority of regions in southern Thailand, but with fishes that require medium-hard water for optimal maintenance and successful breeding in the aquarium. *Betta simplex* from the

Betta simplex ♂

Spawning scenes of Betta simplex

Betta simplex ♂

Betta simplex ♀

Betta simplex, ♂ above, ♀ below

Tham Sra Kaew and the adjoining ponds are short-lived in soft water.

Coordinates: 08° 10. 15 N   098° 40.45 E (LINKE, H.)

(Tham Sra Kaew, behind the village of Ban Nai Sa).

A few hundred metres after the village of Ban Nai Sa, heading north on National Highway 4033 and next to the entrance sign for the "Wangtarntip Garden Resort", a watercourse, on average 8 metres (26 feet) wide, crosses beneath the road. Large groups of various types of *Cryptocoryne* cover large areas of the river bottom. The banks are predominantly overgrown with dense vegetation. The flowing water is slightly cloudy. *Betta simplex* also lives in this biotope, only its coloration is here somewhat different in comparison to that at Tham Sra Kaew. The *Betta simplex* from this site are more brightly coloured and exhibit a bolder blue and brown coloration. On the basis of observations to date they are apparently somewhat smaller in their adult total length.

Coordinates: 08° 09.04 N   098° 47.67 E (LINKE, H.)

("Aquatic plant stream", a few hundred metres after the village of Ban Nai Sa in a northerly direction.)

Biotope data:

Water parameters measured in the watercourse a few metres from the main lake of Tham Sra Kaew (research in 1990: LINKE, H.):

pH: 7.0
Conductivity: 530 μS/cm
Water temperature: 26.9 °C (80.5 °F).
General hardness: 11 °dGH
Carbonate hardness: 11 °dKH

The following water parameters were measured in the main lake Tham Sra Kaew (research in February 2005: LINKE, H & LINKE, M.):

pH 7.0
Conductivity: 541 μS/cm
Water temperature: 27 °C (80.5 °F)
General hardness: 12 °dGH
Carbonate hardness: 7 °dKH

Water parameters in the "aquatic plant stream" a few hundred metres after the village of Ban Nai Sa in a northerly direction (research in February 2005: LINKE, H. & LINKE, M.):

pH: 8.3
Conductivity: 550 μS/cm
Water temperature: 27 °C (80.5 °F)
General hardness: 14 °dGH
Carbonate hardness: 11 °dKH

Reproduction:
Mouthbrooder.

The surrounding area is very hilly, a picturesque karst landscape characterised by white limestone rocks, sometimes covered in plants

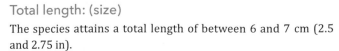

Total length: (size)

The species attains a total length of between 6 and 7 cm (2.5 and 2.75 in).

Remarks: (differences from other species of the genus)

In full coloration male *Betta simplex* exhibit a light red-brown body coloration, which also extends onto the fins. The caudal fin has a dark border followed by a delicate light outer edging. The anal fin is similarly coloured, only here the dark border extends further towards the middle and then becomes light grey. The other half of the fin, next to the body, is then light red-brown in

colour. The anal fin has a narrow white edging. The caudal fin is rounded to straight; the posterior part of the anal fin, likewise that of the dorsal fin, is also rounded. The lower head and the breast area up to the centre of the body are a striking light green. Males have a larger head compared to females.

References:

- NEUGEBAUER, N. 2005: Neue Fundorte vom Krabi-Kampffisch *Betta simplex*,1994. Aquarium life, 2 : 48-51.
- DIEKE, H. 2005: Nicht immer simpel - Der maulbrütende Krabi-Kampffisch *Betta simplex*. Aquarium life, 2: 54-57.

Biotop of *Betta simplex* in northwest of Krabi in South Thailand, spring (small lake) of Tham Sra Kaew and Nine Ponds, behind Ban Nai Sra village

Betta smaragdina LADIGES, 1972

Explanation of the species name:

Latin smaragdinus (feminine smaragdina) = "emerald-like", relating to the coloration of the body and fins.

English name:

Emerald Betta, Blue Betta

Original description:

LADIGES 1972: *Betta smaragdina* nov. spec. Aquar. Terrar., 25 (6): 190-191, Figs.

Systematics:

Betta splendens group

Natural distribution:

The origin of the specimen used for the description is given as Korat/Thailand.

The natural habitat lies in the north-east of Thailand, at the village of Nong Khai on the River Mekong, which forms the border with Laos, close to the capital, Vientiane.

Because the datum of Korat/Thailand for this species led to incorrect interpretation of the locality data, hitherto reports on the precise habitat have been misleading. D. SCHALLER has himself corrected his own data. It is not the region around the town of Nakhon Ratchasima, which was formerly known as Korat, that is the natural habitat of *Betta smaragdina*, but the town of Nong Khai (Nong Chai) on the River Mekong, on the Korat plateau in the north, 305 kilometres north of Nakon Ratschasima (Korat). It was in Nong Khai that in 1970 D. SCHALLER collected the types subsequently determined in 1972 by W. LADIGES.

Biotope data:

Betta smaragdina lives in the water accumulated in the rice fields, in ditches, in small waterholes, and in swamp regions, often in the mud and among dense vegetation. Sometimes also in very small residual pools. At the beginning of the rainy season, when the water level rises and the swamps and residual waters are refilled with water, or when the rice fields are flooded, the fishes return to these larger areas of water and interrupt their otherwise often solitary existence. The water parameters are then predominantly in the neutral zone (pH around 7) and the water has only a low mineral content.

Reproduction:

Bubblenest spawner.

Total length: (size)

Around 6 cm (2.5 in).

Remarks: (differences from other species of the genus)

The males have longer fins compared to females. Both sexes

Betta smaragdina ♂

Displaying male *Betta smaragdina*

are coloured an intense emerald green. During courtship and spawning the female exhibits broad light crossbands on the body. There is only very faint red on the fins, or none at all.

In the region of Nong Khai, Udont Thani, and Khon Kaen *Betta smaragdina* is also used for fish fights. This contradicts the entirely peaceful behaviour of this species generally reported to date. According to the Thai fish-fight organisers in Nong Khai, the most aggressive fishes are caught in the Udon Thani region 53 km further south. Nowadays it is known that the majority of these individuals are no different in their aggressiveness and willingness to fight than the cultivated forms of the wild *Betta splendens* hitherto used elsewhere.

This *Betta smaragdina* from Udon Thani was recently designated *Betta* cf. *smaragdina* 2 on the basis of DNA studies, ie as a variety that differs from the typical form. In addition there is a second form, in this case termed *Betta* cf. *smaragdina* 1, which is distributed around 250 km further south in the area of Nakon Ratschasima (Korat) and to the south of there. These fishes also exhibit red on the membranes of the caudal and anal fins. Hence they are regarded as a sibling taxon to *Betta smaragdina*.

Natural distribution:
One southerly location for *Betta* cf. *smaragdina* 1 lies in Lake Lamchae near Ban Mak south of Nakon Ratschasima. The Lamchae Talee is a large hydro-electric lake. It lies in the province of Korat, only a few kilometres south-west of the town of Khon Buri in the south-east of Thailand. The lake is fed by the rivers Tasa and Lamchae and drains south towards Prachin Buri. While one side of the lake is bordered by a hydro-electric dam, the other three sides have natural borders. The entire area

forms part of a national park. Nowdays *Betta* cf. *smaragdina* 1 predominates in the lake. They are apparently washed into the lake via the affluent rivers, and survive very well and also breed in the shallow shore zones, in and beneath the very copious *Eichhornia crassipes* growing there. Unfortunately the numerous floating plants in the bank zones make collecting very difficult. This can be achieved only using a large dragnet, which is extended beneath the floating plants and then hauled in along with the floating plants and numerous other fishes as well as *Betta* cf. *smaragdina* 1.

The water in the bank area at the village of Ban Maka was clear and almost still. Water parameters were measured as follows (research in January 2005: LINKE, H., LINKE, M., & PUMCHOOSRI, A.):

pH 7.0
Conductivity: 57 µS/cm
Water temperature: 27 °C (80.5 °F)
General hardness: 1 to 2 °dGH
Carbonate hardness: 1 to 2 °dKH

The research took place in the morning hours. The radiation of the sun was already very strong at this time, so that higher water temperatures are to be expected in the afternoon.

Coordinates: 014° 24.67 N   102° 13.66 E (LINKE, H.)
(Lake Lamchae at Ban Maka.)

Reproduction:
Bubblenest spawner.

Total length: (size)
Around 6 cm (2.5 in).

Displaying male *Betta smaragdina*, "Copper" cultivated form

Displaying male *Betta smaragdina*, "Copper" cultivated form

Spawning and brood-care stages of *Betta smaragdina*

The Lamchae Talee is a large reservoir in eastern Thailand. It is home to numerous labyrinthfishes including *Betta smaragdina*. Fishing with large nets is very lucrative there

Remarks: (differences from other species of the genus)

The males have longer fins compared to females. Both sexes are coloured an intense emerald green. The caudal and anal fins also exhibit red on the membranes. During courtship and spawning the female exhibits broad light crossbands on the body.

References:

- ROLOFF, E. 1972: Pflege und Zucht von *Betta smaragdina* Ladiges. Aquar. Terrar., 25 (6): 192-193.

- SRIWATTANAROTHAI, N., PANIJPAN, B. *et al*. 2010: Molecular and morphological evidence

- supports the species status of the Mahachai fighter *Betta* sp. Mahachai and reveals new species of *Betta* from Thailand.

- Journal of Fish Biology. The Fisheries Society of the British Isles.

Water-filled ditch in a rice field in the north-east of Thailand, habitat of *Betta smaragdina*

Residual water at the edge of a rice field

Young rice plants in the fields of north-east Thailand

Fishes from the rice fields after capture

A splendid male *Betta smaragdina* immediately after capture

Explanation of the unofficial name:

The unofficial name relates to the distribution region.

Synonyms:

Betta edithae

Natural distribution:

This previously "unrecognised" *Betta* species is found in the lower area of the Sungai Kapuas. The precise distribution region has not been positively documented so far. The species purportedly lives along the road from Pontianak to Sanggau, the road from Anjungan to Penjuh, and in a small river with a large waterfall that crosses the road near Airplaik, 25 km from Mandor in the direction of Sanggau; all West Kalimantan (Kalimantan Barat), Borneo, Indonesia.

Biotope data:

As far as is known at present, the species lives in the calm zones of various flowing watercourses containing clear, neutral to slightly acid water.

Reproduction:

Mouthbrooder.

Total length: (size)

Around 6 cm (2.5 in).

Remarks: (differences from other species of the genus)

Hitherto this species has often been regarded as *Betta edithae*. The species *Betta edithae* exhibits three parallel longitudinal stripes, the central stripe being somewhat broader and ending with a break followed by a round spot. The species *Betta* sp. Airplaik likewise has three longitudinal stripes, but these are of equal width and the central and lower stripes run together to merge on the tail. It is, however, possible that this fish is merely a local variant with different markings. Hence additional material with precise locality details will be important for further research.

Both species are problem-free in their maintenance and breeding.

Betta sp. Airplaik ♂

The river is larger after the rocky weir

The big weir and waterfall near Airplaik

Norbert Neugebauer fishing in a side channel near Airplaik

Explanation of the unofficial name:

Relates to the collecting site in the Bukit Lawang area.

Natural distribution:

These fishes were imported by Frank GRAMS in 1995 from the Bukit Lawang region, island of Sumatra, Indonesia.

Biotope data:

At present (2009) no details are available regarding the natural habitat.

Reproduction:

Mouthbrooder.

Total length: (size)

Around 9 cm (3.5 in).

Remarks: (differences from other species of the genus)

The fishes exhibit parallels with *Betta pugnax* in their markings and coloration. The species may be identical with the species *Betta* cf. *pugnax* mentioned by TAN & NG, 2005, from the island of Nias, Utara Province, in the west of the island of Sumatra.

Betta sp. Bukit Lawang ♂

Betta sp. Bukit Lawang ♂ with striped pattern

Explanation of the unofficial name:
The name relates to the distribution region.

Natural distribution:
The only specimen known to date was caught in a tributary of the Danau Calak. The site was a large lake-like widening in the river, up to 5 km long with several small affluents, and very rich in fishes. This "lake" lies in the Air Musi river system, close to the village of Bailang, west of Sekayu, north-west of Palembang, Sumatra Selatan Province, Indonesia.

Biotope data:
The water parameters in the "lake" were: water temperature 31.1°C (88 °F), pH 6.3, and conductivity 25 µS/cm. And in the tributary, pH 5.12 and conductivity 12 µS/cm. The water was slightly cloudy and yellowish in colour. (Research in September 2008; LINKE, H.)

Coordinates: 02°57′33 S   103°59′49 E (LINKE, H.)

Reproduction:
Mouthbrooder.

Total length: (size)
Around 7 cm (2.75 in).

Remarks: (differences from other species of the genus)
The only specimen known to date is probably a female. The species lives syntopic with a previously unknown *Parosphromenus* species, inter alia.

Betta sp. Danau Calak ♂

Betta sp. Duc Hua

Explanation of the unofficial name:

The name relates to the distribution region close to the village of Duc Hoa.

Natural distribution:

The collecting site for this *Betta species* lies in a swamp region and along an irrigation canal in an area of scrub and grassland, close to the village of Duc Hoa, to the north-east of Ho Chi Minh City in Vietnam and around 45 km from the border with Cambodia.

Coordinates: 10°54'01 N   106°29'20 E (LINKE, H.)

Biotope data:

The natural habitat is an area of scrubland with a swamp region, and is dominated by vigorous tall grass. The land is used as grazing for cattle and water buffalo. The following water parameters were recorded:

pH value: 3.52
Conductivity: 592 µS/cm
Water temperature: 29 °C (84 °F)
The water was clear and slightly brown in colour. (Research in February 2009: CHIANG, N. & LINKE, H.)

Reproduction:

Mouthbrooder.

Total length: (size)

Around 6 cm (2.5 in).

Remarks: (differences from other species of the genus)

The species *Betta* sp. Duc Hoa was very common in this area and lives syntopic with, inter alia, another previously unknown *Betta* species, very close to *Betta imbellis*. The population density of this species, *Betta* cf. *imbellis*, was very high at the site.

Betta sp. Duc Hoa is sometimes regarded as a variant of *Betta prima* occurring in Vietnam. The fishes from Vietnam are, however, distinguished by having three parallel longitudinal bands on the body and extending onto the caudal peduncle, with the central band ending in a spot on the caudal peduncle.

Betta sp. Duc Hua ♀

Betta sp. Duc Hua ♂

Betta sp. Duc Hua, ♂ in front, ♀ behind

Catching Betta sp. Duc Hua in a swamp

Betta sp. Duc Hua ♂

This is in fact also seen in *Betta prima*, but in *Betta* sp. Duc Hoa the lower longitudinal band, running from the lower part of the mouth to the tail, is more clearly defined, and there is a clear pattern of spots visible on the fin membranes of the caudal and anal fins, absent in *Betta prima*.

Betta sp. Jantur Gemuruh

Explanation of the unofficial name:

Relating to the collecting site in the Jantur Gemuruh area, close to the village of Melak.

Natural distribution:

These fishes were first recorded by DICKMANN, KNORR, & GRAMS in the vicinity of the village of Melak in the area of the Sungai Mahakam-Sungai Barong, west of the town of Samarinda in East Kalimantan (Kalimantan Timur), and imported alive.

Biotope data:

Betta sp. Jantur Gemuruh was caught in the bank region of a reservoir by the Jantur Gemuruh waterfall, on the way to the orchid farm, close to the village of Melak. The bank zones were shallow and covered with fallen leaves, and were described as overgrown with vegetation. The pH was between 5.5 and 6, the general hardness just below 4 °dGH, and the water temperature 26.8 °C (79.7 °F). The species lives in this area with two further mouthbrooding *Betta* species, one in the same biotope and termed *Betta* sp. *aff. balunga* by DICKMANN, and the other, termed *Betta* sp. aff. *unimaculata*, at a higher population density in a very shallow stream nearby, only a metre (40 in) wide. The species may be a local variant of *Betta ocellata*. (Research in 1996; DICKMANN, KNORR, & GRAMS)

Reproduction:

Mouthbrooder.

Total length: (size)

Around 9 cm (3.5 in).

Remarks: (differences from other species of the genus)

In its appearance *Betta* sp. Jantur Gemuruh exhibits parallels with *Betta pulchra* from western Malaysia.

Betta sp. Jantur Gemuruh ♂

Explanation of the unofficial name:

The species was first recorded in the Kubu region.

Natural distribution:

Betta sp. Kubu was first recorded by BAER, NEUGEBAUER, & LINKE in 1990, to the south of the town of Pangkalanbuun, just before the village of Kubu in the south-west of Central Kalimantan (Kalimantan Tengah). This was a specimen with a total length of around 5 cm (2 in), which was preserved in the field. Only when the species was imported alive in 1992 were these fishes observed in the aquarium.

In June 2009 we found *Betta* sp. Kubu in the region 14 km west of Kubu, around 40 km south of Pangkalanbun, in the Sungai Benipah (Sungai Nippa). (Research in 2009: LINKE, H. et al.) Since then it has become known that the species is distributed throughout the south-west region of Kalimantan.

Biotope data:

The habitat of this species was a small blackwater stream with clear, slightly flowing, very soft and very acid water. The general and carbonate hardness were both less than 1° (German), the pH was 5.0 and the electrical conductivity was measured as 14 µS at a water temperature of 23.7 °C (74.7 °F).

Reproduction:

Mouthbrooder.

Total length: (size)

Around 8 cm (3 in).

Remarks: (differences from other species of the genus)

This new species is characterised by a comparatively pointed head form and a large eye. The coloration is not very striking. In addition to soft brown shades and green coloration in the membranes of the fins, the species exhibits a round dark spot on the caudal peduncle. The central spine of the caudal fin is prolonged. The ventral fins are short.

Blackwater biotope near Kubu in south-west Kalimantan Tengah, habitat of *Betta* sp. Kubu

Betta sp. Kubu ♂

Betta sp. Kubu with striped pattern

Explanation of the unofficial name:

Relating to the distribution region in the border region between Kalimantan Barat and Sarawak.

Natural distribution:

According to details from the exporters, Siam Pet Fish Trading in Bangkok, Thailand, the species was found in and imported from the area of Lake Luar (Danau Luar) north of Selimbau in northern Kalimantan (Kalimantan Barat), close to the border with Sarawak (eastern Malaysia).

Reproduction:

Mouthbrooder.

Total length: (size)

The species attains a total length of 8 cm (3 in) when adult.

Remarks: (differences from other species of the genus)

The species exhibits parallels with *Betta stigmosa*, but apparently grows somewhat larger and has no pattern of dots on any of the unpaired fins, and males exhibit an antique green operculum and a striking underhead coloration. There are also parallels with *Betta* sp. Satun in terms of external appearance and coloration.

Betta sp. Lake Luar ♂

Betta sp. Langgam

Explanation of the unofficial name:
The name relates to the distribution region at Langgam in the province of Riau, Island of Sumatra, Indonesia.

English name:
Langgam Betta

Systematics:
Betta pugnax group

Natural distribution:
River and swamp region with streams, in a swampy area of scrub and woodland along the track at Langgam, broad sand track with ferry, 67 km west of the town of Kerincikiri (paper factory) in the province of Riau, central Sumatra, Indonesia.

Coordinates: 01°04.35 N   096°07.59 E

The area is characterised by clear-felling and oil-palm plantations as well as plantations of fast-growing trees for paper manufacture, with only a few undisturbed areas, including regions of scrub and woodland.

Biotope data:
These fishes live in blackwaters and were found in shallow bank zones overgrown with marginal vegetation and scrub, and in shallow wooded waters with gently flowing water. The water parameters were pH 5.25 and conductivity 7 µS/cm at 26.8 °C (80.24 °F). In January (high water period) the water was only slightly brown in colour with a variable (weak to strong) current. The visibility was around 100 cm (40 in). The observations and research took place during the rainy season (high water).

(Research in 2008: SIM, T., CHIANG, N., & LINKE, H.)

Reproduction:
Mouthbrooder.

Total length: (size)
Up to 9 cm (3.5 in).

Remarks: (differences from other species of the genus)
Unfortunately the research in January 2008 recorded only two males.

The species is very close to *Betta fusca*, but in comparison to that species have a smaller number of scales in a longitudinal series, and fewer rays in the dorsal and anal fins.

Betta sp. Langgam ♂

Blackwater biotope in the Langgam area, Riau Province, Sumatra

Sand road for transporting wood through the remaining bushland

Small rivers pass beneath the sand road

The bush and the rainforest are mostly flooded

The elevated position of the huts indicates the water level possible

Betta sp. Langgam after capture

Preliminary examination after capture

Explanation of the unofficial name:

As far as is known to date the species comes from East Kalimantan.

Natural distribution:

At present (2012) no precise details of distribution and the natural habitat are available.

Biotope data:

At present unknown.

Reproduction:

Mouthbrooder.

Total length: (size)

Around 7 cm (2.75 in).

Remarks: (differences from other species of the genus)

These fishes are similar to *Betta ocellata* and *Betta unimaculata* in coloration and appearance, but on the basis of observations to date remain smaller. Additional material is required to clarify identification.

Betta sp. East Kalimantan ♀

Underhead markings of *Betta* sp. East Kalimantan

Betta sp. East Kalimantan ♂

Betta sp. Satun

Explanation of the unofficial name:
Relating to the distribution region in the province of Satun in south-west Thailand.

English name:
Satun Fightingfish.

Systematics:
Betta pugnax group.

Natural distribution:
The natural distribution of *Betta* sp. Satun lies in the area of the province of Satun in the south-west of southern Thailand. K. KUBOTA, Siam Pet Fish Trading in Bangkok, was the first to record these fishes in 2003, in a watercourse at "Thole Bon", close to the National Park (pers. comm.).

Biotope data:
Narrow watercourse in a wooded depression at the approach to the Yao Koi waterfall on National Highway 4184 in the area west of the town of Satun in the direction of the border with Malaysia. The following water parameters have been determined:

pH value: 6.85
Conductivity: 47 µS/cm
Water temperature: 23.8°C (74.8 °F).

The water was flowing and only slightly brown in colour (clearwater). The visibility underwater was around 80 cm (32 in). The on average 6 metres (20 ft) wide watercourse was everywhere shaded by brush and trees. A previously unknown mouthbrooding *Betta* species, not yet scientifically determined and here termed *Betta* sp. Yao Koi, was also recorded during the research in this biotope. (Research in February 2009: KUBOTA, K., PUMCHOOSRI, A. & LINKE, H.)

Coordinates: 06°45'25 N   100°09'15 E (LINKE, H.)

A further collecting site was a narrow watercourse in a valley (cultivated land with buildings) that crosses the road from Khuan Ka Long to Ma Nang, a branch of National Highway 4137. This was a narrow watercourse, on average 20 cm (8 in) deep. Here too the water was slightly brown in colour. (Research in

Betta sp. Satun ♀

Betta sp. Satun ♂

Underhead coloration of *Betta* sp. Satun

February 2009: KUBOTA, K., PUMCHOOSRI, A. & LINKE, H.)

The following water parameters were recorded:

pH value: 6.64
Conductivity: 30 μS/cm
Water temperature: 26.2 °C (79.2 °F).
The water was slightly flowing and clear. The banks were thickly lined with emerse vegetation that almost completely shaded the water.

Coordinates: 06°51'30 N   100°04'18 E (LINKE, H.)

Reproduction:
Mouthbrooder.

The species does not predate on its offspring and hence numerous youngsters can be grown on with the parent fishes in well-planted aquaria. It is beneficial to maintain the species in the ratio of two males per female.

Total length: (size)
Adults of this species attain a total length of around 7 cm (2.75 in).

Remarks: (differences from other species of the genus)
The species exhibits many parallels with *Betta stigmosa*, described in 2005 from near the Sekayu waterfall (Air Terjun Sekayu) in the area south-west of Terengganu in western Malaysia. Another comparable species, termed *Betta* sp. Lake Luar, comes from the area of blackwater lakes in the north of

Kalimantan, in the north-west of Kalimantan Barat. Like *Betta*sp. Satun, this species arrived in the trade via Siam Pet Fish Trading Bangkok (Katsuma KUBOTA).

Betta sp. Satun differs from *Betta stigmosa* mainly by the absence of the pattern of dots on the membranes of the caudal and anal fins.

Betta sp. Satun exhibits a red-brown iris and, in males, bright antique green opercula whose colour extends over the entire under-head region and for long way upwards. The striking green coloration also extends all over the body and is usually underlain with red-brown on the lower part. The fins exhibit the same coloration. Females exhibit similar but less bold coloration, and also a striking underhead pattern.

There are also many parallels with *Betta kuehnei*, but the currently known natural habitats of all the species discussed here are so widely separated that without exception they are probably distinct species. Nevertheless conspecificity between the species *Betta kuehnei* and *Betta* sp. Lake Luar cannot be ruled out.

References:

- LINKE, H., 2007: Betta News spezial. European Anabantoid Club/ A K Labyrinthfische im VDA: 3-30.

- Linke, H., 2008: Türkisgrün Pracht-*Betta* aus Süd-Thailand. Betta News, Journal of the European Anabantoid Club/ Arbeitskreis Labyrinthfische, 2008 (4): 13-14.

- KÜHNE, J., 2008: Den Hartwasserfischen der Gattung *Betta* auf der Spur. Zoologischer Central Anzeiger, 2008 (9):

Spawning scenes and brood-care stages of the mouthbrooding *Betta* sp. Satun

Explanation of the unofficial name:

The species is found in the Sematan area in western Sarawak.

Natural distribution:

Phil DICKMANN discovered the species in the region south of the town of Sematan, north-west of Kuching in Sarawak (eastern Malaysia), and imported a small number of specimens.

The species lives in small, clear, flowing watercourses.

Reproduction:

Mouthbrooder.

Total length: (size)

Adult males of this species attain a total length of up to 14 cm (5.5 in).

Remarks: (differences from other species of the genus)

The males of this species have a slender head and exhibit green scale coloration all over the body, except that on the lower body this green coloration changes to dark brown to black on the lowest two to three series of scales. The membranes of the dorsal fin are without markings, those of the anal fin sometimes exhibit just a few small, dark dots, and those of the caudal fin a pattern of double vertical streaks that produce the effect of a vertical stripe pattern. The ventral fins extend to the middle of the body when folded.

Head close-up of *Betta* sp. Sematan

Betta sp. Sematan

Explanation of the unofficial name:

The name relates to the purported distribution region.

Systematics:

Betta coccina group

Natural distribution:

The species was first exported in 2009 by the company KURNIA AQUARIUM in Palangkaraya, Kalimantan Tengah. In appearance the species is very similar to *Betta foerschi*, but remains smaller. According to the exporter's data, these fishes are found in the Sengalang area, on the road from Palangkaraya to Kualakurun, around 15 km north of the Sungai Kahayan at the edge of the town of Palangkaraya. However, collecting attempts in the area have failed to produce any fishes of this species despite considerable effort, and hence we cannot at present confirm the distributional data of the exporter.

Biotope data:

The purported biotopes in the Sengalang area are blackwater swamp regions along the road to Bukitrawi-Kualakurun, and have a pH value of 3.73 and a conductivity of 31 µS at a water temperature of 29.9 °C (85.8 °F). (Research in 2009: LINKE, H. et al.)

Coordinates: 02°05'08 S   113°57'16 E (LINKE, H.)

Reproduction:

Bubblenest spawner.

Total length: (size)

Adult size is purportedly around 5 cm (2 in).

Remarks: (differences from other species of the genus)

The species is not identical to *Betta uberis* as sometimes stated. It is true that there are many similarities in coloration, but there are also noteworthy morphological differences. These include the notably higher fin-ray count in the dorsal fin in *Betta uberis*, not shared by *Betta* sp. Sengalang.

References:

- LINKE, H. 2009: New *Betta* species from Kalimantan. Betta News, Journal of the European Anabantoid Club/A KL, 2010 (1): 26.

Betta sp. Sengalang ♂

Betta sp. Sengalang ♂

Betta sp. Sengalang ♀

Betta sp. Sengalang ♀

Explanation of the unofficial name:

The name Sungai Dareh relates to the distribution region.

Natural distribution:

According to information from Thomas G. K. SIM at Sindo Aquarium in Jambi, this *Betta* species is found in the area of the Sungai (River) Dareh in the south-east of the province of Padang, island of Sumatra, Indonesia.

The Sungai (River) Dareh flows through a hilly to mountainous landscape at the eastern edge of the neighbouring province of Padang to the west of Jambi. *Betta* sp. Sungai Dareh is found along the Sungai Dareh south of the village of the same name, where it lives in small, densely vegetated side valleys in both larger and very small flowing watercourses that empty into the Sungai Dareh after only a few hundred metres. These watercourses are habitats for numerous well-known aquarium fishes and as yet unknown species, including a second *Betta* species that we were able to identify as *Betta fusca*.

Biotope data:

Both species were found in a narrow watercourse with clear, slightly brown-stained water with little current. The margins were overgrown with emerse plants and the watercourse lay in the half-dark of the woodland. Only rarely did sunlight reach the water's surface for a short time. The water temperature measured 26 °C (79 °F). The bottom consisted of light sand and dead leaves. There were lots of dead trees and branches lying in and around the margins.

Reproduction:

Mouthbrooder.

Total length: (size)

Apparently up to 8 cm (3 in).

Remarks: (differences from other species of the genus)

Unfortunately it would appear that so far only a single male of *Betta* sp. Sungai Dareh has been exported to Europe.

Head close-up of *Betta* sp. Sungai Dareh

Betta sp. Sungai Dareh Male

Betta sp. Tana Merah

Explanation of the unofficial name:
Relating to the Tana Merah waterfall, the collecting site for the species.

Natural distribution:
The species was collected by DICKMANN, KNORR & GRAMS 1996 at the waterfall, in a stream several metres wide, upstream of a large pool used as a bathing place and with a large car park. The collecting site lies only a little north of the town of Samarinda in East Kalimantan (Kalimantan Timur).

Biotope data:
The fishes were to be found predominantly in the bank region, beneath scrub and roots spreading into the water, as well as in large residual pools that usually contained just one adult specimen and several juveniles.

Reproduction:
Mouthbrooder.

Total length: (size)
Around 10 cm (4 in).

Remarks: (differences from other species of the genus)
The species undoubtedly belongs to the *Betta unimaculata* group. The eight dark crossbands on the body are species-typical. In addition, the large round caudal fin is a striking feature. The ventral fins are short. The body colour is light brown to orange. The lower part of the head is light in colour. The fishes exhibit a dark mouth coloration and a dark streak on the lower snout.

The species exhibits parallels with *Betta patoti*, but the number of crossbands on the body is smaller. In *Betta patoti* there are 10 to 12, in *Betta* sp. Tana Merah only eight.

Betta sp. Tana Merah ♂

Betta sp. Tideng Pale

Explanation of the unofficial name:

The name Tideng Pale relates to the collecting site.

Natural distribution:

This comparatively large species was caught in the area of the Sungei Mentarang, east of the village of Malinau, in the north of East Kalimantan (Kalimantan Timur). The species was first recorded by DICKMANN, KNORR and GRAMS.

Biotope data:

These fishes live between Malinau and Tarakan in small, deep black, moorland ditches filled with water. According to the collection data the water was almost stagnant, with a pH value of 4.5-5.0 and a general hardness of less than 3 °dGH. The water temperature measured 29 °C (84 °F). (Research in April-May 1996; DICKMANN, KNORR & GRAMS.)

Reproduction:

Mouthbrooder.

Total length: (size)

Around 10 cm (4 in).

Remarks: (differences from other species of the genus)

Males exhibit an anal fin that is much prolonged and pointed posteriorly, and a lancet-shaped caudal fin. In fright coloration there is a broad, dark longitudinal band on the upper part of the lower half of the body. This is composed of three narrow bands. The ventral fins are short and when folded extend to somewhat past the centre of the anal fin. The lower part of the head is a pale golden colour and exhibits dark spots.

These fishes are peaceful. On the basis of appearance the species apparently belongs to the *Betta akarensis* group.

Betta sp. Tideng Pale ♂

Betta sp. Waeng

Explanation of the unofficial name:
Relating to the distribution in the region around the village of Waeng.

English name:
Waeng Betta

Systematics:
Betta pugnax group

Natural distribution:
According to information from Siam Pet Fish Trading Bangkok, the fishes were caught in the Waeng region, around 25 km south-west of the town of Sungai Kolok in the south of the province of Narathiwat in the east of southern Thailand, near to the border with Malaysia. The fishes thus live in the distribution region of *Betta apollon*, *Betta pallida*, and *Betta pi*. This is, astonishingly, the greatest concentration of mouthbrooding *Betta* species in Thailand, and confined to a comparatively small area, naturally bounded to the east by the sea. With one exception all the species are endemic there. A very interesting region from an ichthyological viewpoint. Unfortunately further field observations and research are currently impossible for political reasons.

Reproduction:
Mouthbrooder.

Total length: (size)
Around 7.5 cm (3 in).

Remarks: (differences from other species of the genus)
Males are characterised by a bright antique green operculum and body-scale margins. This impressive coloration is also sometimes seen in the unpaired fins. The anal fin is bordered with bluish to faint black. The body colour is grey to reddish brown. The caudal-fin membranes exhibit an irregular spot pattern.

Betta sp. Waeng ♂

Betta sp. Waeng ♀

Underhead pattern of Betta sp. Waeng ♀

Betta sp. Yao Koi

Explanation of the unofficial name:
The name relates to the distribution region.

Natural distribution:
The species has to date been recorded in the region of the Yao Koi waterfall in the Thale Ban National Park and in the adjacent watercourses and lakes by National Highway 4184 and north thereof in the province of Satun in south-west Thailand.

Biotope data:
Narrow watercourse in a wooded valley at the approach to the Yao Koi waterfall on National Highway 4184 in the area west of the town of Satun in the direction of the border with Malaysia. The following water parameters were determined:

pH value: 6.85
Conductivity: 47 µS/cm
Water temperature: 23.8 °C (75 °F).

The water was flowing and only slightly brown in colour (clearwater). The visibility underwater was around 80 cm (32 in). The on average 6 metres (20 ft) wide watercourse was everywhere shaded by brush and trees. (Research in February 2009: KUBOTA, K., PUMCHOOSRI, A., & LINKE, H.)

Coordinates: 06°45'25 N   100°09'15 E (LINKE, H.)

Reproduction:
Mouthbrooder.

Total length: (size)
Around 7 cm (2.75 in).

Remarks: (differences from other species of the genus)
In terms of coloration the species exhibits many parallels to *Betta pugnax from* Penang, Malaysia, but the fishes from south-west Thailand remain smaller, and on the basis of observations to date are full-grown at around 7 cm (2.75 in).

Watercourse in the Thale Ban National Park

Betta sp. Yao Koi ♀

Betta sp. Yao Koi, ♀ above, ♂ below

Road in the Yao Koi area

Inspecting *Betta* sp. Yao Koi after capture

Biotope in the Thale Ban National Park, Satun Province, south Thailand, habitat of *Betta* sp. Yao Koi

Biotope of *Betta* sp. Yao Koi

Biotope in the Thale Ban National Park, Satun Province, south Thailand, habitat of *Betta* sp. Yao Koi

Different kinds of *Betta* sp.

Betta sp. aff. coccina

Betta sp. Riau Red

Photo by Nathan Chiang

Different kinds of *Betta* sp.

Betta sp. *aff. obscura* Kalimantan Tengah

Betta sp. South Thailand

Betta sp. Jenga Danum Kalimantan Timur

Different kinds of *Betta* sp.

Betta sp.

Betta sp. Antuta from north Kalimantan Timur

Betta sp. Tarantang west Kalimantan Tengah

Different kinds of *Betta* sp.

Betta sp. Melak area Kalimantan Timur

Betta sp. middle East Kalimantan Timur

Betta sp. Skrang Sarawak

Different kinds of *Betta* sp.

Betta sp. Kalimantan Timur

Betta sp. Barbugus east Kalimantan Tengah

Betta spilotogena NG & KOTTELAT, 1994

Explanation of the species name:
Latinised Greek *spilotogena* = "with spotted cheeks", referring to the dark spots on the head (and lower body).

Original description:
KOTTELAT, M. & NG, P.K.L., 1994: Revision of the *Betta waseri* species group (Teleostei: Belontiidae) Raffles Bull. Zool. 42, (3): 606-607.

Systematics:
Betta waseri group

Natural distribution:
The holotype was caught in a small watercourse in the north of the island of Bintan. Additional fishes of this species have been recorded in "freshwater swamp, Tanjong Bintan, northeastern Pulau Bintan".

Coordinates:
01°09'34.6" N 104°31'57" E (SIVASOTHI, N. et al., 1993)

Biotope data:
Betta spilotogena has been recorded in the inundation zones, swamp regions, and small watercourses in the north of Pulau (= island) Bintan, as well as at other sites, on the road to Uban Pinang, kilometre marker 56, near Gunung Demit.

Reproduction:
Mouthbrooder.

Total length: (size)
Up to 12 cm (4.75 in).

Remarks: (differences from other species of the genus)
The species is one of the peaceful fightingfishes. *Betta spilotogena* is usually light brown in colour and has no markings on the fin membranes. The underhead marking is an important character for differentiation from the other species of the group. The species also differs in the pattern on the head, body, and fins as well as by having a different head shape. The pattern of dots on the posterior operculum is noteworthy.

Underhead pattern of *Betta spilotogena* ♂

Betta spilotogena ♂

Explanation of the species name:
splendens = "brilliant", "splendid".

English name:
Siamese Fightingfish, Siamese Fighter

Synonyms:
Betta pugnax. var. *trifasciata*
Betta pugnax. var. *rubra*

Original description:
REGAN, C.T., 1910: The Asiatic fishes of the family Anabantidae, Proc. Zool. Soc.of London, page 767-787.

Systematics:
Betta splendens group

Natural distribution:

In the scientific description by REGAN in 1910 the distribution of this fish was given as the Menam River in Siam, nowadays Thailand, and Pinang in the Malay Peninsula, now western Malaysia. Nowadays we know that the reference to Pinang must relate to the distribution of *Betta imbellis* and not that of *Betta splendens*.

For this reason a neotype has been scientifically designated from a locality in Thailand.

But when speaking of the home of this fish its distribution cannot be limited to Thailand, at least not in the case of the "fighters". Because in the past there have been close, albeit not always friendly, links between Thailand and the neighbouring country of Cambodia, it is not possible to exclude a historic connection with the Khmer in the matter of *Betta splendens*

as well. However, at present there is no direct evidence of any natural occurrence in Cambodia. The *Betta splendens*-like fishes recorded from the land of the Khmer may be a naturalised cultivated form of *Betta splendens*, or similar fishes that exhibit the typical coloration of *Betta splendens* except that they lack the species-typical green-blue colour to the gill-cover seen in *Betta imbellis*. The same applies to the fishes recorded from southern Vietnam.

According to my own frequent studies in Vietnam, Cambodia, Myanmar, and western Malaysia, the natural occurrence of *Betta splendens* is restricted to Thailand. They are found from Chiang Mai/Lampang in the north to Surat Thani/Phangnga in the south. Further to the south the preferred habitat of *Betta splendens* is instead occupied by *Betta imbellis*. In the west the distribution of *Betta splendens* ends at the foot of the mountains that also form the boundary with Myanmar, and in the east the habitat is occupied by *Betta smaragdina* to the east of a north-south line from Pak Chong to Nakhon Ratchasima (Korat). It is no longer possible to ascertain how far this is its natural distribution or whether introduced "fighters" are also involved. The distribution region cited here also includes small areas in which *Betta splendens* has not been recorded to date.

Biotope data:

Natural habitats include rice fields, the irrigation ditches supplying them, areas of residual water, and swampy areas. In all cases the water is standing or has only a very slight current, and there is typically a vigorous growth of sheltering plants,

Betta splendens Crowntail

Betta splendens Crowntail, male

Betta splendens Crowntail, male

Betta splendens Crowntail, male

Betta splendens Crowntail, male

Betta splendens Crowntail, male

Betta splendens, male

Betta splendens Crowntail, male

usually also growing emerse. There are also large areas of water covered in floating plants such as *Eichhornia* and *Pistia* where the fishes can find shelter from predators. This floating vegetation often also acts as a "protective coating" that prevents the water from over-heating. Rice fields are sometimes a deadly habitat, as they dry up in the course of the development of the rice and often there is no possibility of the fishes escaping to areas where there is still water.

Betta splendens always lives in areas of shallow water, sometimes barely 3 cm (1.25 in) deep, filled with a vigorous growth of plants, where the temperature of the water can quickly rise to 35 °C (95 °F) in the rays of the sun. Contrary to popular opinion, *Betta splendens* usually lives in neutral to slightly acid, soft water in the wild. All the natural waters contain a smaller or larger percentage of humic substances, and although these natural waters sometimes appear heavily polluted, they often have a lower germ count than water in the aquarium does on occasion. As regards water parameters, the hydrogen-ion concentration (pH) is usually in the range 6.0 to 7.0. Sometimes, however, the pH value is considerably lower. A pH value above 7 is rare and not to be recommended. The water is generally mineral-poor and the water temperature lies between 22 and 33 °C (71.5 and 91.5 °F) depending on the latitude and the time of year. In the northern parts of Thailand, at higher altitudes, the water temperature in the winter months can drop to as low as 10 °C (50 °F) at night, but rises again by day to 22 to 25 °C (71.5 to 77 °F).

The correct maintenance of these labyrinthfishes should thus always take place in water that is as low in germs as possible. The water parameters in the natural habitats include a pH value of 6.0 to 7.0 (the lower value of 6.0 will suit the fishes better and permit a lower germ density) and an electrical conductivity of between 50 and 200 µS/cm, i.e. relatively mineral-poor and hence soft water. For optimal maintenance *Betta splendens* should be kept in biologically clean, soft, slightly acid water with a water temperature varying between 23 and 27 °C (73.5 and 80.5 °F). Higher water temperatures for long periods will shorten the lifespan. Given optimal maintenance, the species has an average life expectancy of from 14 to 18 months.

Reproduction:

Bubblenest spawner

Betta splendens is one of the bubblenest-building species – in other words the male takes in atmospheric air and then coats this with an oral secretion to create small bubbles of air, which en masse produce a foam-like material which he uses to construct a bubblenest at the water's surface. The pair then embrace and spawn beneath this nest. This embrace involves the male wrapping his body around the female and at the same time turning her onto her back. During this "union" of their bodies the female releases the eggs which are immediately fertilised by the male. The eggs, which then start to sink, are next collected up – generally by both partners – and then carried to the bubblenest. Because the oral mucus used to construct the bubblenest is inimical to bacteria, the eggs are able to develop well in an environment low in germs.

This mating procedure takes place over a period of on average 60 to 90 minutes. Thereafter the male alone assumes the care of the brood. In a suitably large aquarium the female will guard the breeding territory against predators. In small tanks, however, the female should be removed and the brood care left to the male alone.

The development period of the brood to free-swimming is around four days. After they become free-swimming the fry should be fed the smallest possible pond foods, such as infusorians (slipper animalcules or rotifers). Usually, however, the newly free-swimming fry will also immediately take tiny, freshly-hatched *Artemia* nauplii. But they must be the nauplii of a small *Artemia* species. These nauplii are a very good first food, but they should always be used during the first 12 hours after they hatch as thereafter their nutritional value decreases. Immediately after hatching *Artemia* nauplii still have a nutrient-rich yolk sac of their own. This yolk sac contains important nutrients, including for young fry of *Betta splendens*. If you wait longer and use the nauplii over a period of two to three days, you will be feeding only the bodies of the nauplii, the nutrient-poor "shells" that remain, as they will have consumed the yolk sac themselves. Unless the nauplii are themselves fed they will no longer have any nutritional value and hence are of no value to the tiny fish fry. For this reason you should always use two *Artemia* cultures, started at an interval of 24 hours, for the optimal rearing of fish fry. Assuming the nauplii require 36 hours to hatch completely, I have found that if I set the cysts (eggs) to hatch in the evening then the nauplii will be ready to use in the morning and evening of the day after the next. In the interim I start the second culture 24 hours after the first. This ensures controlled, nutritious feeding for the first few days.

After the fry become free-swimming the male should also be removed from the breeding tank, although the majority of fathers do not normally attack their offspring, regarding them as food.

Given good feeding and constant very good water quality then the offspring will grow very well and the sexes will be distinguishable after around 8 weeks. By this time they may well have already attained a total length of 4 cm (1.5 in). In a large maintenance tank with plants, then provided the population density isn't too great both sexes (or even just the males) can be kept together until they reach an age of 5 to 6 months, without battles and injuries resulting. However, a prerequisite for this is that the fishes are maintained without disturbance. No fishes should be taken out and then put back again. In addition, too much "tinkering" in the aquarium can destroy the "carefully cultivated peace" and lead to outbreaks of aggression, putting at risk the possibility of the fishes continuing to live together.

Total length: (size)
Around 6.5 cm (2.5 in)

Remarks: (differences from other species of the genus)
Once upon a time the Siamese Fightingfish, *Betta splendens*, was just a small fish of little note in the tropical rice fields. Before long, however, humans started using its aggressive behaviour towards conspecifics to stage competitions and maintaining

Betta splendens Crowntail, male

Betta splendens Crowntail, pair

Betta splendens Crowntail, male

Betta splendens Crowntail, male

it in captivity for its bright colours. In its native Thailand it rapidly developed a career as the Pla Kat, the biting and tearing fish (pla (Thai) = fish and kat (Thai) = bite, tear). And as so often happens its natural aggressiveness wasn't enough for us humans and fishes with a greater degree of belligerence were bred. But people were also enchanted by its varied colours. Another reason for targeted breeding. Then came the idea of altering the size of the fins and hence their form. Nowadays we have not only the original short-finned wild form of *Betta splendens* but also numerous variants with different fin forms and above all unusual colours.

Betta splendens was scientifically described in an issue of the Proc. Zool. Soc. London dated 1909, but whose publication was

Betta splendens Doubletail

Betta splendens Doubletail, male

Betta splendens Doubletail, male

Betta splendens Doubletail, pair

Betta splendens Doubletail, female

Betta splendens Doubletail, male

Betta splendens Doubletail, male

Betta splendens Doubletail, male

Betta splendens Doubletail, pair

Betta splendens Doubletail, male

cultivated forms than in the Pla Kat Lug Tun, the fishes of the fields, and even more in the Pla Kat Lug Moh, the fishes of the pot, the short-finned *Betta splendens* bred for fighting.

The reproduction of the latter is no different to that of the wild-caught fishes or of the colourful long-finned forms. The best and most aggressive fighters are used for breeding. Their offspring are grown on as juveniles in wide (around 2 metres (80 in) in diameter), up to 1 m (40 in) deep ceramic pots. The surface of the water is covered with large *Eichhornia crassipes* and *Betta splendens*. These provide good hiding-places and in addition improve the water quality by virtue of their high nutrient requirement. Given optimal maintenance and feeding, the Pla Kat grow on into good fighters in 6 to 7 months in these ceramic pots. But until they are separated they live together peacefully apart from minor squabbles. Without battles and without fin damage or other injuries. Only after they are separated does their aggressiveness increase. After separation, or even a short period kept singly, they can no longer be put together.

The Pla Kat are next tested for their suitability as fighters and the selected fishes are kept in water rich in humic substances for around two weeks in preparation for a fighting event;

delayed until the spring of 1910, which is the reason why many authors regard the year of the description as 1910. But the question of 1909 or 1910 has not damaged the popularity of the Pla Kat. Far from it. Today *Betta splendens* is one of the best-known and most popular aquarium fishes. But the people of Thailand are less interested in the bright colours of the various

Betta splendens Halfmoon

Betta splendens Halfmoon, male

Betta splendens Halfmoon, male

Betta splendens Halfmoon, pair

Betta splendens Halfmoon, male

Betta splendens Halfmoon, male

Betta splendens Halfmoon, male

Betta splendens Halfmoon, pair

Betta splendens Halfmoon, male

Betta splendens Halfmoon, male

Betta splendens Halfmoon, male

Betta splendens Halfmoon, male

according to the Thai breeders this is to harden their scales so that they are better able to withstand their opponent in battle. The same method is also used successfully after the fight in order to heal any injuries.

There are only a very small number of breeders of fighters in comparison to the numbers breeding *Betta splendens* for decorative purposes, i.e. as ornamental fishes. Hundreds of thousands of individuals of the Pla Kat are bred in practically every colour combination imaginable as beautiful, colourful ornamental fishes. Often they are reared in flat bottles that formerly contained brandy and with a volume of 0.5 litres (1 pint) apiece, several thousands of which are stood on a level concrete surface, each containing a *Betta splendens*. Here they generally grow on into splendid fishes in the course of 5 to 6

Betta splendens Halfmoon, male

Betta splendens Halfmoon, male

Betta splendens Halfmoon, male

Betta splendens Halfmoon, male

Betta splendens Halfmoon, male

Betta splendens Halfmoon, male

Betta splendens Halfmoon, male

Betta splendens Halfmoon, pair

Betta splendens Halfmoon, pair

Betta splendens Halfmoon, male

Betta splendens Halfmoon, male

Betta splendens Halfmoon, pair

Betta splendens Halfmoon, male

Betta splendens Halfmoon, male

Scenes of the *Betta splendens* Halfmoon Red mating

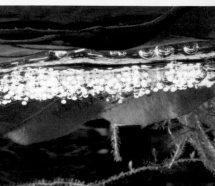

Betta splendens Halfmoon, male

Betta splendens Halfmoon, pair

Betta splendens Halfmoon, male

Betta splendens Halfmoon, male

Betta splendens Halfmoon, male

Betta splendens Halfmoon, 2 males

Betta splendens Elephant-Ear, male

Betta splendens Elephant-Ear, male

months. The larger fish farms may have up to 50,000 bottles. These bottles stand packed closely together and this permits the necessary maintenance work to be performed on them – in this way water changes and feeding can be effected without problem. Water changes generally take place twice weekly, and feeding daily. Until a few years ago the only item of food on the menu was black mosquito larvae. Because of the risks to health from feeding these mosquito larvae, nowadays the fishes are fed mainly on manufactured foods and/or alternative live foods such as *Tubifex*, water fleas, or adult,.

The majority of breeders have devoted themselves to the breeding of veil-finned fightingfishes, and only small groups are involved with the cultivated forms Crown-Tail (zigzag with foreshortened membranes between the fin rays), Half-Moon (the fins in a half-moon form), and Short-Tail (short-finned, in various colours); and most recently an as yet very small group with the breeding of Jumbo or Giant *Betta splendens*, a cultivated form that when adult is more than two and a half times the size of a normal *Betta splendens*. Until recently these Jumbo Bettas were available only as the Short-Tail form, but now also as Half-Moon.

While these large Pla Kat, as well as the small forms, are nowadays available in practically every colour, the colour yellow was formerly very rare among the cultivated forms of *Betta splendens* available in the trade. There is a special reason for this. In Thailand yellow is the colour of the king, and this colour was reserved for and permitted only to the royal house. Even today those of the population who are loyal to the king dress in yellow. Whether or not it was the wish of the king to breed yellow fightingfishes as well, is no more than speculation. But they were for a long time a privilege of the royal household. Nowadays these yellow fishes may not be commonplace but they do repeatedly crop up in the trade. Apparently it is not easy to breed a clear yellow colour. The situation is quite different with regard to the other colours. Nowadays *Betta splendens* exists in numerous colour "creations" that were unimaginable a few years ago. Which is one reason for the worldwide popularity of these unusual fishes.

It is no exaggeration to state that they are regarded as special fishes in Thailand. The great markets such as the Yatuchak, held every weekend in Bangkok, or the weekly Sunday market in Tombori, provide proof enough with their numerous stalls

Betta splendens Jumbo

Betta splendens Jumbo or Giant Short tail, male

Betta splendens Jumbo or Giant Short tail, male

Betta splendens Jumbo or Giant Halfmoon, male

Betta splendens Jumbo or Giant Short tail male, normal *Betta splendens* Short tail male in front

Betta splendens Jumbo or Giant Short tail, male

Betta splendens Jumbo or Giant Short tail, male

Betta splendens Jumbo or Giant Short tail, pair, female in front

Betta splendens Jumbo or Giant Short tail, male

Betta splendens Jumbo or Giant Short tail, pair, female in front. Laby

Betta splendens Jumbo or Giant Short tail, male

Betta splendens Jumbo or Giant Short tail male, normal *Betta splendens* Short tail male in front

Betta splendens Jumbo Competition in Singapore 2009

Betta competition at Aquarama 2009 in Singapore

The largest Betta show in the world takes place every two years at Aquarama in Singapore

A red-green Jumbo (Giant) *Betta* in the competition

Green Jumbo *Betta* with the beginnings of halfmoon fins

Jumbo *Betta* are increasingly being bred in bright colours

The short tail is also now showing the beginning of elephant ears category

Jumbo (Giant) *Betta* are still rare in competitions

Betta splendens Jumbo or Giant Betta Farm 1

Young Jumbo (Giant) Bettas are also grown on in individual bottles

Here a multitude of bottles standing on the floor, each with a single Betta growing on

Concrete vats of varying sizes are used as breeding containers for bettas

Pairs are put to breed in these concrete vats covered with a coarse mesh

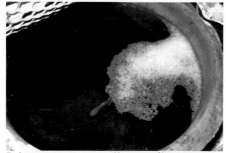

Each concrete vat is occupied by a pair of bettas. Here the males build their bubblenests and spawning takes place

The interior furnishing of these dark concrete vats is limited to a large Sea Almond leaf to improve water quality and provide a retreat for the female

Jumbo *Betta* male with wild-form coloration

Shorttail Jumbo *Betta* male growing on

Sales stand for Jumbo and other *Bettas* at the market in Thomburi.

Betta splendens Jumbo or Giant Betta Farm 2

Water change using a "water rake" for Jumbo *Betta* males growing on in separate containers

Large vats of water for the communal rearing of *Betta* females

Breeding set-up for *Betta*

Large concrete vats for growing on larger Jumbo (Giant) *Betta* males

selling Bettas. Here they are offered for sale ready-packed in plastic bags as well as in special, tastefully decorated and carefully tended containers. At the same time the passion of the vendors for these fishes is evident. Often the actual sale plays only a secondary role to the discussion about the Pla Kat on display. Any lover of the Siamese Fightingfish will find everything his heart desires here. All over the country there are Betta shops, and there are even stalls selling *Betta splendens* at the edge of the various main roads, next to those offering fruit and drinks.

Nowadays the Siamese Fightingfish is also an important export item for Thailand, and obviously quality is increasingly an important issue. But *Betta splendens* are bred not only in Thailand but also in Indonesia. The Indonesian fishes are often hardier in the aquarium and usually just as colourful. And the Pla Kat "bug" is also widespread among the fans of *Betta splendens* in the neighbouring Asian countries, and it is possible to buy fightingfishes in all Asian countries. Only the presentation is different. While the majority of dealers prefer to display them in small glass containers holding up to a litre (2

Betta splendens longfin

Betta splendens Longfin, male

Betta splendens Longfin, male

Betta splendens Longfin, male

Betta splendens Longfin, male

Betta splendens Longfin, male

Betta splendens Longfin, male

Betta splendens Longfin, 2 males

Betta splendens Longfin, male

Betta splendens Longfin, male

Betta splendens Longfin, male

Betta splendens Longfin, female

Betta splendens Longfin, 2 males

Betta splendens Longfin, male

Betta splendens Longfin, pair, female in front

Betta splendens Longfin, male

Betta splendens Longfin, male

Breeding scene of *Betta splendens* Longfin Red

pints), there are also very variable practices. In the tourist city of Hue in Vietnam, for example, in the aquarium-hobby shops by the Song Huong Giang (the "perfumed river"), they are displayed to view in sealed, shallow bags, suspended in rows on a board, and among the bicycle-riding dealers in Saigon they are offered for sale in small bags, together with other ornamental fishes, at the edge of the streets of the "city of millions". And in Hong Kong's Fish Street, in the Kowloon District, where there are aquarium-hobby shops packed with fishes and equipment along both sides of the road for around 400 metres (1/4 mile), *Betta splendens*, along with many other fish species, are displayed in bags on the "sales wall" next to the street or suspended at the entrance to the store.

But there are also *Betta splendens* that are sold as special fishes, in carefully tended sales establishments and as quality fishes

Betta splendens Short tail

Betta splendens Short tail, male

Betta splendens Short tail, male

Betta splendens Short tail, male

Betta splendens Short tail, male

Betta splendens Short tail, male

Betta splendens Short tail, male

Betta splendens Short tail, male

Betta splendens Short tail, male

Betta splendens Short tail, male

Betta splendens Short tail, male

Betta splendens Short tail, male

Betta splendens Short tail, male

Betta splendens Short tail, male

Betta splendens Short tail, male

Betta splendens Short tail, male

Betta splendens Short tail, male

Betta splendens Short tail, male

Betta splendens Short tail, male

Betta splendens Short tail, male

Betta splendens Short tail, male

Betta splendens Short tail, male

Betta splendens Short tail, pair, female in front

Betta splendens Short tail, male

Betta splendens Short tail, pair, female in front

Betta splendens Short tail, male

Betta splendens Short tail, male

Betta splendens Short tail, male

Betta splendens Short tail, male

Betta splendens Short tail, male

Betta splendens Short tail, pair, female in front

for exclusive enthusiasts with a serious interest. In Guangzhou in southern China there are Betta shops in which *Betta splendens* are sold in transparent plastic boxes barely 10 cm (4 in) in diameter and only 5 cm (2 in) deep. Several thousand of these boxes are stacked high on pallets and on the sales shelves. Sorted according to colour and fin form. In America, where supermarkets often include a small pet department, there are sometimes separate *Betta splendens* counters as well, with copious information and equipment for problem-free maintenance. In larger aquarium-hobby shops there is often even a special "Betta Lounge" with special wares for the so-called optimal maintenance of *Betta splendens*. These include so-called "Betta water", decanted into bottles. The fishes themselves are sold in clear plastic boxes with a sealed cover. Sorted and labelled by sex and colour. Special equipment, designed for the maintenance of *Betta splendens*, is also offered for sale. Even going as far as little "fish bowls" with separate lighting. Apparently *Betta* enthusiasts in America have a different mindset.

And yet no aquarium fish has so great a fan club by virtue of its behaviour, its reproductive biology, its pleasing coloration, and its variety of appearance. This is also demonstrated by the *Betta splendens* shows held worldwide, where it is not their fighting ability but their beauty that wins recognition and trophies. The largest events of this kind include the exhibitions in Asia, for example the Aquarama in Singapore or the Aquaria held in Guangzhou in southern China until a few years ago. Here there are usually 400 to 500 containers with *Betta splendens* competing for prizes and trophies. And the winners? These are invariably different colours and forms from event to event. Often fishes of unimaginable perfection in their finnage and their beautiful coloration.

References:

LINKE, H. 2004: *Betta splendens* - Ein Fisch mit vielen Gesichtern. Aquarium live, Bede Verlag, Ruhmannsfelden, 3: 55-59.

Betta splendens Wildform

Betta splendens wild form, male

Betta splendens wild form, male

Betta splendens wild form, male

Betta splendens wild form, male

Field of *Betta splendens* Wild in south Thailand

Habitat of *Betta splendens* in southern Thailand

Rice field in southern Thailand

Area of residual water at the edge of a rice field, habitat of *Betta splendens*

Betta splendens Competition in Guangzhou China

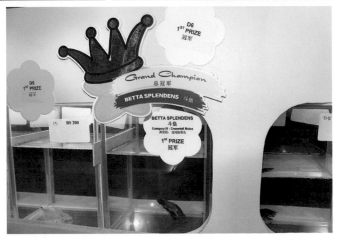

Betta splendens EAC Competition in Germany

Betta splendens Competition in Singapore 2005

Betta splendens Competition in Singapore 2007

Betta splendens Competition in Singapore 2009

Exhibition in Dresden Germany

Betta splendens Farm : Straits Aquariums Singapore

Betta splendens Thailand Farm 1

Betta splendens Thailand Farm 2

Atison *Betta splendens* Farm Thailand

Betta splendens Farm in Thailand

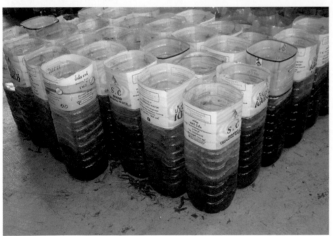

Field of *Betta splendens* in Chaiya Thailand

Field of *Betta splendens* in Khao Yai Thailand

Field of *Betta splendens* in north Thailand

Field of *Betta splendens* winter time north Thailand

Betta splendens fighting game on old central Thailand

Betta splendens fighting game on south Thailand

Betta splendens market in Chatuchak Bangkok Thailand

Ricefield of *Betta splendens* Thailand

Betta splendens shops in Saigon, in Vietnam

World of *Betta splendens* - Beijing (Peking) China

World of *Betta splendens* - United States

World of *Betta splendens* - Hong Kong

World of *Betta splendens* - Thailand

Betta stigmosa TAN & NG, 2006

Explanation of the species name:

Latin *stigmosus* (feminine *stigmosa*) = "spotted", referring to the spots on the fin membranes, the body, and the underhead region.

Original description:

TAN, H.H. and NG, P.K.L. 2005: The fighting fishes (Teleostei: Osphronemidae: Genus *Betta*) of Singapore, Malaysia and Brunei. The Raffles Bulletin of Zoology 2005 Supplement 13: 64-65.

Systematics:

Betta pugnax group

Natural distribution:

The species is thought to be endemic to the region of the Sekayu waterfall (Air Terjun Sekayu) in the area south-west of Kuala Berang in the province of Terengganu, around 53 km south-west of Kuala Terengganu in the north-west of western Malaysia.

Coordinates:

04° 57' 40" N 102° 57' 14" E (TAN, H.H. & NG, P.K.L., 2006)

Biotope data:

The *Betta stigmosa* pictured here were caught in a slow-flowing watercourse, up to 5 metres (20 ft) wide and clear and almost colourless, at the edge of the car park at the waterfall. The bottom consisted of light brownish sand. The fishes were provided with shelter and hiding-places from predators by large, rounded stones and in places the foliage of emerse bank vegetation trailing in the water, as well as by dead leaves and wood. The research site overall was an eroded gulley, 2 to 3 metres (6.5-10 ft) deep. The entire biotope was up to 90 % overgrown and hence heavily shaded. The water temperature was between 25 and 26 °C (77 and 79 °F) and the pH value 6.6. The species was comparatively numerous. (Research in August 2007: LINKE, H.)

Reproduction:

Mouthbrooder.

Total length: (size)

Around 8 cm (3 in).

Remarks: (differences from other species of the genus)

The small spots on the membranes of the fins, particularly the caudal fin, are regarded as species-typical for *Betta stigmosa*.

Adult males of this species exhibit a bold antique green on the opercula and sometimes also on the breast, and have no or only a very faint underhead pattern.

By contrast, females of this species exhibit a bold, dark underhead pattern but lack any antique green coloration.

Betta stigmosa ♂

Betta stigmosa ♂

Head close-up of *Betta stigmosa* ♂

Entrance to the Sekayu Waterfall (Air Terjun Sekayu)

Watercourse at the Sekayu Waterfall, habitat of *Betta stigmosa*

Biotope in the watercourse at the Sekayu Waterfall

The water in the biotope is only a few centimetres deep and very stony

In places large dead leaves cover the bottom in the biotope

Betta stigmosa after capture in the biotope.

Explanation of the species name:

Greek *stiktos* = "spotted", referring to the round dots on the unpaired fins.

English name:

Spotted Green Fighter.

Local Name:

Trey kroem omboke

Original description:

TAN, H.H. and NG, P.K.L. 2005: The fighting fishes (Teleostei: Osphronemidae: Genus *Betta*) of Singapore, Malaysia and Brunei. The Raffles Bulletin of Zoology 2005 Supplement 13: 95.

Systematics:

Betta splendens group

Natural distribution:

The specimens used for the description were collected on 16th February 1994 by T. R. ROBERTS in the area east of the town of Stung Treng, between Stung Treng and Banlung in north-eastern Cambodia.

Cambodia: Mekong basin, small swampy stream from Stung Treng to Ban Lung (ca. 3/4 to bridge over Tonle Srepok) (ca. 13° 30'N 106° 30'E).

L. HERMANN and H. LINKE recorded the species on 16th February 2007 in a slow-flowing forest stream that crosses the road (sand track) beneath a bridge 38 km from Stung Treng (Stoeng Treng), i.e. around 19 km from the fork in the road to Stung Treng-Kratie (National Highway 7) in the direction of Banlung, around 5 km before the village of Chop in the direction of the bridge over the Tonle Srepok. This watercourse is called the Oleang Krous River by the local people. It is a slow-flowing watercourse that leaves the forest close to the bridge and flows north through a bush landscape.

The watercourse at this spot is used by the local people as a washing place and water source.

Betta stiktos ♀

Betta stiktos ♂

Biotope data:

The water is clear and slightly brownish in colour. The water depth was on average 50 to 80 cm (20-32 in). The margins were densely overgrown with plants. The areas of submerse plants were the habitat of numerous fish species including *Betta stiktos*. The species was relatively numerous here. The water temperature at 10.00 am was 24.6 °C (76.3 °F).

Coordinates: 13° 24.08 N   106° 13.82 E

(Research in February 2007: HERMANN, L. & LINKE, H.)

Reproduction:

Like the other members of the *Betta splendens* group, *Betta stiktos* is a bubblenest spawner. Breeding is problem-free in soft, slightly acid water. The male alone tends the brood. At the same time it is possible that after only a few days further spawning activity may be seen beneath the bubblenest. The male can tend eggs and larvae simultaneously without problem. Once any fry have become free-swimming they continue to be tolerated in the nest or in the vicinity of the nest. They are comparatively small and should be provided with infusoria and/or rotifers as first food. Newly-hatched *Artemia* nauplii can be taken only if they derive from small *Artemia* species.

Total length: (size)
Around 6 cm (2.5 in).

This water course is called the Oleang Krous River by the local people. It is the habitat of *Betta stiktos*

The old road bridge over the Oleang Krous River

Breeding scenes of *Betta stiktos*

Betta stiktos larva

The species is very similar to *Betta smaragdina* in appearance and may even be identical with that species. Species-typical features are the round dots on the membranes of the caudal and anal fins as well as the vivid green markings on the opercula.

View from the old road bridge on the Oleang Krous River

Widening in the Oleang Krous River after it leaves the forest

The widening in the stream is used as a washing place

The Oleang Krous River is a narrower, flowing stream beneath the road bridge

Male *Betta stiktos* immediately after capture

Explanation of the species name:

Dedication in honour of Pater Heinz STROH (MSF).

English name:

Pater Stroh's Fightingfish

Original description:

SCHALLER, D. & KOTTELAT, M. 1990: *Betta strohi* sp. n., ein neuer Kampffisch aus Südborneo (Osteichthyes: Belontiidae), DATZ, Stuttgart, 43 (1): 31-37.

Systematics:

Betta foerschi group

Natural distribution:

According to the data in the original scientific description, *Betta strohi* was described from the Nataik Sedawak area, around 30 km south of Sukamara, Central Kalimantan (Kalimantan Tengah).

Biotope data:

The fishes we brought back came from the Nataisedawak region, around 4 km south of Sukamara in the south-west of Central Kalimantan (Kalimantan Tengah). We were able to find these fishes without problem in the local springs. The water was clear with a slight current and a strong brown coloration.

Reproduction:

Mouthbrooder.

Total length: (size)

Around 6 cm (2.5 in).

Remarks: (differences from other species of the genus)

Betta strohi is without question very closely related to *Betta foerschi*. Nevertheless, as SCHALLER & KOTTELAT state in the original description, there are also clear differences in morphological characteristics. These may include the prolonged central portion of the caudal fin, extended into a point, so far

Betta strohi ♀

Betta strohi ♂

Betta strohi ♂, nocturnal coloration

recorded in adult males. In addition the females exhibit a different pattern in their fully expressed and courtship coloration, where bold dark crossbands are visible on the body and clearly extending onto the anal fin.

References:

LINKE, H: 2007: Betta News spezial. European Anabantoid Club/AK Labyrinthfische im VDA: 3-30.

Biotope of *Betta strohi* in south of Sukamara

The main street of Sukamara

The harbour in Sukamara

The domestic well in the Nataisedawak area where Pater STROH discovered the *Betta*

The population density of the *Betta* was very high in the well

Explanation of the species name:

Latin *taeniatus* (feminine *taeniata*) = "striped".

English name:

Borneo Betta
Banded Betta
Striped Betta

Original description:

REGAN, C.T. 1910: The Asiatic Fishes of the Family Anabantidae. Proc. Zool. Soc. London, 1909, 54: 781, Pl. 78 (fig. 1).

Systematics:

Betta picta group

Natural distribution:

The holotype originated from the River Senah in Sarawak, eastern Malaysia.

Additional locality: Sungai Kuhas, 6-9 km left of Tebelu Tebakong turnoff, 5.8 km into right trail (Sungai Riih), Sarawak, Borneo.

Coordinates:

01°09'10.0 N,   110°29'22.7 E (Research in January 1996: TAN, H.H.).

The distribution of the species is restricted to the western part of Sarawak in eastern Malaysia (northern Borneo). The habitat is flowing watercourses, large and small. We recorded

Betta taeniata at various sites investigated. One of these was at Sutong, close to the turn-off from the main road from Sibu to Kuching, in the Sungai Raja south of Sri Aman. Also at Kampong Lanchang, around 6 km along a small road in the direction of Tebedu (7 km), after turning off the road from Kuching to Sibu precisely 60 km from Kuching. The fishes here were living in a small watercourse at the edge of the hill near the village of Kampong Lanchang as well as in the larger Sungai Kuhas at the edge of the village.

Biotope data:

In the small watercourse at the edge of the village of Kampung Lanchang the water temperature was 25.3°C (77.5 °F), the pH

Betta taeniata ♀

Betta taeniata ♂

Head close-up of *Betta taeniata* ♂

value 6.62, and the electrical conductivity 25 µS/cm. After a long period of rain the water was clouded and yellowish-brown in colour, with a fast current. The fishes were found only in areas of low current at the margins. The population density was very small.

In the Sungai Kuhas we found these fishes in the vegetated bank zones and among dead leaves and plants in marginal zones with weaker current. The water temperature here was 24°C (75 °F). The water was clear and without coloration.

(Research in August 2008: YAP, P., LO, M., CHIANG, N., & LINKE, H.)

Reproduction:
Mouthbrooder.

Because the small number of specimens that are occasionally available are predominantly wild-caught, it is probable that successful development of the eggs will be possible only in water matching that of the natural habitat. This means soft, mineral-poor, acid water with added humic substances.

The fishes usually pair beneath a large sheltering aquatic plant-leaf or an overhanging piece of wood, only a few centimetres above the bottom. This spawning site is closely guarded, mainly by the female. Other fishes are driven away.

The spawning is preceded by several "dummy runs", in which the female swims around the slightly curved male and touches the centre of his body with her mouth at the level of the base of the dorsal fin. The male then wraps himself around the female in a brief embrace, forming a U-shape around her body with his head and caudal fin pointing upwards. The embrace lasts for around 5 seconds, then the pair separate. Eventually eggs are laid during one of these embraces. The female picks up the eggs just laid in her mouth, collecting them from where they lie on the anal fin of the male. Only after several minutes of vigorous

Sign in the village of Kampong Lanchang

The Sungai Kuhas at the edge of the village of Kampong Lanchang

"churning" of the eggs by the female does she spit them, one at a time, immediately in front of the mouth of the male, and if the latter doesn't snap them up – as often happens in the beginning – she catches them up again herself. This often results in the pair playing a game of "who can catch the egg first". Usually there is no renewed spawning until all the eggs from a spawning pass (around 5 to 15) have been passed to the male. Throughout the process the territory is guarded, with the female being the more active partner in this, as during pairing. Additional dummy runs may take place throughout the whole spawning process. The individual spawning passes take place at intervals of 10 to 20 minutes and the entire spawning may last for several hours. Thereafter, when the male has taken all the eggs from the female, he retires, with much inflated gills and frequent vigorous chewing motions, to a quiet, sheltered area of the tank where he is initially still defended against other fishes by the female.

The fry are released from the mouth of the male after around ten days of development. They grow relatively quickly. The parents do not usually hunt down their offspring, but a good covering of floating plants in the aquarium is nevertheless advisable to provide shelter for the fry.

Total length: (size)
Around 7 cm (2.75 in).

The broad Sungai Kuhas is also a habitat of *Betta taeniata*

Betta taeniata lives mainly in shallow marginal zones

Remarks: (differences from other species of the genus)

Males of this species exhibit a very intensely coloured blue-green operculum, sometimes with delicate green scale margins on the body and a bold blue-green coloration on the marginal zones of the anal and caudal fins, bordered by a broad black outer edging.

References:

- SMITS, H.M., 1945: The freshwater fishes of Siam, or Thailand. Bull. US. Nat Mus., 188: 455.

- VIERKE, J. 1984: *Betta taeniata* REGAN, 1910 and *Betta edithae* spec. nov., zwei Kampffische von südlichen Borneo. Das Aquarium, 176: 58-63.

- LINKE, H: 2007: Betta News spezial. European Anabantoid Club/A K Labyrinthfische im VDA: 3-30.

The old wooden houses in Kampong Lanchang are worthy of seeing

The broad Sungai Kuhas is fast-flowing in places, especially at low water, and shows its stony bottom

Betta taeniata after capture

The small streams at the edge of the village of Lanchang are also habitats of *Betta taeniata*

Fishing in a small shallow stream near Lanchang

The margins of the stream were lined with predominantly emerse vegetation

In places the little stream was heavily overgrown with plants

In places the widenings in the stream contained dense clumps of aquatic plants

Here too commercial usage is influencing the habitat of *Betta taeniata*

Explanation of the species name:

Dedication in honour of Professor Tom Lam Toong Jin

Original description:

KOTTELAT, M. & NG, P. K. L., 1994: Revision of the *Betta waseri* species group (Teleostei: Belontiidae) Raffles Bull. Zool. 42 (3): 603-606.

Systematics:

Betta waseri group

Natural distribution:

The holotype was caught in the inundation forest around 15 km outside Kota Tinggi, on the road to Mersing in the area of the Mupor River in the south-east of the Malayan peninsula (western Malaysia).

Biotope data:

Betta tomi is a species that lives by preference in blackwaters. The water in the natural habitat is very soft and acid with a pH value of 5.5.

Reproduction:

Mouthbrooder

Total length: (size)

Around 11 cm (4.25 in).

Remarks: (differences from other species of the genus)

According to NG & KOTTELAT the *Betta waseri* group contains eight species that can be distinguished mainly by differences in the patterning of the fin membranes, the head proportions, the scale counts, the under-head markings and the patterns on the body and head. In addition to *Betta waseri*, the group comprises *Betta hipposideros*, *Betta spilotogena*, *Betta tomi*, *Betta renata*, *Betta pi*, *Betta chloropharynx*, and *Betta pardalotus*.

These fishes can be maintained without problem in large aquaria with a wealth of plants. Robust foods, including terrestrial insects, are highly advisable. As well as a dark, species-typical under-head marking, *Betta tomi* has a dark outer margin to the anal fin.

Head close-up of *Betta tomi* ♂

Betta tomi ♂

Explanation of the species name:

Dedication in honour of Mrs. Tussy NAGY de Felsö Gör.

English name:

Tussy's Fightingfish

Original description:

SCHALLER, D.1985: *Betta tussyae* spec. nov., ein neuer Kampffisch aus Malaysia. (vorläufige comm.) Aquar. Terrar. 38 (8): 348-350.

Systematics:

Betta coccina group

Natural distribution:

Type locality: Asian Highway Route No. 18, 16 kilometres by road south of Kuantan, as well as about 77 km south of Kuantan on road parallel to east coast, 17 km south of Pekan, Pahang State, Eastern Malayan Peninsula, western Malaysia.

The natural distribution is restricted to the area north and south of the town of Kuantan on the east coast of western Malaysia.

On the basis of personal knowledge, the apparently northernmost confirmed location for this species lies in a wooded swamp, around 300 m (1000 feet) from the Sungai Bungus, 30 km south-west of Kijal on National Highway 3 (Dungun-Kuantan) on the east coast of western Malaysia.

Reached from Kijal via the T6 road and then turning left onto the T13 in the direction of Kampong Kedai Binjai and Kampong Bungkus. This locality is a region of wooded blackwater swamp, with a pH value of 5.6 after heavy rainfall. (Research in August 2006: LINKE, H. & ZAHAR, Z.)

Coordinates: 04° 18. 13 N   103° 21.37 E (LINKE, H.)

A further confirmed location north of Kuantan lies west of the village of Kampong Cerating, on the right-hand (inland) side of National Highway 3 in the direction of Kuantan. After passing several businesses, and crossing the large, modern, wide-spanning bridge over the railway, on the left-hand side there is a small blackwater stream, which crosses the road beneath

Betta tussyae ♂ in fright coloration

Betta tussyae ♂

Courting pair of *Betta tussyae*

Betta tussyae ♂

m (8-80 in) wide and 0.5-2.0 m (20-80 in) deep, stagnant or with a slight current. These biotope data given by KRUMENACHER and WASER were recorded during the low-water months.

In the light of this the results of my own research in January 1987 – that is, at high water a month after the main rainy period – may be of interest. On the Kuantan-Mersing-Johor Baharu road, between kilometre stones 16 and 17 to Kuantan or 313 and 312 to Johor Baharu, a small, up to 8 m (26 feet) wide river, with bays on its east side up to 20 m (66 feet) wide, crosses the road beneath a concrete bridge around 12 m long. In places it flows through densely wooded zones which, however, afford little shade. The water was very clear and heavily stained brown. There was a strong current in the middle part of the river. An astonishingly large number of *Betta tussyae* and above all *Parosphromenus nagyi* were living in the numerous small, up to 50 cm (20 in) deep bays in the bank. Hence in around 30 minutes I was able to catch around 30 specimens of the latter species in four small bays, mainly in water 20 to 40 cm (8 to 16 in) deep among dead leaves and emerse plants. Average size was between 2 and 3 cm (0.75-1.25 in). *Betta tussyae* was also relatively numerous here and again measured between 2 and 3 cm (0.75-1.25 in) long. The water in these bays was almost motionless and had a temperature of 23.9 °C (75 °F). The pH value was 3.62, but dropped to 3.28 in the still zones between plants and dead leaves. The electrical conductivity measured 49 μS/cm. General and carbonate hardness were both less than one degree (German). The area was in the main exposed all day to the sun. The research was conducted towards 14.30 hours. Interestingly the numbers of juvenile *Betta tussyae* found were relatively high. A sweep of the net in the immediate vicinity of the bank produced up to five fishes of this species at a time. The fishes were always found among dense plants, leaves, or similar in the shallow (up to 20 cm (8 in) deep) water in the still bank zones. The dark red-brown colour of the water was astonishing. The bottom consisted of red-brown sand containing clay and laterite, producing a very slippery mixture on which it was impossible to walk.

a small bridge around 100 metres (330 feet) after the railway bridge. It was densely vegetated in places but readily accessible. At the end of August the residual water depth was mainly only 10 to 30 cm (4-12 in). The stream was a blackwater with a slight current and a pH value of 3.8. (Research in August 2006: LINKE, H. et al.)

Coordinates: 04° 07. 58 N   103° 22.14 E (LINKE, H.)

A confirmed location to the south lies at the edge of National Highway 3 from Kuantan to Pekan (N3 Kuantan-Johor Bahru). This is a narrow blackwater ditch alongside National Highway 3 between kilometre stones 305 and 306 heading for Kota Bahru or kilometre stones 22 and 23 heading for Kuantan, at the edge of dense rainforest.

Coordinates: 03° 39. 74 N   103° 17.77 E (LINKE, H.)

Biotope data:

The species usually lives in small streams and shallow, slow-flowing bodies of water with a depth of only a few centimetres in the dry season. Here the fishes often lurk among dead leaves or in dense aggregations of aquatic plants. Renè KRUMMENACHER (1985), who, along with Alfred WASER, recorded this species at several collecting sites in this area, gives the following data: Kuantan-Segamat road, 3.6 km before the sign indicating 141 km to Segamat or 32.4 km before Kuantan, right of the road. Water pH 5.5, conductivity 15 μS/cm, tea-coloured water, 0.2-2.0

Reproduction:

Bubblenest spawner.

Total length: (size)

The species attains a total length of around 5 cm (2 in).

Remarks: (differences from other species of the genus)

The differentiation of the sexes is not always easy. Males exhibit somewhat longer dorsal, anal, and ventral fins. In addition they are more boldly coloured by comparison.

Betta tussyae are typical blackwater fishes. For optimal maintenance it is important to use clean, mineral-poor, very acid water with a pH value between 4.5 and 5.0, a temperature between 24 and 27 °C (75 and 80.5 °F), with added humic substances. Like the other small red fightingfishes this species is at risk from Velvet Disease (*Oodinium*). By and large, only the best water quality can prevent this, and hence regular partial water changes are obligatory.

Betta tussyae biotope : North Kampong Cherating West-Malaysia

Betta tussyae biotope : North Kampong Cherating West-Malaysia

Explanation of the species name:

Latin *uber* (feminine uberis) = "plentiful", "abundant", referring to the comparatively high number of rays in the dorsal fin.

English name:

Red Pangkalanbun Fightingfish

Original description:

TAN, H.H. and NG, P.K.L. 2006: Six new species of fighting fish (Teleostei: Osphronemidae: *Betta*) from Borneo. Ichthyol. Explor. Freshwaters 17 (2): 105-108.

Systematics:

Betta coccina group

Natural distribution:

The specimen used for the description was collected by Patric YAP, an exporter from Singapore, in the Sungai Arut area in the vicinity of the town of Pangkalanbun, also known as Pangkalanbuun or Pangkalanbun Raya. Further specimens have come from Kalimati in the Palankanbun area and from a region 5 km north of Sukadana in southern West Kalimantan (Kalimantan Tengah), Borneo. Norbert NEUGEBAUER recorded the species as long ago as 1993 in blackwater streams south of the village of Kubu in the south of Kumai, in the region where the Sungai Kumai empties into the Java Sea.

In June 2009 we recorded *Betta uberis* in the Sungai Benipah (Sungai Nippa), in the region 14 km west of Kubu, around 40 km south of Pangkalanbun. (Research in 2009: LINKE, H. et al.)

Biotope data:

Research took place in 1993 in swamp woodland in the vicinity of the small port of Kubu, south-west of Kumai, where the Sungai Kumai enters the Java Sea. The fishes live there in blackwaters, in very soft, very acid water with a conductivity of 40 µS/cm and a pH of 4.0. (Research in January 1993; NEUGEBAUER, N.)

The 2009 research was in a small blackwater river, the Sungai Benipah, which crosses the road beneath a small bridge.

Coordinates: 02°56'11 S   111°37'45 E (LINKE, H.)

Apart from three small wooden houses, there was nothing but

Betta uberis ♀

Betta uberis ♂

Betta uberis ♂

scrub in the surrounding area. There were no longer any taller trees to provide shade. The water had a slight current and was dark red-brown in colour. The banks were heavily vegetated. The following water parameters were recorded:

pH value: 3.91
Conductivity: 12 µS/cm

Water temperature: 28.9 °C (84 °F)
(Research in June 2009: LINKE, H.)

Reproduction:
Bubblenest spawner.

Betta uberis in Sungai Benipah in the area 14 km west of Kubu

Plants in the Sungai Benipah

The black water of the Sungai Benipah with clumps of plants

Fishing in the Sungai Benipah was very hard work and *Betta uberis* weren't common

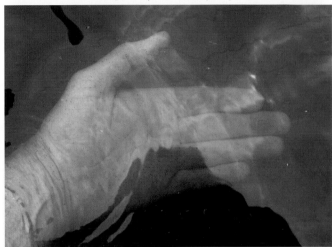
The black water of the Sungai Benipah was a deep red-brown colour

Milestone by the Sungai Benipah (Sungai Nippa)

Sphaerichthys selatanensis shortly after capture in the Sungai Benipah

Total length: (size)

The species attains a total length of up to 4.5 cm (1.75 in).

Remarks: (differences from other species of the genus)

Betta uberis has 0-l, 13-17 rays in the dorsal fin and can be readily distinguished from the other "small reds" by this higher dorsal-fin ray count (wider dorsal fin). *Betta uberis* does not always exhibit a turquoise spot on the flank, but sometimes also has a solid turquoise scale coloration on the sides of the body. In 2009 two similar species, termed *Betta* sp. Palangka and the somewhat smaller *Betta* sp. Senggalang, arrived in the trade from the Palangkaraya area, around 300 km west of Pangkalanbun. These are sometimes referred to as "*Betta uberis* from the Palangkaraya area", but it is clear from a comparison of

the adult sizes and the number of rays in the dorsal fin that we are dealing with different species.

Like the other small red fightingfishes this species is at risk from Velvet Disease (*Oodinium*). By and large, only the best water quality can prevent this, and hence regular partial water changes are obligatory.

References:

- NEUGEBAUER, N. 1993: Neuer *Betta* aus Kalimantan. Aquarium Heute 11 (2/93): 271.
- LINKE, H., 2009: New (?) *Betta* species from Kalimantan. Betta News, Journal of the EAC/AKL, 2009 (3): 22.
- LINKE, H., 2010: The new *Betta* from Kalimantan. Betta News, Journal of the EAC/AKL, 2010 (1): 26.

Explanation of the species name:

Latin *unimaculatus* (feminine *unimaculata*) = "one-spotted", "having one spot".

English name:

Howong Betta
One-Spot Betta

Synonyms:

Parophiocephalus unimaculata
Betta ocellata

Original description:

POPTA, C. M. L., 1905: Suite des descriptions préliminaires des nouvelles espèces de poissons recueillies au Bornéo central par M. le Dr. A. W. NIEUWENHUISEN 1898 et en 1900. Notes from Leyden Mus., 25:171-186

Systematics:

Betta unimaculatus group

Natural distribution:

POPTA 1905 gives the natural range as Howong and Kayan River, Kalimantan Timur, Borneo.

F. INGER & P. K. CHIN 1962 give the natural range as the Tawau District, Sabah, eastern Malaysia.

TAN & NG 2005 report the species' occurrence in the Sungai Kayan basin as well as around Howong in the upper Sungai Mahakam region, including the affluents Sungai Nap, Sungai Seba, Sungai Bako, and others, around 300 km south-west of the town of Malinau. (First imported by LINKE, H. 1980.)

Biotope data:

The author was able to record the species in 1980 in the region around Tawau, a town on the Celebes Sea, close to the border with the Indonesian part of the island of Borneo, Kalimantan Timur. The area around Tawau consists almost exclusively of oil-palm, rubber, and cocoa plantations. The local watercourses are predominantly under the influence of human activity. Because

Light Blue cultivated form of *Betta unimaculata*

Betta unimaculata from Kampong-Imam/Tawau, East-Malaysia

Betta unimaculata mating

After spawning the fertilised eggs lie on the anal fin of the male

Betta unimaculata after capture

all the watercourses are flowing, the water here is constantly renewed, and frequent rainfall accelerates this effect. *Betta unimaculata* is very numerous in this region. The fishes live under cover, usually as solitary individuals, among plants and bank scrub in the water margins. These are usually medium-sized individuals, often with a total length of up to 9 cm (3.5 in). In the side-arms of the on average 2 to 3 metres (6.5 to10 feet) wide watercourses there are often small, up to 3 cm (1.125 in) long, juveniles living among the leaf litter lying in the water, which often has a depth of just 2 cm (0.75 in).

Reproduction:
Mouthbrooder.

Breeding this species is not difficult. The procedure differs only slightly from that in *Betta pugnax*. In almost all mouthbrooding *Betta species* the female takes the eggs into her mouth after they have been fertilised, and then transfers them to the male by spitting them out in front of his mouth so he can in turn snap them up. But in *Betta unimaculata* the male collects up the eggs himself after fertilisation and there is no intermediate stage involving transfer by the female.

The brooding of the offspring by the male takes around 15 days at a water temperature of 25 °C (77 °F). The around 7 mm (0.25 in) long fry are then released from the mouth of their father. They can immediately take freshly-hatched *Artemia* nauplii as their first food. There is no longer any relationship between the father fish and the offspring after they have been released from his mouth, but the male does not normally hunt down the fry. Given a rich and varied diet as well as good water quality the young grow very rapidly. After around two months the young *Betta unimaculata* may already measure around 5 cm (2 in) long.

Total length: (size)
Around 11 cm (4.25 in).

Remarks: (differences from other species of the genus)
The sexes can be distinguished even in half-grown individuals. Males are more colourful, have a larger head, and exhibit bright green opercula.

The species prefers robust food and is in general peaceful even towards smaller fishes. These fishes present no problems in their maintenance and have no special requirements regarding water chemistry. It should nevertheless be noted that these are not fishes for very soft, and especially not for very acid water. Medium-hard, neutral to slightly alkaline water is beneficial for maintenance.

The identification of the species *Betta unimaculata* is apparently not absolutely clear-cut. It is possible that there was confusion with *Betta ocellata* even by WEBER & de BEAUFORT (1922) and INGER & CHIN (1962), and the very dark-coloured fishes from the Tawau area have been termed *Betta unimaculata* in the aquarium hobby since 1980.

Biotope of *Betta unimaculata* in Kampong Imam

Small stream in an oil palm plantation in the area of Kampong Imam - Biotope of *Betta unimaculata*

Small stream in an oil palm plantation in the area of Kampong Imam/Tawau

Head close-up of *Betta unimaculata*

References:

- INGER, R. F. & CHIN, P. K., 1962: The fresh-water fishes of North Borneo. Chicago Natural History Museum, USA: 158-161

- TAN, H.H. & NG, P.K.L., 2005: Fighting fishes (Teleostei: Osphronemidae: Genus *Betta*) of Singapore, Malaysia and Brunei. The Raffles Bulletin of Zoology, 13: 75-76.

Betta waseri KRUMMENACHER, 1986

Explanation of the species name:
Dedication in honour of Alfred WASER.

English name:
Waser's Betta

Original description:
KRUMMENACHER, R., 1986: *Betta waseri* spec. nov.. Aquaria (St. Gallen), 33 (12): 177-181.

Systematics:
Betta waseri group

Natural distribution:
The natural distribution of *Betta waseri* lies west and south of the town of Kuantan in western Malaysia.

The holotype was collected in a small rainforest stream that crosses the Kuantan - Kuala Lumpur road, 500 metres before kilometre-stone 232 heading for Kuala Lumpur. Heading for Kuantan, 22.5 km before Kuantan, in the deepest part of a depression. (Research in August 1985: WASER, A. & KRUMMENACHER, R.)

Biotope data:
In the natural habitat these fishes live in clear, flowing, very soft and very acid blackwaters, sometimes with dense areas of plants. Here they are found syntopic with, inter alia, *Parosphromenus nagyi, Betta pugnax, Betta bellica, richogaster trichopterus, and Betta tussyae.*

Water parameters:
pH value: 5.7
Conductivity: 30 µS/cm
Water temperature: 27 °C (80.5 °F)

These data are for a small watercourse around 1 m (40 in) wide, with a water depth of around 5 cm (2 in) but up to 1.5 m (5 feet) deep in occasional water-hollowed places. The water was clear with a slight current. (Research in August 1985: WASER, A. & KRUMMENACHER, R.)

Reproduction:
Mouthbrooder.

Total length: (size)
Up to 14.5 cm (5.75 in).

Remarks: (differences from other species of the genus)
This species is undoubtedly one of the largest mouthbrooding *Betta species* and on capture was already 13.5 cm (5.25 in) in total length (collected by Walter SCHÄR). The fishes in question attained a total length of 14.5 cm (5.75 in) in the aquarium and hence this is the largest fightingfish known to date in this

Betta waseri ♂

Underhead markings of *Betta waseri* ♂

group. The coloration described by KRUMMENACHER is remarkable. According to him the species, when in full colour, is brown-red, becoming lighter ventrally. The fins are likewise brown-red in colour. However, the most striking characters are the deep black lips and the lower tip of the mouth with its species-typical pattern on the underside of the head, which is likewise coloured deep black. In fright coloration *Betta waseri* exhibits a broad, dark, longitudinal band beginning behind the eye and ending in a dark spot on the caudal peduncle.

Betta waseri juvenile

The gnus *Colisa*

The genus name *Colisa* appeared for the first time when it was used for *Colisa lalius* and *Colisa unicolor* by CUVIER & VALENCIENNES in 1831. All the other species were described in one of the two genera *Trichopodus* or *Trichogaster*. In 1923 MYERS revised the nomenclature of the labyrinthfishes, including the gouramis, and the name *Colisa* was used for the species previously included in the genera *Trichogaster* and *Trichopodus*. This recommendation has been followed since 1923 without any dissent, except that very recently a few people have tried to change things back.

The name *Colisa* (= *kholisha*) for these gouramis comes from the everyday language of the people of Bengal and the Assam region.

There have been duplicate descriptions several times in the history of the "dwarf gouramis", sometimes also known as the "western gouramis" – in other words male fishes received different names to conspecific females and were assigned to their own separate genera. Thus BUCH.-HAM 1822 describes the female of one species as *Trichopodus chuna* and the male as *Trichopodus sota*, and the female of another species as *Trichopodus colisa* and the male as *Trichopodus bejeus*. There is also confusion about who was responsible for the descriptions. Thus numerous species were described by Dr. Francis BUCHANAN, who subsequently changed his name to HAMILTON for family reasons. Hence the name HAMILTON and the abbreviation BUCH.-HAM are often used for scientific purposes.

The genus *Colisa* currently contains five described species and a number as yet not examined scientifically. All are ideal aquarium fishes and well suited to maintenance in the community aquarium. They have no special requirements regarding water and food – they are omnivores that can be maintained without problem in moderately hard and slightly alkaline water. All *Colisa* species have long, threadlike ventral fins which are equipped with cells that sense by touch and taste.

When it comes to the scientific classification of the species there are still a lot of unanswered questions regarding *Colisa labiosa* and *Colisa fasciata* and their distribution. In particular their occurrence in Myanmar remains little studied and hence there is a need for additional collections. In 2004, 2005, and 2010 I made collections at various places in Myanmar and in so doing found *Colisa labiosa* and a species comparable to *Colisa fasciata* as well as species that cannot as yet be unequivocally classified scientifically. A small selection of photos of these fishes, with shots of the species in the field and in the aquarium in display coloration, may serve to provide some idea of the variety:

Wild caught photo of *Colisa* from the irrigation ditches in the rice fields at Yenatha on National Highway 3 from Mandalay to Mogok.

Aquarium photo of *Colisa* from the irrigation ditches in the rice fields at Yenatha on National Highway 3 from Mandalay to Mogok.

Wild caught photo of *Colisa* from the rice fields between Kamaby and Paung, on the road from Bago to Mawlamyine (Molmein).

Aquarium photo of *Colisa* from the rice fields between Kamaby and Paung, on the road from Bago to Mawlamyine (Molmein).

Wild caught photo of *Colisa* from Lake Inle near the village of MAING THAUK on the eastern side of the lake.

Aquarium photo of *Colisa* from Lake Inle near the village of MAING THAUK on the eastern side of the lake.

Aquarium photo of *Colisa* from the water-filled ditches along the National Highway from Toungoo to Mandalay.

Aquarium photo of *Colisa* from the water-filled ditches along the National Highway from Toungoo to Mandalay.

Aquarium photo of *Colisa* from the flooded fields along the road from Bago to Yangon at Intagwa, around 8 km after Bago.

Aquarium photo of *Colisa* from the flooded fields along the road from Bago to Yangon at Intagwa, around 8 km after Bago.

Wild caught photo of *Colisa* from the residual lagoons near the village of TA LET KHER on the road from Pathein to Ngwe Saung.

Wild caught photo of *Colisa* from the rice fields at NIGET TA THUNG on the road from Yangon to Pathein.

Aquarium photo of *Colisa* from the residual lagoons near the village of TA LET KHER on the road from Pathein to Ngwe Saung.

Aquarium photo of *Colisa* from the rice fields at NIGET TA THUNG on the road from Yangon to Pathein.

Aquarium photo of *Colisa* from the residual lagoons near the village of TA LET KHER on the road from Pathein to Ngwe Saung.

Aquarium photo of *Colisa* from the rice fields at NIGET TA THUNG on the road from Yangon to Pathein.

Wild caught photo of *Colisa* from the water-filled ditches along the road from Pathei to Yangon, around 31 miles before Yangon.

Colisa bejeus (HAMILTON, 1822)

Explanation of the species name:
From the language of the local people in Bengal, north-eastern India.

English name:
Banded Gourami

Synonyms:
Trichopodus colisa

Original description:
HAMILTON, F., 1822: A account of the fish in the river Ganges and its branches. Archibald Constable & Co, Edinburgh, & Hurst, Robinson & Co., Cheapside, London: 118-119, 372.

Natural distribution:
The natural distribution of the species is probably restricted to north-eastern India, along with the Assam region, Bangladesh, and possibly north-western Myanmar.

Biotope data:
The species lives in rice fields and associated irrigation ditches, in swamps and small rivers and streams. The water parameters are pH neutral to slightly alkaline, and soft to more mineral-rich. Like all other *Colisa* species, the species prefers dense areas of plants providing cover, and marginal undergrowth in the water.

Reproduction:
Bubblenest spawner.

Total length: (size)
Up to 8 cm (3 in).

Remarks: (differences from other species of the genus)
In his work of 1822 HAMILTON described the female as *Trichopodus colisa* and the male as *Trichopodus bejeus*. Frank SCHÄFER, currently the best informed about the genus and its problems, has declared the species *Colisa bejeus* valid on this basis.

Colisa bejeus ♀

Colisa bejeus ♂

Colisa bejeus pair, ♀ above, ♂ below

References:

- SCHÄFER, F. 2003: Wieviele *Colisa*-Arten gibt es?, Aquaristisches Fachmagazin Berlin-Velten, 173: 12-16.

- SCHÄFER, F.2008: Fadenfische der Gattung *Colisa*-eine aktuelle Übersicht / Gouramis of the genus *Colisa*-an actual overview, Betta News, Journal des European Anabantoid Club mit Arbeitskreis Labyrinthfische im VDA, 2: 7-14.

- SCHÄFER, F. 2009: *Colisa-Trichogaster-Trichopodus*, Betta News, Journal des European Anabantoid Club mit Arbeitskreis Labyrinthfische im VDA, 1: 6-8.

Colisa bejeus ♂

Colisa chuna (HAMILTON,1822)

Explanation of the species name:
Derived from *chuna kholisha*, the name used for the species in north-eastern Bengal in north-eastern India.

English name:
Honey Gourami

Synonyms:
Trichopodus sota
Colisa sota
Polyacanthus sota
Trichopodus chuna
Trichogaster chuna
Trichogaster sota
Polyacanthus chuna

Original description:
HAMILTON, F., 1822: A account of the fish in the river Ganges and its branches. Archibald Constable & Co, Edinburgh, & Hurst, Robinson & Co., Cheapside, London: 119, 372.

Natural distribution:
In eastern India and Bangladesh, in the areas around the rivers Brahmaputra, Ganges, and Jamuna-Ganga. The species is found in small watercourses, swamps, and rice fields, as well as among floating plants in areas of residual water, large and small.

Biotope data:
Colisa chuna lives by preference among dense vegetation in shallow, still areas of water.

We recorded the species in small bodies of water and also in large flooded areas and rice paddies near Dhamrai in Bangladesh, west of Dhaka and halfway along the road to Aricha, near to the great Jamuna-Ganga river. The numerous flooded fields and residual pools along the road were largely covered in a dense layer of Water Hyacinth. The water temperature in this layer of floating plants, around 10 cm (4 in) beneath the water's surface, was around 22.2 °C (72 °F), but only 20 °C (68 °F) at around a metre (40 in) of depth. The water was only slightly

Colisa chuna ♀

Colisa chuna ♂

Habitat of *Colisa chuna* in Bangladesh, west of Dacca

cloudy and light brownish in colour. The smaller fishes such as the *Colisa* species lived mainly in the covering of vegetation. The research site was permanently exposed to the sun. The maximum air temperature at the time (January, and hence winter) measured 22 to 23 °C (72 to 73.5 °F). During the rainy period (March to June/July) the temperature falls by a few degrees and the water level is sometimes 2 to 3 metres (6.5 to 10 feet) higher, and at that time only higher areas of land and the roads are free of water. In the subsequent hot dry season the temperature often rises above 30 °C (86 °F).

According to the local fishermen, the spawning season for the majority of fish species is in March, with the onset of the rains. The dried-up, harvested rice fields and the desiccated grass and scrub form a good nutrient base for the development of micro-organisms in the water and hence an important food supply for fish fry.

In this area *Colisa chuna* lives in the company of *Colisa lalia* and *Colisa fasciata*, infrequently also with *Ctenops nobilis*, *Badis badis*, and numerous other species. All the fishes living there are very frequently caught for food.

Colisa chuna ♂ , orange cultivated form

Colisa chuna ♀ , orange cultivated form

Water parameters:

pH value: 6.0
Conductivity: 182 μS/cm
Water temperature: 22.6 °C (72.7 °F)
General hardness: 4 °dGH
Carbonate hardness: 4 °dKH
(Research in January 1987: DERWANZ, K. & LINKE, H.)

Reproduction:

Bubblenest spawner.

The male builds a small bubblenest among plants and spawning takes place during an embrace. The male performs the brood care until the fry are free-swimming.

Total length: (size)

Up to 5 cm (2 in).

Remarks: (differences from other species of the genus)

When adult, males develop a bright orange-red body including the caudal fin. The gill surfaces beneath the eyes, the breast, the lower part of the belly, and the spinous part of the anal fin are deep blue to almost black in colour. The spinous dorsal and the upper area of the soft-rayed part of the fin are lemon yellow.

Female are uniform grey with a dark brown longitudinal band which sometimes fades. Unfortunately subdominant and frightened males also exhibit this female coloration, making it often difficult to distinguish the sexes.

Recently *Colisa chuna* has also been available in various cultivated colour forms.

The species is sometimes offered in the trade as *Colisa sota* and at present also as *Trichogaster chuna*. Both names are synonyms.

The parent fishes should be removed from the breeding tank once the fry have become free-swimming. The fry are amongst the smallest of all freshwater fishes and correspondingly difficult to rear, but breeding is simple otherwise.

References:

- SCHÄFER, F. 2003: Wieviele *Colisa*-Arten gibt es?, Aquaristisches Fachmagazin, Berlin-Velten, 173: 12-16.

- SCHÄFER, F.2008: Fadenfische der Gattung *Colisa*-eine aktuelle Übersicht / Gouramis of the genus *Colisa*-an actual overview, Betta News, Journal des European Anabantoid Club mit Arbeitskreis Labyrinthfische im VD*A*, 2: 7-14.

- SCHÄFER, F. 2009: *Colisa-Trichogaster-Trichopodus*, Betta News, Journal des European Anabantoid Club mit Arbeitskreis Labyrinthfische im VDA, 1: 6-8.

Rice fields in Bangladesh, habitat of *Colisa chuna*

Rice fields in Bangladesh, habitat of various *Colisa chuna*

The dense floating vegetation is a preferred habitat of *Colisa chuna*

Explanation of the species name:

Latin adjective *fasciatus* (feminine *fasciata*) = "striped", "banded".

English name:

Banded Gourami

Synonyms:

Colisa cotra
Colisa vulgaris
Colisa ponticerianer
Polycanthus fasciatus
Trichogaster fasciatus var.
Trichogaster fasciatus
Trichopodus fasciatus

Original description:

BLOCH, M. E. & SCHNEIDER, J. G., 1801: M. E. Blochii, *Systema Ichthyologiae iconibus ex illustratum. Post obitum auctoris opus inchoatum absolvit, correxit, interpolavit Jo. Gottlob Schneider.* Saxo. Berolini. Sumtibus Austoris Impressum et Bibliopolio Sanderiano Commissum. i-lx + 1-584 pp.

Natural distribution:

The distribution of the species purportedly extends from northern Myanmar (northern Burma) across Bangladesh to north-eastern India. The species inhabits large swampy expanses of water with dense areas of floating plants, small streams, and the margins of small rivers, as well as rice fields and associated irrigation ditches.

Biotope data:

The species lives among dense aggregations of plants with open areas in places. Water parameters are of secondary importance. These fishes live mainly in soft to moderately hard, neutral to slightly alkaline water with water temperatures that sometimes rise to 30 °C (86 °F) during the day but can also drop to 20 °C (68 °F) at night. Because the natural habitats lie in subtropical to moderate climatic zones, however, the water temperature during the winter months can drop to as low as 15 °C (59 °F) by night, but thanks to the sunshine usually warms up again to more than 20 °C (68 °F) by day. (See also research by DERWANZ & LINKE on *Colisa chuna* in 1987)

Reproduction:

Bubblenest spawner.

The male builds a large bubblenest among plants or even in areas of open water. The incorporation of pieces of plant is rare.

Total length: (size)

Around 7 cm (2.75 in).

Colisa fasciata ♂

Rice fields in central Myanmar, habitat of *Colisa fasciata*

Remarks: (differences from other species of the genus)

Colisa fasciata is one of the peaceful members of the genus. The male is very attractive and colourful, while the female possesses a grey base colour. Unfortunately the coloration of the male develops only relatively late on, at around 5 cm (2 in) in length, so differentiating the sexes is difficult at the juvenile stage.

Males of this species have a prolonged, pointed dorsal fin with a white edging, and a prolonged, pointed anal fin with a red margin. *Colisa fasciata* is often confused with *Colisa labiosa*, but *Colisa fasciata* remains smaller and the male has a prolonged, pointed anal fin while that of *Colisa labiosa* males is rounded. Males of *Colisa fasciata* exhibit a striking blue coloration on the breast and belly regions and on the anal fin.

References:

- DAY, F., 1878: The fishes of India; being a natural history of the fishes known to inhabit the seas and fresh water of India, Burma and Ceylon. Taylor & Francis, London: 374.

- SCHÄFER, F. 2003: Wieviele *Colisa*-Arten gibt es? Aquaristisches Fachmagazin, Berlin-Velten, 173: 12-16.

- SCHÄFER, F.2008: Fadenfische der Gattung *Colisa*-eine aktuelle Übersicht / Gouramis of the genus *Colisa*-an actual overview, Betta News, Journal des European Anabantoid Club mit Arbeitskreis Labyrinthfische im VDA, 2: 7-14.

The fishes are mainly to be found in the small water-filled ditches at the edges of the rice fields

Fishes from the small water-filled ditches after capture

Colisa fasciata pair, ♂ above, ♀ below

Colisa fasciata ♂

Colisa fasciata ♀

Colisa labiosa (DAY, 1877)

Explanation of the species name:

Latin adjective *labiosus* (feminine labiosa) = "large-lipped", "thick-lipped".

English name:

Thicklip Gourami

Synonyms:

Colisa labiosus
Trichogaster labiosa
Trichogaster labiosus

Original description:

DAY, F., 1877: The fishes of India; being a natural history of the fishes known to inhabit the seas and fresh water of India, Burma and Ceylon. Taylor & Francis, London: part 4, 553-779, Pls. 139-195.

Natural distribution:

The natural distribution of the species sold as "*Colisa labiosa*" in the aquarium hobby (see Remarks: below) is restricted to central and southern areas of central Myanmar (Burma).

Biotope data:

These fishes live in rice fields and associated irrigation ditches, as well as in expanses of water, small rivers, and swamps.

We recorded "*Colisa labiosa*" in the rice fields west of Yangon (Rangoon) in the south of central Myanmar, where they were living at a high population density among aquatic plants and in the marginal scrub of channels and rice fields by the road from Yangon to Pathein (Bassain). Our study site here was near Niget Ta Thung, around 2 miles west of the bridge over the River Ayeyarwady (Irrawaddy).

Coordinates: 17° 02'05 N   095° 31'47 E.

Water parameters:

pH value: 6.48
Conductivity: 263 µS/cm

Colisa labiosa ♀

Colisa labiosa ♂

Colisa labiosa ♂, red cultivated form

Colisa labiosa ♀, red cultivated form

Colisa labiosa ♀,
red-brown cultivated form

Colisa labiosa ♂, red-brown cultivated form

Field of *Colisa labiosa* near Toungoo to Mandalay in Myanmar

Water temperature: 29 °C (84.2 °F).

The water was brownish, murky, and still. The water depth was mainly 80 to120 cm (32 to 48 in). The landscape was flat and cultivated, with rice fields, vegetable crops, and residual swamps.

(Research in March 2010; WASER, G. & A., HERRMANN, L., BAYER, H., & LINKE, H.)

Reproduction:
Bubblenest spawner.

Total length: (size)
Up to 10 cm (4 in).

Remarks: (differences from other species of the genus)

At present the distribution of the species *Colisa labiosa*, described in 1822, can be stated only provisionally, because the precise identity of the species is also in question due to the loss of the holotype.

Males of this species have a rounded anal fin compared to *Colisa fasciata*.

References:

- SCHÄFER, F. 2003: Wieviele *Colisa* Arten gibt es?, Aquaristisches Fachmagazin, Berlin-Velten 173: 12-16.
- SCHÄFER, F.2008: Fadenfische der Gattung *Colisa* - eine aktuelle Übersicht / Gouramis of the genus *Colisa* - an actual overview, Betta News, Journal of the European Anabantoid Club mit Arbeitskreis Labyrinthfische im VDA, 2008 (2): 7-14.
- SCHÄFER, F. 2009: *Colisa-Trichogaster-Trichopodus*. Betta News, Journal of the European Anabantoid Club mit Arbeitskreis Labyrinthfische im VDA, 2009 (1): 6-8.

Field of *Colisa labiosa* near Pathein in Myanmar

Field of *Colisa labiosa* at east Yangoon in Myanmar

Field of *Colisa labiosa* between Yangon to Pathein in Myanmar

Field of *Colisa labiosa* near Mawlamyine east Myanmar

Explanation of the species name:

Latin adjective *lalius* (feminine *lalia*), derived from the name used for the species in Bengal in north-eastern India.

English name:

Dwarf Gourami or Lalia

Synonyms:

Trichopodus lalius
Trichogaster lalius
Polyacanthus lalius
Colisa lalius
Colisha lalius
Colisa unicolor
Trichogaster unicolor

Original description:

HAMILTON, F., 1822: A account of the fish in the river Ganges and its branches. Archibald Constable & Co, Edinburgh, & Hurst, Robinson & Co., Cheapside, London: 120, 372.

Natural distribution:

Colisa lalia is to date known from Bangladesh, from the Assam region to west of Calcutta. It has yet to be established whether the *Colisa* species found in Pakistan is perhaps also *C. lalia*.

Biotope data:

The species lives in small streams, ditches, and rivers, as well as in rice fields and swamps. We were able to catch *Colisa lalia* in small accumulations of water and also in large residual lakes and rice fields. Our research site was by the road from Dhaka (Dacca) to Mymensingh, shortly before the village of Trisal in Bangladesh. The site was a lake with grassy banks, and the surrounding area was used for cultivating rice. There were small groups of floating plants in places on the surface of the water. The lake measured around 100 x 200 m (330 x 660 feet) and was only around 2 m (6.5 feet) deep at the time (dry season, in January). The water was clouded and slightly mud-coloured.

Colisa lalia ♀

Colisa lalia ♂

Mating in *Colisa lalia*, red cultivated form. At the end of the brief embrace the transparent eggs sink briefly and then rise to the water's surface

There was no water movement. The water temperature at the water surface measured 21.9 °C (71.4 °F), and only 19.1 °C (66.4 °F) at 80 cm (32 in) of depth. The air temperature warmed up to around 25 °C (77 °F) during the midday period. The lake wasn't shaded. All the fishes living there were caught as food.

Water parameters:

pH value: 7.4
Conductivity: 103 µS/cm
Water temperature: 21.9 °C (71.4 °F)
General hardness: 2 °dGH
Carbonate hardness: 2 °dKH
(Research in January 1987: DERWANZ, K. & LINKE, H.)

Reproduction:

Bubblenest spawner.

Males of *Colisa lalia* are real little master builders when it comes to constructing their bubblenest, which includes pieces of the leaves of various aquatic plants, especially floating plants, and forms a large but compact work of art up to 6 cm (2.5 in) in

diameter and often protruding for up to 2 cm (0.75 in) above the water's surface. They are among the few labyrinthfishes that build their nests with a lot of care.

Once a male *Colisa lalia* has finished constructing his nest he goes looking for a female who is willing to spawn. He courts his chosen one with fins spread and display movements and them swims back to the nest, stopping several times on the way to make sure that the female is following. Once the two fishes are beneath the nest, the male swims around his partner and displays in his finest colours. Often the female is frightened by the ardour of her mate and takes flight. This ritual is repeated after a few minutes and so mating takes place only after several dummy runs. The pair now circle one another and the female then touches the belly to tail region of the male with the tip of her mouth. This contact is the trigger for the male to embrace his partner, wrapping himself around the female and turning her onto her side, and then a few seconds later onto her back. At this point the eggs are expelled and simultaneously fertilised. The pair separate after a few seconds and the eggs are released to rise to the surface of the water. The female then leaves the spawning site and the male immediately collects up the eggs with his mouth and puts them into the nest. Only when all the eggs have been collected and secured with additional air bubbles does the next mating episode take place. If there are several females in the aquarium then the male may spawn with different partners during a single spawning episode.

After around two hours the male starts to tend the eggs, a task he performs alone. At a water temperature of 27 °C (80.5 °F), the first larvae rupture their eggshells around 12 to 14 hours after spawning. After around 24 hours the larvae already exhibit a distinct head form and the embryonic spinal column begins to differentiate into the spinal column. In addition the first protective pigmentation becomes visible on the body. After around 33 hours the eyes already possess a gold-black iris, which is clearly defined and already capable of movement after 52 hours. 10 hours later, that is 62 hours after spawning, the larvae exhibit good protective pigmentation all over the body. The fins are completely visible, the yolk sac is exhausted, and the internal organs can be seen to be functional.

At this point in time, that is 2 days and 14 hours after spawning, the fry become free-swimming and leave their father's nest. At first the fry try to remain together, but a few hours later they are spread all over the tank. The parents should be removed from the breeding tank at the point where the offspring become free-swimming. The fry can now be fed powdered food, and if possible slipper animalcules as first food, and an airstone, operating at a moderate rate, should now be suspended in the aquarium. After 5 to 7 days *Artemia* nauplii can be fed.

Total length: (size)
Around 5.5 cm (2.25 in).

Remarks: (differences from other species of the genus)

The differences between the sexes are easy to detect. The males are very much more colourful and exhibit a striking stripe pattern on the body. The females are usually plain grey in colour and sometimes exhibit a blue sheen on the body. They also often remain somewhat smaller by comparison, and are plumper in the breast/belly region.

References:
• DAY, F., 1878: The fishes of India; being a natural history of the fishes known to inhabit the seas and fresh water of India, Burma and Ceylon. Taylor & Francis, London: 375, plate LXXIX, fig. 5

Colisa lalia pair, red cultivated form, approaching to mate

Embrace during the mating of *Colisa lalia*

Colisa lalia beneath the nest after mating. The eggs are still
floating free in the water and will only later rise into the nest

The pair embrace during mating

Mating beneath the bubblenest

Release from the embrace
after mating

Colisa lalia ♂

Colisa lalia ♀

Colisa lalia ♂ , blue cultivated form

Colisa lalia ♀ , red cultivated form

Colisa lalia ♂ , red cultivated form

Colisa lalia, red cultivated form, ♂ above, ♀ below

Colisa lalia ♂ , wild-caught

Colisa lalia ♂ , wild form in aquarium good condition keeping

Colisa lalia ♀ , wild form

Field of *Colisa lalia*

Rice-field scenery in Bangladesh, habitat of *Colisa lalia*

Setting out young rice plants in Bangladesh

Fishes from the rice fields in Bangladesh at the market

Breeding Farm of *Colisa lalia*

Breeding set-up for *Colisa* at a farm in Malaysia

Breeding containers at a breeding farm for *Colisa lalia* in Malaysia

Red *Colisa lalia* being counted prior to despatch

Red *Colisa lalia* in a net for sorting

Market :
Colisa lalia in Fishstreet of Kowloon Hongkong

Aquarium fishes (including *Colisa*) in bags for sale in Hong Kong

Colisa lalia being sold in bags in the Fish Street in China.

Colisa sp. Lake Inle ♂

Colisa sp. Lake Inle ♀

Biotope of *Colisa* sp. Lake Inle in Myanamr

The genus *Ctenops*

One of the most unusual of the labyrinthfishes was first imported for the aquarium in 1912. But only for a short time, and we then had to wait for many years for a further importation. For a long time details of maintenance and breeding remained unknown and were largely a matter of supposition. The genus remains monotypic and little studied to the present day. The much elongated head and the pointed snout with its large mouth, together with the dorsal fin set well back on the body and the rounded caudal fin, give this, the only member of the genus, a very unusual appearance. The species is sometimes also assigned to the genus *Trichopsis* due to the incorrect classification by BLEEKER. It remains unclear whether and to what extent the two genera are related.

COLISA LALIA (HAMILTON, 1822)   339

Explanation of the species name:

Latin adjective *nobilis* = "noble", "splendid", "distinguished".

English name:

Pointed-head Gourami.

There are various local names used in different geographical regions - Modumala, Nakteceuda, Naktecolischa, Nagdani, and Chibcilli.

Synonyms:

Trichopodus nobilis
Trichopsis nobilis

Original description:

McCLELLAND, J. 1845: Description of four species of fishes from the rivers at the foot of the Boutan Mountains. J. Nat. Hist. Calcutta 5 (18): 274-282.

Systematics:

Ctenops group (monotypic)

Natural distribution:

Type locality: rivers in north-eastern Bengal and Assam, in India. The species is found in small rivers and swamp regions from the plains to the mountains in north-east India and Bangladesh. According to fishermen in Bangladesh its habitat and distribution is mainly the south-western and western areas, from Dhaka to the vicinity of Calcutta.

Biotope data:

In general the occurrence of these fishes is restricted to small pools of water and still, shallow zones of streams and rivers, where they live in the upper layers of the water among dense vegetation. The composition of the water generally encountered in this region is interesting, however. The water is soft to moderately hard with a pH around the neutral point. In January 1987 I was able to record the following water parameters near Dhamrai, in an area of water covered with water hyacinths by the road to Aricha on the Jamuna River, where the Jamuna (Bramaputra) and the Ganges meet: general and carbonate hardness were both 4 degrees (German), the pH was 6.76, and the conductivity 182 µS/cm at a water temperature of 22.6 °C (72.7 °F). (Research in January 1987: DERWANZ, K. & LINKE, H.)

Reproduction:

Paternal mouthbrooder

In 1988 Sven BITSCH reported on the breeding of *Ctenops nobilis*. According to him this species is a maternal mouthbrooder. More than 150 fry, measuring 5-6 mm (0.2 in) in length were released from the mouth of the female after a long period of brooding and immediately took freshly hatched *Artemia* nauplii as their first food. But in 1992 BRITZ reported, on the basis of his own observations, that *Ctenops nobilis* was unequivocally a paternal mouthbrooder.

Ctenops nobilis

River in Bangladesh, home of *Ctenops nobilis*

Total length: (size)

Up to 9 cm (3.5 in).

Remarks: (differences from other species of the genus)

Large aquaria are recommended for maintenance, with corresponding amounts of décor items such as bogwood and a thick covering of vegetation, several centimetres deep, below the surface of the water.

This to date only very rarely imported species first arrived in Europe in large numbers in the middle of 1986. The small number of specimens of *Ctenops nobilis* imported in 1912 and 1955/56 aroused little attention. Even today there are gaps in our knowledge as regards maintenance requirements and reproductive biology. Even the differentiation of the sexes is largely a matter of supposition. On the basis of experience to date *Ctenops nobilis* is delicate and tricky in its maintenance. It remains unknown whether these fishes – which apparently occupy a systematic position between the genera *Sphaerichthys* and *Trichopsis* – are trophic specialists.

In the aquarium these fishes inhabit the upper layers of the water. They prefer densely vegetated areas and those with a covering of floating plants. On the basis of experience to date they are not predatory fishes. Their preferred foods, which are taken with relish, are all sorts of insects and insect larvae, with black mosquito larvae apparently particularly attractive. It is unknown whether the strikingly prolonged mouth has any particular significance. The species is still regarded as delicate and difficult in its maintenance, but it is unclear whether – given that to date only wild-caught stocks have been available – this is the result of damage during transportation or whether we still lack the knowledge required to maintain these fishes correctly. Perhaps they require subtropical maintenance conditions, i.e. cooler water temperatures.

Ctenops nobilis also inhabits the irrigation channels in the rice fields

Motor boats are still very uncommon on the rivers of Bangladesh

Ctenops nobilis after capture

References:

- SCHMIDT, J. 1987 Getarnte Räuber. Auch unter Labyrinthern gibt es Blattfische. Das Aquarium 21 (7), 338-342.

- KROKOSCHA, M. 1988: *Ctenops-* und *Sphaerichthys*-Zucht und systematische Stellung. DATZ 41 (8), 300-302.

- BRITZ, R.1992: Bemerkungen zur Pflege und Zucht von *Ctenops nobilis*, DATZ 45 (11), 692-694.

- ENDE, H.-J., 2004: *Ctenops nobilis* - der Spitzkopfgurami, eine Momentaufnahme. Der Macropode (26) 9/10: 202-204.

The genus *Helostoma*

The fishes known as Kissing Gouramis are at present assigned to a monotypic genus; in other words there is just one species in this genus. Unfortunately it is no longer possible to ascertain their natural distribution nowadays. The species was originally found in Thailand, Malaysia, Sumatra, Java, and Borneo. There is a grey, silvery-green form of this species and a cultivated "pink"-coloured variety, with the latter being the better known. These fishes can attain a total length of up to 25 cm (10 in). They are farmed as food fishes and nowadays this practice is no longer limited to South-East Asia.

Because they are greatly prized as food fishes they are bred in many places for this purpose. And because in the warm tropical zones a still-living fish fetches more money in the market than a dead one, the species, being an air-breathing labyrinthfish, is better able to survive transportation to the customer than a fish that breathes only via its gills. In addition large *Helostoma temminckii* tip the scales at around a kilogram (2.2 lbs), and hence they are regarded as a lucrative commodity.

Fish breeders in Thailand keep Kissing Gouramies in open-air pools measuring around 8 x 15 m (26 x 50 feet) with a water depth of around 80 cm (31 in). The water temperature can rise to 30 °C (86 °F) in the heat of the sun. The high rate of evaporation during the dry season is compensated by the addition of spring water, while during the rainy season Nature maintains a balance. Young fishes measuring 4 to 5 cm (1.5 to 2 in) long are placed in these large pools of water and in the course of a year grow on into splendid large specimens. Feeding consists of whatever food occurs naturally in the open-air pools plus fresh pig manure. The latter contains semi-digested, not fully utilised pig food and its high vegetable component constitutes a very vitamin-rich food. The uneaten residues then encourage the proliferation of plankton and algae in the water, and the "well-fertilised" water attracts insects to lay their eggs, producing an increased population of their larvae. Nevertheless the self-purification of the water is astonishing. The fishes grow on well in this environment.

Particularly large, strong fishes are preferred as broodstock. Before the ponds are pumped empty in order to harvest the Kissing Gouramis, a cast net is used to catch out several specimens for inspection. Those fishes that are suitable for breeding are sorted out and the sexes placed in separate plastic nets that are suspended in the ponds to create floating containers with an area of 2 to 3 m2 (20 to 30 sq.feet). Round concrete tanks with a diameter of around 80 cm (31 in) are used for breeding. These are cleaned thoroughly and filled with fresh water. The water is around 40 cm (16 in) deep. A pair of *Helostoma temminckii* is placed in each tank. If the females are full of eggs as the result of the long separation of the sexes, mating takes place after just a few hours. The eggs are expelled as the pair embrace, and rise to the surface of the water. Spawning takes place in the open water and no nest is constructed. No plants or other material are placed on the water's surface to protect the eggs. After spawning the fishes are removed from the breeding container.

If the eggs are numerous (large specimens can release more than 10,000) they are removed and placed in large concrete tanks with an area of 6 to 8 m2 (60 to 80 sq.feet) with around 20 cm (7 in) of water. The very small fry receive the tiniest water fleas as their first food, supplemented for the first three days with crumbled hard-boiled egg yolk. With daily water changes the fry grow on very rapidly to strong, healthy specimens in these concrete containers, and at a length of around 4 to 5 cm (1.5 to 2 in) they are exported as aquarium fishes or transferred to the big outdoor pools to be grown on further as food fishes.

Because Kissing Gouramis take their food predominantly from the water's surface, it is very important to feed with a floating food (such as flake, for example) with a high component of pure vegetable material and vitamin-rich raw ingredients, and this is also an important measure for successful growing-on. Only well-fed individuals will thrive and grow into splendid specimens. The "grazing" of algae-covered surfaces between meals is evidence of their considerable requirement for vegetable food and this also provides them with the requisite intake of micro-organisms.

The water temperature should be around 25 °C (77 °F), and efficient filtration of the water will lead to improved well-being.

Explanation of the species name:

Dedication in honour of the Dutch doctor and naturalist C. J. TEMMINCK.

English name:

Kissing Gourami

Synonyms:

Helostoma temmincki
Helostoma tambakkan
Helostoma oligacanthum
Helostoma servus
Helostoma xanthoristi
Helostoma rudolfi.

Original description:

CUVIER, G. & VALENCIENNES, A., 1829: Hist. Nat. Poiss., VII: 342.

Natural distribution:

These fishes are distributed all over South-East Asia, in part as the result of breeding as food fishes, and in part representing in the natural habitats of the original wild form. Natural distribution regions are known from Thailand, Malaysia, Sumatra, Java, and Borneo.

Biotope data:

The species prefers slow-flowing rivers and lakes and has a preference for areas of dense plants. Habitats include large rivers, swamp regions, and also smaller lake-like aggregations of water.

Biotope data from the island of Sumatra:

Lake with several small affluents (Danau Calak). A very fish-rich large lake up to 5 km long, close to the village of Bailang, west of Sekayu, north-west of Palembang, province of Sumatra Selantan, island of Sumatra.

Coordinates:

02° 57' 33 S   103° 59' 49 E (H. LINKE)

The lake is a murky, loam-brown, mixed water with a water temperature of 31.3 °C (88.3 °F), a pH of 6.3, and a conductivity of 25 µS/cm. (Research in September 2008: H. LINKE et al.)

Reproduction:

The breeding of this species is somewhat more difficult than that of other labyrinthfish species. Two important factors are prerequisites for breeding success. Firstly the fishes must be at least 12 to14 cm (4.75 to 5.5 in) long, and secondly the tank must be at least 100 cm (40 in) long (200 cm (80 in) is even better) and at least 60 cm (24 in) deep. The surface of the water should be covered as far as possible with numerous aquatic plants. In places dried grass with stiff stems should be laid on the surface to protect the ascending eggs from the often greedy appetites of the parents. This also produces an important side benefit, as the dried grass or hay will encourage the proliferation of infusorians such as slipper animalcules (*Paramecium*) and other unicellular organisms that will then be available to the fry as first food.

Helostoma temminckii after capture

Helostoma are a popular food fish sold at the markets in Sumatra

Breeding farm for *Helostoma temminckii*. The fishes are grown on in large ponds

Selected Helostoma are kept in nets

Most of the *Helostoma* bred on the farms are the pink cultivated form

Sexing is performed by testing for the emission of sperm

Water parameters play a subordinate role in breeding. The male will involuntarily produce a few air bubbles that rise to the water's surface, but doesn't construct a bubblenest. The pair embrace in the open water and the eggs are released at random. Because this species is very productive, large individuals can produce up to 10,000 eggs per spawning. The parents should be removed after spawning. The eggs will float at the water's surface or stick to the plants. The fry swim free after around four days at a water temperature of 30 °C (86 °F) and it is difficult to satisfy their appetites. They require the smallest of foods. Given good and abundant feeding, the little *Helostoma temminckii* will grow on rapidly. After around four weeks, however, they must survive a critical stage, as it is now that the respiratory labyrinth becomes functional and the fry begin to also take atmospheric air at the water's surface, so the air needs to be at an appropriate temperature. The air temperature, like that of the water, should be 30 °C (86 °F), which means that the breeding tank must be tightly covered.

Total length: (size)

In the wild between 15 and 20 cm (6 and 8 in), captive-breds up to 25 cm (10 in), rarely larger.

Remarks: (differences from other species of the genus)

Differentiating the sexes is very difficult as there are no external distinguishing characteristics. Only larger specimens, around 12 cm (4.75 in) long, permit accurate sexing – when the fishes are viewed from above the body form of the female looks broader. This is one possibility, for use where the female is not yet ripe and recognisable by swelling of the breast and belly region. The second possibility is to identify the males. In order to do this each fish must be removed from the water, held in the hand, and turned onto its back. If the belly region is then massaged gently, evenly, and very carefully on both sides of the body, then after a short time white sperm will be visible in the genital opening of the ripe male. An interesting method of determining the sexes, long practised by fish breeders in Thailand.

Catching *Helostoma* in Calak Lake in Sumatra

The population density of *Helostoma temminckii* is very high in Calak Lake

The fishermen at Calak Lake live in floating huts that can be moved very quickly if required

The species is a planktivore and thus a trophic specialist. These fishes prefer water fleas and *Cyclops* plus *Artemia* nauplii. Vegetable food is also very important. Vegetable dry foods are greatly enjoyed.

These fishes are particularly famed for their "kissing" behaviour during courtship and display. Aquaria with a minimum length of 120 cm (48 in) should be used for maintenance.

The genus *Luciocephalus*

These unusual predators are sometimes known as pikeheads. They are labyrinthfishes and mouthbrooders. The first species description was that of *Luciocephalus pulcher* (as *Diplopterus pulcher*) in 1830 by J. E. GRAY, but the genus description of *Luciocephalus* by BLEEKER in 1851 is regarded as valid. The name is derived from lucius, referring to the European Pike, *Esox lucius*, and the Greek *kephale* = head.

The pikeheads are interesting predators. Their bodies are slender and elongate. The mouth is likewise elongate and, along with the head, represents about a third of the body length. In the wild these fishes live mainly in standing to slightly flowing water, and almost exclusively in the upper layers of the water. When hunting food they usually position themselves among plants at the water's surface and lie in wait for prey. The latter includes insects, their larvae, and above all fish fry, and in the case of adult pikeheads small slender fishes with a total length of up to 5 cm (2 in).

There are currently two species in the genus, both rare in the aquarium hobby. Their maintenance is not without problem. They are soft-water fishes that require soft, acid water for optimal maintenance as well as a diet exclusively of clean, healthy live foods.

Explanation of the species name:

Latin *aura* (noun in apposition) = a "gleam" or "sheen", referring to the metallic spots and stripe on the flanks.

English name:

Peppermint Crocodile Fish

Original description:

TAN, H.H. & NG, P.K.L. 2005: The Labyrinth Fishes (Teleostei: Anabantoidei; Channoidei) of Sumatra, Indonesia. The Raffles Bulletin of Zoology 2005 Supplement 13: 129-130.

Natural distribution:

The natural habitat of *Luciocephalus aura* lies in the south-east of the island of Sumatra as well as large parts of Kalimantan Tengah in the south of the island of Borneo. We found the species in the Soak Putat, a widening in the course of the River Pijoan, near to the village of Pijoan on the road from Kota (= town) Jambi to Muarabulian in the province of Jambi; as well as in the area of Sentang in the north of the province of Sumatra Selantan, which borders the province of Jambi to the south; both sites in south-eastern Sumatra.

Biotope data:

Pijoan/Jambi Province:

We found the species in the lateral lagoons and inundation zones of the Sungai Pijoan, in shallow-water areas along the banks and in the rainforest. The water was slightly cloudy and yellowish in colour, with a pH of 5.0 and a water temperature of 27.0 °C (80.5 °F).

Sentang/Sumatra Selatan Province:

The village of Sentang lies on the road from Kota Jambi to Kota Palembang in the province of Sumatra Selatan, island of Sumatra, Indonesia. The natural habitat is reached via a fork in the road heading east in the village of Sentang. A large part of the surrounding terrain was being used for the cultivation of rubber trees.

Luciocephalus aura, the protrusible tip of the snout is clearly visible

Luciocephalus aura ♂

photo / Andreas Hartl

Luciocephalus aura photographed in the field

Luciocephalus aura after capture in Kalimantan

Blackwater biotope in central Kalimantan, habitat of *Luciocephalus aura*

Coordinates: 01°56.07 S   103°42.53 E (LINKE, H.)

In the biotopes east of Sentang these fishes were found in and around groups of plants in areas of little current. The water temperature measured 27.2 °C (81 °F) and the air temperature around 30 °C (86 °), dropping to around 25 °C (77 °F) at night; the pH was 5.0 and the conductivity 3 µS/cm. These were blackwater habitats. The clear, brown-stained water was still to slightly flowing, and the underwater visibility was around a metre (40 in).

Habitat of *Luciocephalus aura* in central Kalimantan

Luciocephalus aura after capture in central Kalimantan

Freshly-caught *Luciocephalus aura* with total lengths of 13 to 15 cm

Breeding episode in *Luciocephalus aura*

The pair seek to embrace

The fishes maintain constant body contact as they swim around each other

A breeding episode can last for up to eight hours

Completely embraces are sometimes to be seen

Male *Luciocephalus aura*

Luciocephalus aura juvenile measuring 17 mm total length immediately after release from the mouth

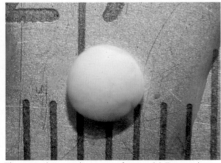
The eggs have a diameter of around 2.7 mm when laid

The eggs of *Luciocephalus aura* are yolk-coloured rather than transparent

Reproduction:

Mouthbrooder.

Luciocephalus aura is also one of the most interesting labyrinthfishes in terms of its reproductive behaviour. My *Luciocephalus aura*, at the time measuring around 11 cm (4.5 in) in length, were maintained in a 200-litre aquarium. Their only tankmates in this aquarium were an adult pair of *Betta obscura*. As for all the blackwater fishes that I keep, the water was conditioned in a large glass container and, after a long period of "biological maturation", also used for water changes in the *Luciocephalus aura* aquarium. Its parameters were a conductivity varying from around 10 to 20 microSiemens at a water temperature of 28° C (82.5 °F) and a pH of 4.0 to 4.3. The water was enriched by the addition of natural humic substances. The wild-caught fishes soon settled in with good feeding with small fish fry, glassworms, whiteworms (*Enchytraeus albus*), and adult *Artemia*. Only healthy live foods were used.

The first mating was seen after around four weeks. I thus established that the female was somewhat slimmer than the male and exhibited less colour on the lower part of the body.

Luciocephalus aura pair, above ♀ , below ♂

Mating took place in the lower part of the aquarium, usually just above the bottom, and lasted for more than eight hours. The female swam, almost snake-like, around her partner and repeatedly contacted his dorsal and head regions with her head. The courtship repertoire also included parallel positioning next to the body of the male as well as stroking and sinuous touching of the male by the body of the female. There were also traces of the embracing usual in labyrinthfishes. The eggs were expelled in small numbers (6 to 8) at long intervals and then picked up by the male.

The male brooded the eggs in his mouth. With an average size of 2.7 mm they were very large compared to those of other labyrinthfish species. Most of the time the male remained among dense vegetation. After a brooding period of 23 or 24 days he released the fry. They already had a total length of around 17 mm (0.6 in) and were fully developed, and in the following days took freshly-hatched *Artemia* nauplii as their first food without problem. With good feeding and optimal water quality, the little *Luciocephalus aura* grew very rapidly and after around three months were already 5 to 6 cm (2-2.5 in) long on average. From a very early stage the diet of the young *L. aura* again included small fish fry of suitable size. At a total length of

around 2.5 cm (1 in) they already exhibited the coloration of the parent fishes.

Total length: (size)
Up to 15 cm (6 in).

Remarks: (differences from other species of the genus)
Pathogen-free, soft, acid water is very important for optimal maintenance, and very soft, very acid water is an important prerequisite for successful breeding. The addition of humic substances is essential. The food should always consist of fresh, healthy food organisms as otherwise the fishes will react very badly. Unhealthy, dirty, and dead, partially fungussed food organisms can very rapidly lead to the death of the pikeheads.

References:
- HARTL, A. 2005: *Luciocephalus pulcher* – ein ebenso heikler wie faszinierender Labyrinther. Der Makropode, 27 (1/2): 5-11.
- NELLES, R. 2010: Meine Zucht von *Luciocephalus aura*. Internet Forum der IGL
- LINKE, H. 2012: Faszinierende *Labyrinthfische*/Fascinating labyrinth fish, *Betta* News 4/2012: 6-8

Explanation of the species name:

Latin adjective *pulcher* = "beautiful".

English names:

Pikehead

Brown Pikehead

Synonyms:

Diplopterus pulcher

Original description:

GRAY, J. E. 1830: Illustrations of Indian zoology; chiefly selected from the collection of Major-General Hardwicke, F.R.S. No page number, Pl. 87 (fig. 1).

Natural distribution:

The following locality is cited for the neotype designated by TAN & NG, 2005 : Mersing area, watercourse around 66 km from Kluang or 166 km from Batu Pahat, Johor Province, West Malaysia. However, *Luciocephalus pulcher* is also known from southern Thailand, Borneo, Sumatra, and the islands of Bangka and Belitung, as well as various islands in the Riau Archipelago.

Malaysia:

We were able to find the species in the surviving rainforest in the drainage of the Sungai Tambang in the Gahang Nitar region, west of Jemaluang in the area of Mersing in the south-east of West Malaysia.

Coordinates: 02° 21.23 N   103° 41.65 E

Kalimantan:

We recorded the species in the area south-west of Pundu in Central Kalimantan (Kalimantan Tengah, Borneo).

Coordinates: 01° 57.85 S   113° 02.48 E

And in a blackwater swamp region by the road from Palangkaraya to Kuala Kurun, around 15 km after the large bridge at the edge of Palangkaraya heading in the direction of Kuala Kurun.

Head close-up of *Luciocephalus pulcher*

Photo / Arend van den Nieuwenhuizen

Luciocephalus pulcher ♂

Photo / Arend van den Nieuwenhuizen

Coordinates:
02° 05.08 S   113° 57.16 E

Sumatra:

We found the species in the Soak
Putat, a widening in the course of
the River Pijoan, near to the village
of Pijoan on the road from Kota (=
town) Jambi to Muarabulian in the
province of Jambi.

Biotope data:

Malaysia:

Our researches in the rainforest area
in the Gahang Nitar region produced
interesting variations. The water
parameters in the Sungai Tambang
were pH 5.41 and conductivity
43 µS/cm at a water temperature
of 29.9 °C (86 °F), while in the
adjacent, flooded forest area the pH
was 4.76 and the conductivity 10µS/
cm at a water temperature of 28.5
°C (83.3 °F). The water was clear
and only slightly brown in colour.

Kalimantan:

Here the species was recorded
in a blackwater stream around a
kilometre west of the bridge in
Pundu. The water was very soft
and acid. The pH was 4.2 and the
conductivity measured 29 µS/cm
at a water temperature of 27.3 °C
(81.1 °F).

Sumatra:

We found the species in the lateral
lagoons and inundation zones of
the Sungai Pijoan, in shallow-water
areas along the banks and in the
rainforest. The water was slightly
cloudy and yellowish in colour, with
a pH of 5.0 and a water temperature

Accumulations of leaves at the water surface are preferred lurking places of *Luciocephalus*

Biotope of *Luciocephalus pulcher* in western Malaysia

Female *Luciocephalus pulcher* after capture

Luciocephalus pulcher photographed in the field

The *Luciocephalus pulcher* measured 11 to 13 cm total length on capture

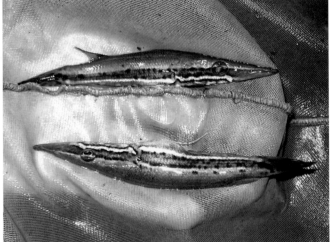

Luciocephalus pulcher after capture in western Malaysia

of 27.0 °C (80.5 °F). We also found *Luciocephalus aura* in the same area. *Luciocephalus pulcher* is also known to occur in numerous other parts of the island of Sumatra, and on the islands of Bangka and Belitung.

Reproduction:
Mouthbrooder.

Remarks: (differences from other species of the genus)
Pathogen-free, soft, acid water is very important for optimal maintenance. The notes on the feeding and reproduction of *Luciocephalus aura* apply equally here.

References:

- HARTL, A. 2005: *Luciocephalus pulcher* – ein ebenso heikler wie faszinierenden Labyrinther. Der Makropode, 27 (1/2): 5-11.

- TAN, H.H. & NG, P.K.L. 2005: The Labyrinthfishes (Teleostei: Anabantoidei, Channoidei) of Sumatra, Indonesia. The Raffles Bulletin of Zoology 2005 Supplement No. 13: 115-138.

- NELLES, R. 2010: Meine Zucht von *Luciocephalus aura* –Internet Forum of the IGL

- LINKE, H. 2012: Faszinierende / Labyrinthfische/Fascinating labyrinth fish, Betta News 4/2012: 6-8

The genus *Macropodus*

This genus, erected in 1801, contains the oldest scientifically described, so-called ornamental fishes. That includes not only tropical and subtropical species but also those that live in temperate zones. Its natural distribution extends from the cold regions of northern China across Japan to tropical central Vietnam. The regions north of the Truong Son Mountains and the Deo Hai Van, the so-called Pass of Clouds, form a natural boundary to the south. In fact nowadays *Macropodus* – chiefly *Macropodus opercularis* – are found in many places all over the world. But this is almost always the result of intentional or unintentional release of fishes from aquarium populations into the wild. Their spread has been favoured by numerous factors, with fish farms and the aquarium-fish trade in particular, but also the many fans of these fishes, playing a part.

While, depending on their origins, *Macropodus ocellatus* are fishes that can tolerate very low water temperatures and in live in temperate climatic zones, *Macropodus opercularis* are subtropical to tropical fishes. Although they too can, depending on their origins, tolerate water temperatures as low as 10 °C (50 °F) for short periods, they prefer higher temperatures. All other *Macropodus* species are warmth-lovers and "children of the tropics" to a greater or lesser degree.

For many years just three species were assigned to the genus: *Macropodus ocellatus*, *Macropodus opercularis*, and *Macropodus spechti*. In 2002 the species *Macropodus hongkongensis* and *Macropodus erythropterus* were scientifically described, but in 2008 the latter was declared a synonym of *Macropodus spechti*. In 2005 four new species, apparently closely related to the species *Macropodus opercularis*, were described in Vietnam, but unfortunately there were no descriptions or illustrations of their appearance in life. Comparisons with other, existing species were also lacking. The species scientifically described in 2005 are *Macropodus baviensis*, *Macropodus lineatus*, *Macropodus oligolepis*, and *Macropodus phongnhaensis*.

Explanation of the species name:

Latin adjective *baviensis* = "of Ba Vi" - the species lives in the area around the village of Ba Vi in North Vietnam.

English name:

None

Original description:

NGUYEN, H. D. & NGUYEN, V. H. in NGUYEN 2005: Ca Nuoc Ngot Viet Nam, Tap. III, BO THUY SAN TRUNG TAM TIN HOC, Bien soan: KS. Nguyen Van Hao, NHA XUAT BAN NONG NGHIEP (Agriculture Publinshing House), pp. 644-646

Natural distribution:

In flooded fields and rice paddies and associated irrigation ditches, and in ponds, in the area around the village of Ba Vi, south of the River Hong, north-west of the city of Hanoi in North Vietnam.

Biotope data:

No data are available at present.

Reproduction:

Probably a bubblenest spawner.

Total length: (size)

Purportedly around 7 cm (2.75 in).

Remarks: (differences from other species of the genus)

Males of this species are more colourful than females. There are 7-8 red and blue vertical bars on the body. The anal fin is green in colour. The caudal and anal fins exhibit numerous black spots. The ventral fins are partly pink in colour.

Biotope of *Macropodus* in Vietnam

Explanation of the species name:

Latin adjective *hongkongensis* = "of Hongkong", referring to the distribution region in and to the north of the city state of Hongkong in southern China.

English name:

Hongkong Paradisefish

Synonyms:

None

Original description:

FREYHOF, J. & HERDER, F., 2002: Review of the paradise fishes of the genus *Macropodus* in Vietnam, with description of two new species from Vietnam and southern China (Perciformes: Osphronemidae) Ichthyol. Explor. Freshwaters 13 (2): 147-167.

Natural distribution:

To date *Macropodus hongkongensis* has been recorded in the border zone to the north of Hongkong and in the north and east of the neighbouring province of Guangdong to the north (Jieyang, Shanwei, and Huizhou). Purported reports of the species from Taiwan have not been confirmed.

Biotope data:

These fishes live mainly in areas of little current in streams flowing down from areas of higher ground and in watercourses with bank vegetation and aquatic plants. The species has also been found in rice fields. It is supposedly sometimes found living syntopic with *Macropodus opercularis* and crosses between the two species have been recorded. All locations known to date have been in the subtropics and experience high summer temperatures and moderate to cool temperatures in the winter.

Reproduction:

Bubblenest spawner.

Total length: (size)

Up to 9 cm (3.5 in).

Remarks: (differences from other species of the genus)

The species is comparatively aggressive and is only very rarely available in the aquarium hobby. The yellow colour of the iris supposedly distinguishes it from *Macropodus opercularis* and *Macropodus spechti*, in which the iris is red.

References:

- SCHINDLER, I., 2008: Die Arten der Gattung *Macropodus*, The species of the genus *Macropodus*, Betta News, Journal des European Anabantoid Club mit AK Labyrinthfische im VDA, 2008 (3): 14–19.
- ZHOU HANG, 2008: Verbreitungsgebiet von *Macropodus hongkongensis* auf dem chinesischen Festland.- Distribution of *Macropodus hongkongensis* in mainland China, Betta News, Journal des European Anabantoid Club mit AK Labyrinthfische im VDA, 2008 (4): 27–29.

Macropodus hongkongensis ♂

Explanation of the species name:
Latin adjective *lineatus* = "lined", "with lines", referring to the longitudinal lines on the body.

English name:
None

Synonyms:
None

Original description:
NGUYEN, V. H., NGO, S. V. & NGUYEN, H. D.in NGUYEN 2005: Ca Nuoc Ngot Viet Nam, Tap. III, BO THUY SAN TRUNG TAM TIN HOC, Bien soan: KS. Nguyen Van Hao, NHA XUAT BAN NONG NGHIEP, pp. 641-643.

Natural distribution:
The holotype originated from the area around Mon Chink in the limestone hills in Phong Nha, Quang Binh Province, around 250 km north of the Trung Son Mountains and the Doe Hai Van (Pass of Clouds) in the north of central Vietnam.

Reproduction:
Probably a bubblenest spawner

Total length: (size)
Total length is currently unknown, but is probably around 7 cm (2.75 in).

Remarks: (differences from other species of the genus)
According to the original description, the colour of the body is dark blue without vertical barring but with 5 to 6 longitudinal lines. A large part of the ventral fins is white in colour. At certain times of the year the males are very much more colourful than the females. The species exhibits 8 to 9 lines on the body.

Small Town in Vietnam

Macropodus ocellatus CANTOR, 1842

Explanation of the species name:

Latin adjective *ocellatus* = "with an ocellus" (eye-spot), referring to the opercular spot.

English name:

Roundtail Paradisefish

Synonyms:

Chaetodon chinensis
Macropodus ctenpsoides
Macropodus opercularis
Polyacanthus chinensis
Macropodus chinensis
Polycanthus paludosus
Polyacanthus opercularis

Original description:

CANTOR, T.E., 1842: General features of Chusan, with remarks on the flora and fauna of that island. Ann.Mag. Nat. Hist. (N.S.) 9 (60): 484-493.

Natural distribution:

The species is distributed in central China to north of Shanghai on the east coast, on the Korean peninsula, and in central and southern Japan, though it has not been confirmed as originally native to Japan. These fishes are found south to the subtropics and north to the boundary region between the temperate and subarctic climatic zones. This means that fluctuations in water temperature are not uncommon, with up to 30 °C (86 °F) in the warm seasons of the year and down to freezing point in the cold months; this must always be taken into consideration if maintenance in the aquarium is to be successful.

Biotope data:

Macropodus ocellatus lives in irrigation ditches associated with rice fields, in swamps, and in streams and small rivers, by preference among areas of dense vegetation. Water parameters are of secondary importance.

Macropodus ocellatus ♀

Macropodus ocellatus ♂

Macropodus ocellatus pair, ♀ in front

Distribution region of *Macropodus ocellatus* in the south Shanghai, China

Distribution region of *Macropodus ocellatus* in the west Shanghai, China

Reproduction:

Bubblenest spawner.

These fishes spawn beneath a small bubblenest, by preference among dense vegetation. The developing brood is tended by the male.

Total length: (size)

Males up to 8 cm (3.125 in); females remain somewhat smaller and have smaller fins compared to males.

Macropodus ocellatus ♂

Remarks: (differences from other species of the genus)

In general these fishes can be described as "grey mice". Only in display coloration does the male exhibit an impressive pattern of markings and striking colours. The female changes colour mainly at spawning time, when she becomes very light, almost cream-coloured. The sexes can be readily told apart. Males, even when only half-grown, have a more prolonged dorsal and anal fin. Females are plumper in the belly region when filled with eggs. In addition the sexes can be identified by holding them up to a light, as described for *Trichopsis pumila*.

For successful maintenance the water temperature in the aquarium should if possible be lowered to 6 to 10 °C (43 to 50 °F) for an extended period during the winter months (i.e. the cold part of the year), and can be raised to up to 30 °C (86 °F) in the summer. Without this temporary drop in temperature these fishes are very short-lived and/or susceptible to disease.

The species is comparatively rare and was not available in the aquarium hobby for many decades. Unlike all other *Macropodus* species these fishes have a rounded rather than a forked caudal fin.

References:

• PAEPKE, H.-J., 1994: Die Paradiesfische. Die neue Brehm-Bücherei, Vol. 616, Westarp Wissenschaften, Magdeburg, Germany.

• FREYHOF, J. & HERDER, F., 2002: Review of the paradise fishes of the genus *Macropodus* in Vietnam, with description of two new species from Vietnam and southern China (Perciformes: Osphronemidae) Ichthyol. Explor. Freshwaters,13 (2): 147-167.

• KUBOTA, K. 2009: Wildlebende Labyrinthfische in Japan, About the wild labyrinth fish seen in Japan Betta News, Journal of the European Anabantoid Club mit AK Labyrinthfische im VDA, 2009 (3): 6-8.

Explanation of the species name:

Latinised Greek *oligolepis* = "with few scales", referring to the smaller than usual number of scales in the lateral line.

Original description:

NGUYEN, V. H., NGO, S. V.& NGUYEN, H. D. in NGUYEN 2005: Ca Nuoc Ngot Viet Nam, Tap. III, BO THUY SAN TRUNG TAM TIN HOC, Bien soan: KS. Nguyen Van Hao, NHA XUAT BAN NONG NGHIEP, pp. 644

Natural distribution:

The holotype originated from the area of Sohn Trach, Phong Nha-Ke Bang, Quang Binh Province, around 250 km north of the Trung Son Mountains and the Doe Hai Van (Pass of Clouds) in the north of central Vietnam.

Biotope data:

No details are available at present, but because the species lives in a limestone region then it is reasonable to assume mineral-rich, alkaline waters.

Reproduction:

This species is probably a bubblenest spawner.

Total length: (size)

Possibly around 8 cm (3.125 in).

Remarks: (differences from other species of the genus)

According to the original description, the body of *Macropodus oligolepis* is light yellow in colour. There are 7-8 lines running along the body, and numerous irregularly distributed black spots. The fins are grey and slightly yellow at their margins.

Cycling ornamental fish shops in Vietnam

Macropodus opercularis LINNAEUS, 1758

Explanation of the species name:

Latin *opercularis* = "having an opercular spot", referring to the spot on the posterior edge of the gill-cover.

English name:

Paradisefish

Synonyms:

Labrus opercularis
Labrus operculatus
Macropodus filamentosa
Macropodus ocellatus
Macropodus venustus
Macropodus viridi-auratus
Platypodus furca
Polyacanthus opercularis
Polyacanthus viridi-auratus

Original description:

LINNAEUS, C. 1758: Systema Naturae, 10th edition: 283.

Natural distribution:

The species is widely distributed. The natural habitat extends from central China and the southern islands of Japan in the north south to the tropical regions of North Vietnam. The Trung Son Mountains in central Vietnam appear to be the natural boundary in the south. These fishes have also been recorded further south, but this probably doesn't represent their natural distribution but intentional or unintentional introduction.

Biotope data:

The species inhabits streams, small rivers, and ponds, as well as swamps and rice fields, predominantly in lowland areas. *Macropodus opercularis* are also found in higher areas, but only rarely.

Macropodus opercularis pair, ♀ in front, ♂ behind.

Macropodus opercularis ♂

One such exception is a hilly region with an altitude of around 420 m (1400 feet) in the Taipei district in the north of Taiwan, which is home to a very colourful variant of the species, notable for its particularly striking red coloration. We found these fishes in a large upland pool in mid February. The water in the biotope studied had a temperature of 18.5 °C (65.3 °F), a conductivity of 30 µS/cm, and a pH of 6.3. It was a large body of standing water with dense vegetation along the banks as well as stands of aquatic plants in the areas of open water. The water was yellow-brownish in colour and cloudy. The population density was high. The fishes captured were half-grown to adult. The lowest water temperature in the course of the year is around 10 °C (50 °F) in December, and the highest around 28 to 30 °C (82.5 to 86 °F) in July/August. The breeding season is in March/April. (Research in February 2009: N. CHIANG & H. LINKE.)

Reproduction:

Macropodus opercularis is a bubblenest spawner that constructs its nest among floating plants at the water's surface. These fishes spawn in pairs, and the eggs are released while the pair embrace beneath the bubblenest. Both fishes participate in the retrieval of the upward-floating eggs and place them in the nest. When spawning is over the female is driven away from the nest and only the male tends the brood. It is generally possible to have several successive broods growing on in the same aquarium, as if the parent fishes are well fed they will rarely predate on their offspring. A water temperature of 25 to 27 °C (77 to 80.5 °F) is recommended as suitable for breeding. The fry swim free after around 4 days. Tiny pond foods and infusorians are recommended as first foods for the very tiny fry.

Total length: (size)

Males around 10 cm (4 in), females around 8 cm (3.125 in).

Remarks: (differences from other species of the genus)

Macropodus opercularis was the first warm-water fish species to be kept in captivity and has been maintained continuously in the aquarium since its first importation to Europe (1869 to France and 1876 to Germany). Its attraction can be ascribed mainly to the red crossbands and the large fins, especially the impressively forked caudal fin.

References:

- PAEPKE, H.-J., 1994: Die Paradiesfische. Die neue Brehm-Bücherei, Vol. 616, Westarp Wissenschaften, Magdeburg, Germany.
- FREYHOF, J. & HERDER, F., 2002: Review of the paradise fishes of the genus *Macropodus* in Vietnam, with description of two new species from Vietnam and southern China (Perciformes: Osphronemidae) Ichthyol. Explor. Freshwaters, 13 (2): 147-167.

Macropodus opercularis, silver cultivated form

Macropodus opercularis, albino cultivated form

Macropodus opercularis, blue cultivated form

Macropodus opercularis during mating

Macropodus opercularis ♂ , brood care beneath the bubblenest

Embracing the female beneath the bubblenest

Macropodus opercularis at the start of the embrace

• SCHINDLER, I., 2008: Die Arten der Gattung *Macropodus*, The species of the genus *Macropodus,* Betta News, Journal des European Anabantoid Club mit AK Labyrinthfische im VDA, 2008 (3): 14–19.

• KUBOTA, K. 2009: Wildlebende Labyrinthfische in Japan, About the wild labyrinth fish seen in Japan Betta News, Journal of the European Anabantoid Club mit AK Labyrinthfische im VDA, 2009 (3): 6-8.

Field of *Macropodus opercularis* New Taipei City in Taiwan

Habitat of *Macropodus opercularis* on the island of Taiwan

Mountain lake to the north of Taipei, biotope of *Macropodus opercularis*

Mountain region in the Taipei district of Taiwan

Catching *Macropodus opercularis* in a mountain lake in the Taipei district

Macropodus opercularis after capture. The species is very colourful here

Field of Macropodus opercularis near Da Nang in south Vietnam

Rice fields south of the Trung Son mountains in central Vietnam

Macropodus opercularis often inhabits the rice fields

Rice fields west of Da Nang

Swamp in the north of Vietnam

Da Nang, the last stop before the Pass of Clouds into North Vietnam

Explanation of the species name:

Latin adjective *phongnhaensis* = "of Phong Nha", referring to the occurrence of the species in the area of Phong Nhe-Ke Bang (Quang Binh Province).

Original description:

NGO, S. V., NGUYEN, V. H.& NGUYEN, H. D. in NGUYEN 2005: Ca Nuoc Ngot Viet Nam, Tap. III, BO THUY SAN TRUNG TAM TIN HOC, Bien soan: KS. Nguyen Van Hao, NHA XUAT BAN NONG NGHIEP, pp. 640-641

Natural distribution:

The holotype originated from the area around Sohn Trach, Phong Nhe-Ke Bang, Quang Binh Province, around 250 km north of the Trung Son Mountains and the Doe Hai Van (Pass of Clouds) in North Vietnam.

Biotope data:

Currently unknown.

Reproduction:

Probably a bubblenest spawner.

Total length: (size)

Probably up to 7 cm (2.75 in).

Remarks: (differences from other species of the genus)

In the original description the coloration of the body is given as brown with 8 vertical stripes on the body. The fins are dark grey.

Market for food fish in Vietnam

Explanation of the species name:

Dedication in honour of the fish enthusiast SPECHT, who obtained some of these fishes in France and took them back to Germany.

English name:

Black Paradisefish

Synonyms:

Macropodus concolor
Macropodus opercularis concolor
Macropodus opercularis var. *spechti*
Macropodus erythropterus

Original description:

SCHREITMÜLLER, W., 1936: Ein neuer Makropode(?), *Macropodus opercularis* L. var. *spechti*. Das Aquarium, 10: 181-182

Natural distribution:

The distribution region is apparently not very large and restricted to the area north of the Trung Son Mountains and the Doe Hai Van (the Pass of Clouds) in central Vietnam, extending only to about 100 km north of the old imperial city of Hue. Occasional records south of the Trung Son Mountains are probably not natural habitats. The species lives in still and gently flowing waters as well as in swampy fields, and sometimes also in rice fields and associated irrigation ditches. According to research by HERTEL, *Macropodus spechti* also lives in ditches and pools in and around the old royal palace in Hue. HERDER and FREYHOF recorded the species in the vicinity of the River Huong (which flows into the sea near Hue) in Thua Thien Hue Province, and in numerous ditches, swampy fields, and streams, as well as a small number of specimens to the south in Quang Nam Province.

Biotope data:

We found the species in rice fields and swamps as well as inundation zones by National Highway 1 from Hue to Da Nang, south of the Doe Hai Van, around 13 km before L. CO, 51 km before Da Nang, in Vietnam.

The water was predominantly still, clear, and slightly brownish in colour. The water temperature measured around 30 °C (86 °F) (air temperature around 33 °C (91.5 °F)).

Coordinates: 16° 16. 19 N   107° 58. 00 E
(Research in February 2009: N. CHIANG & H. LINKE.)

The population density of *Macropodus spechti* was not very high. The species was living syntopic with *Trichogaster trichopterus* and *Trichopsis vittata*.

Reproduction:

Bubblenest spawner.

Macropodus spechti ♂ , black variant.

Macropodus spechti mating beneath the bubblenest

playing male *Macropodus spechti*

Courting pair of *Macropodus spechti*

Macropodus spechti ♂
during brood care

Swamp in the Hue region in Vietnam

Fishing with a smile

The Hai Van Pass, the Pass of Clouds, the climatic boundary between North and South Vietnam.

Swamp in the Hue region, habitat of *Macropodus spechti*

Macropodus spechti after capture

Catching *Macropodus spechti* using electrodes

Macropodus spechti ♂ , red-backed variant
Synonym: *Macropodus erythropterus*

The species is very productive. It appears that an excess of males among the offspring is the rule. The differences between the sexes are often very late to appear.

Total length: (size)

Males to 9 cm (3.5 in); females remain somewhat smaller.

Remarks: (differences from other species of the genus)

In general *Macropodus spechti* exhibits a grey body colour, but this turns to almost black at higher water temperatures or during threat and courtship display. The fins are then edged with white and the male exhibits bright red ventral fins. In addition, in this species the female has shorter fins and remains somewhat smaller.

This fish was described by AHL as *Macropodus opercularis concolor* in 1937. Later it was classed as a distinct species and given the name *Macropodus concolor*. But because Wilhelm SCHREITMÜLLER had previously termed this species *Macropodus opercularis* var. *spechti* in an article in 1936, in 2002 HERDER & FREYHOF changed the specific name to *Macropodus spechti* in accordance with the Principle of Priority in the rules of zoological nomenclature. Despite a lot of objections and discussion as well as appeals, in 2008 the change of name was confirmed by the International Commission for Zoological Nomenclature .

Macropodus erythropterus, the Redbacked Paradisefish, a species described by FREYHOF & HERDER in 2002, was declared a synonym of *Macropodus spechti* by WINSTANLEY & CLEMENTS in 2008.

References:

- PAEPKE, H.-J., 1994: Die Paradiesfische. Die neue Brehm-Bücherei, Vol. 616, Westarp Wissenschaften, Magdeburg, Germany.

- FREYHOF, J. & HERDER, F., 2002: Review of the paradise fishes of the genus *Macropodus* in Vietnam, with description of two new species from Vietnam and southern China (Perciformes: Osphronemidae) Ichthyol. Explor. Freshwaters,13 (2): 147-167.

- ICZN Opinion 2145 (Case 3255) 2006: *Macropodus spechti* SCHREITMÜLLER, 1936 (Osteichthyes, Perciformes): priority maintained, Bulletin of Zoological Nomenclature 63 (1) March 2006.

- SCHINDLER, I., 2008: Die Arten der Gattung *Macropodus*, The species of the genus *Macropodus*, Betta News, Journal des European Anabantoid Club mit AK Labyrinthfische im VDA, 2008 (3): 14 –19.

- WINSTANLEY, T. & CLEMENTS, K. D., 2008: Morphological re-examination and taxonomy of the genus *Macropodus* (Perciformes, Osphronemidae), Zootaxa 1908: 1-27

The genus *Malpulutta*

This genus, erected in 1937 by the Ceylon scientist DERANIYAGALA, is monotypic – in other words there is only one species in the genus. In fact the describer of the nominate form also described a subspecies, *Malpulutta kretseri minor*, but nowadays this taxon is regarded as just a variant. *Malpulutta kretseri* is variable in its coloration. The nominate form from the area of the north-western province of Dandegamuva in Sri Lanka exhibits not only a marbled grey base coloration on the body but also lots of red and red-brown colour elements. The form described as the subspecies *Malpulutta kretseri minor*, supposedly remains smaller and exhibits more blue and black on a comparable body base colour. Today *Malpulutta kretseri* is again regarded as a rarity in the aquarium hobby. They are uncommon even in the natural habitat and found only in small numbers. Their exportation has been prohibited for some years.

Explanation of the species name:

Dedication in honour of de KRETSER.

English name:

Kretser's Paradisefish

Synonym:

Malpulutta kretser minor

Original description:

DERANIYAGALA, P. E. P. 1937: *Malpulutta kretseri* - a new genus and species of fish from Ceylon. Ceylon Journal Sci., SECT: b: Zool: geol. 20 (3): 351-353. Natural distribution:

Malpulutta kretseri ♂ in normal coloration

The north-west and south-west of the island of Sri Lanka. The describer, DERANIYAGALA, gives the natural habitat as Nikaväratiya and Hettipola in the north-west of the island and Gilimalé in the province of Sabaragamuva. There is a further location in the south-west of the island, around 15 miles north-east of the town of Galle, in a rainforest area surrounded by rice fields, whose watercourses constitute the southern tributaries (in the broadest sense) of the great Ging-Gang River.

Biotope data:

The region known as the Kottawa Forest Reserve, in the south-west of the island, is a small area of rainforest through which narrow watercourses, 1 to 2 m (40 to 80 in) wide and up to 40 cm (16 in) deep, still wend their way, in most cases completely

Aggressive male of *Malpulutta kretseri* often exhibit blue coloration. This sometimes also denotes a local variant

Malpulutta kretseri ♂ when aroused

Sequence of shots showing mating of *Malpulutta kretser*

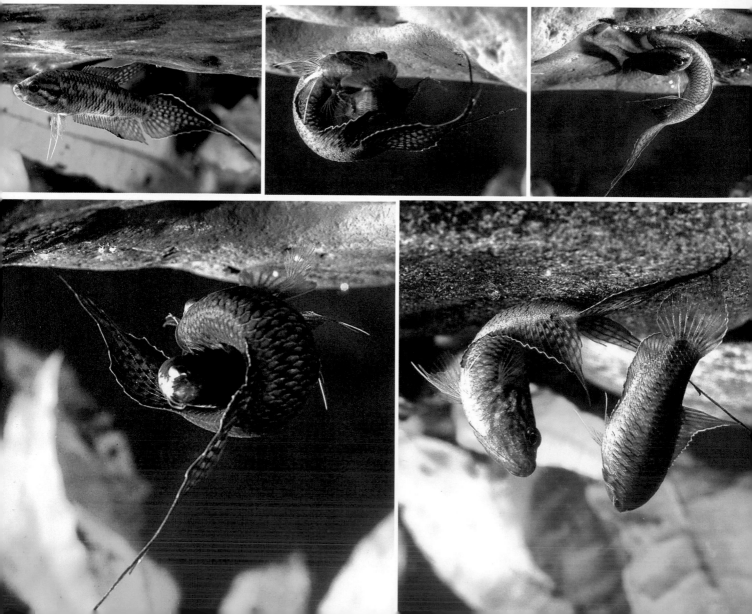

Rice fields in Sri Lanka

The rice-field irrigation channels in Sri Lanka are home of numerous labyrinthfishes

The swamps and small watercourses of Sri Lanka are also habitats of *Malpulutta kretseri*

overgrown by the forest greenery. The rays of the sun rarely reach the surface of the water, so semi-darkness permanently prevails. The bottom consists of sand and is yellow-brownish in colour. The bank zones are largely bordered by overhanging trees and scrub. Numerous dead leaves from this bank vegetation cover the bottom and provide excellent cover. They are favourite haunts of these fishes. There is simultaneously a slight but very important enrichment of the water with humic substances.

Reproduction:

Bubblenest spawner.

Unfortunately this species isn't always easy to breed. Aquaria with a length of up to 50 cm (20 in) are perfectly adequate for the purpose. They like to construct their relatively small nests only a few centimetres above the bottom in caves, or beneath rocks or broad leaves. After a peaceful courtship spawning takes place beneath the nest, with the male wrapping himself around the female and turning her so her genital zone points upwards. After five seconds of this tight embrace, during which the eggs are expelled and fertilised, the male releases himself from the female and the pair sink to the bottom in a form of spawning paralysis. After several seconds the female awakens and collects up the eggs in her mouth, then carries them to the bubblenest. After a few seconds more the male awakens and immediately begins guarding the territory. If there are no other fishes in the aquarium this defensive behaviour isn't seen, and the male immediately devotes himself to collecting up the eggs and tending them. But in this species the female can also be clearly observed to actively collect up the eggs.

The spawning ends after around five hours and the male alone assumes responsibility for the care of the clutch. The larvae break through their shells after 55 to 60 hours at a water temperature of 27 °C (80.5 °F). Because their water tolerance range is unfortunately not very great, approximation of the natural water parameters is important for successful development of the offspring. Only 15 to 25 fry swim free after around five days; a very small brood numerically. Perhaps this is the reason for the species' limited occurrence in the wild.

Malpulutta kretseri is regarded as one of the more delicate and sensitive labyrinthfishes. Aquaria with a length of between 60 and 80 cm (24 and 31 in) are adequate for maintenance.

The large elephant often drives away the small one

Sri Lanka still exudes a special atmopshere

Smiling girls, again while fishing

Expanses of *Cryptocoryne* in the watercourses are likewise habitats of *Malpulutta kretseri*

Large rivers criss-cross the island

Abundant planting, plus caves made from rocks and coconut shells, should be included in the decor. Biologically clean, low-nitrite, well-filtered water is an important prerequisite for successful maintenance. The container should be tightly covered as these fishes are good jumpers and will escape through the tiniest of openings.

Total length: (size)

Malpulutta kretseri is a small, dainty-looking species, in which females are full-grown at 4.5 cm (1.75 in) and males at just 6 cm (2.4 in) total length. This doesn't include the approximately 2-cm (0.75-in) long prolongation of the central caudal-fin ray in males.

Remarks: (differences from other species of the genus)

Malpulutta kretseri are very variable in their coloration. Normally a grey brown base colour is the main feature of their appearance. When they are slightly aroused this changes to a dark marbled pattern and when they are very excited this pattern becomes very dark. The nominate form is brown in colour with red elements and grows somewhat larger than a second variant, still occasionally termed *Malpulutta kretseri* minor, which supposedly remains somewhat smaller. The fins exhibit a blue-black coloration. In males of both variants the caudal fin is greatly prolonged and lanceolate, and the dorsal, anal, and caudal fins are bordered with white. Recommended foods include small live "pond foods" and/or cultivated live foods, plus frozen and flake foods.

The genus *Osphronemus*

The fishes in this genus are only exceptionally maintained as aquarium fishes and continue to be known mainly as food fishes in their native lands. Because, depending on the species, they can attain a total length of up to 70 cm (27 in) or more, they are unsuitable for maintenance in the "small" hobby aquarium and their size limits them to appropriately large tanks, usually in public aquaria. These fishes are often exported as juveniles. The overwhelming majority come from fish farms and are sold as food fishes for human consumption in tropical countries. Occasionally captive-bred colour sports of these species appear on the market, for example "pink", "white", "gold", and unfortunately they are also offered injected with various dyes to create colour patterns on the body.

The genus, erected in 1802 by LACEPÈDE, today contains four species that differ in appearance in terms of coloration and size. In the natural habitat the fishes of the genus live predominantly in larger lakes and in the bank zones of rivers of all sizes where there are plenty of plants. Water parameters are usually of secondary importance.

Explanation of the species name:

Latinised Greek *exodon* = "external tooth", referring to the rows of outward-pointing teeth on the lips.

English name:

Elephant Gourami

Cambodian name:

Trey romeas

Original description:

ROBERTS, T. 1994: *Osphronemus exodon*. a new species of giant gouramy with extraordinary dentition from the Mekong, Nat. Hist.Bull Siam Soc.v.42, pp. 67-77 Fig.1-6 Stung Treng market, Cambodia.

Natural distribution:

To date known only from the Mekong area in southern Laos, close to the border with Cambodia, and in northern Cambodia in the Stung Treng region, possibly south to Kratie.

Biotope data:

Apparently mainly in larger lakes and rivers with abundant vegetation.

Reproduction:

Probably a bubblenest spawner like *Osphronemus goramy*.

Total length: (size)

Up to 70 cm (27.5 in) total length.

Remarks: (differences from other species of the genus)

The species differs from *Osphronemus goramy* through the strikingly large, protruding lips set with rows of teeth, a larger number of hard rays in the dorsal and anal fins, and longer ventral fins.

References:

- RAINBOTH, W. J. 1996: Fishes of the Cambodian Mekong – FAO Species Identification Field
- Guide for Fishery Purposes. Food and Agriculture
- Organization of the United Nations: 218.

Because of its size *Osphronemus exodon* is also a popular food fish

Osphronemus exodon at the market in Stung Treng in north-eastern Cambodia

Osphronemus exodon attains a total length of around 70 cm

The striking rows of teeth on the lips are typical of the species

Explanation of the species name:

The name *goramy* is the local term used for this fish in its native distribution.

English name:

Giant Gourami

Original description:

LACEPÈDE, B. G. E.,1801: Hist. Nat. Poiss. vol.3: 116-117, Pl. 8 (fig. 2).

Synonyms:

Osphromenus gourami
Osphromenus notatus
Osphromenus olfax
Osphromenus satyrus
Osphronemus gorami
Osphronemus gourami
Osphronemus gouramy
Trichopode mentonnier
Trichopodus mentum
Trichopus goramy
Trichopus satyrus

Natural distribution:

Found in many countries in South-East Asia, and also in Sri Lanka and China through cultivation.

It is difficult to determine the actual natural distribution because nowadays the species is widely distributed as a food fish in large parts of subtropical and tropical Asia.

Biotope data:

The species lives in large lakes and slow-flowing rivers. These fishes are vegetarians that feed on emerse and submerse plants, fruits, seeds, and larger algae.

Maintenance in Asia:

The Giant Gourami is maintained chiefly as a food fish in artificial open-air ponds. Its flesh is very delicate in flavour. Adult specimens measure around 70 cm (27.5 in) in length and weigh around 7 kg (15.5 lbs). Because under normal conditions they "keep" longer at warm temperatures at the markets, their culture as food fishes is interesting. They are maintained in artificial open-air ponds around 10 x 20 m (33 x 66 feet) and about 1 m (40 in) deep. In this "biotope" they grow to large, saleable-size fishes within 30 to 36 months. They are often fed entirely on pig manure. The pigsties are constructed

Juvenile *Osphronemus goramy* around 15 cm long

Gold-orange cultivated colour form of *Osphronemus goramy*

Approximately 50 cm long *Osphronemus goramy* in natural coloration

Approximately 20 cm long *Osphronemus goramy* in natural coloration

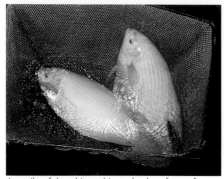

Juvenile of the white cultivated colour form of *Osphronemus goramy*

Approximately 50 cm long specimen of the white cultivated colour form of *Osphronemus goramy*

Adult male of the gold-orange cultivated colour form of *Osphronemus goramy*

immediately above the water, on stilts, for the purpose. The floor consists of a wooden grid on which the pigs live and their dung thus falls straight through into the water. There are two benefits to this: firstly the floor of the sty needs only to be sprayed with water to clean it, and secondly the fishes receive a very vitamin-rich food. The same process is used in the rearing of several other food-fish species. Young Giant Gouramis can attain a length of 25 cm (10 in) in just a year with this type of "maintenance".

Reproduction:

Reproduction takes place just once per year. Giant Gouramis need to be over two years old to breed and do so only at the beginning of the rainy season, perhaps because of the huge influx of fresh water and the concomitant cooling. The male constructs a hemispherical nest for the purpose, using pieces of plants. Only the flat, bottom part has a few "pearls" of foam in the centre. In order to construct the nest the male will even tear off pieces of plant growing close above the surface of the water. The nest itself is constructed in the well-vegetated bank region and has a diameter of around 50 cm (20 in) and a depth of about 25 cm (10 cm). The highest part of the hemispherical nest lies immediately beneath the water's surface and it then extends downwards, with a flat surface at the bottom. The centre of this vegetable hemisphere is filled with soft and dead pieces of leaf. The fishes then spawn during an embrace beneath this vegetable umbrella.

Full-grown Giant Gouramis are very productive.

Fish breeders report that it is not unusual for them to produce 20,000 eggs, often even more. After spawning the male tends his nest. For optimised production the heart of the nest containing the spawn is transferred to a large concrete tank for controlled rearing. At a length of 3 to 4 cm (1.25 to 1.5 in), i.e. at an age of three to four weeks, the fry are transferred to specially prepared open-air pools. Because Giant Gouramis are very greedy eaters and will eat not only pig manure but also waste vegetables, rice, and fruit, they rapidly grow into splendid specimens.

Total length: (size)

This labyrinthfish, nowadays maintained as a food fish in large parts of Asia, is the largest member of its family. It grows to up

Approximately 7 cm long juvenile of *Osphronemus goramy*

Artificially coloured *Osphronemus goramy* with red pattern

Concrete containers for rearing *Osphronemus goramy*

Artificially coloured *Osphronemus goramy* with blue pattern

Tanks measuring around 5 x 5 m for breeding *Osphronemus goramy*

to 70 cm (27.5 in) long and hence is suitable for maintenance only in very large aquaria.

Remarks: (differences from other species of the genus)

The somewhat plump appearance, the round head form, and the large, protruding mouth give this fish an almost sympathetic look. In the juvenile stage, up to around 15 cm (6 in) in length, these fishes have a completely different appearance. The pointed head and dark grey body coloration with 8 to 10 dark cross-bands don't look at all like *Osphronemus goramy*. Only with further growth do the head form and coloration change.

The species is now also offered in various cultivated colours, including white, pink, yellow, and gold. Sometimes white and yellow fishes are also injected with dyes to create a patterned appearance. This is very wrong and undesirable from an animal welfare viewpoint. Anyone considering the maintenance of juveniles should first of all establish whether it will subsequently be possible to pass them on to a public aquarium or zoological garden, as only such institutions have suitably large tanks for adults.

It should also be noted that the fry are extremely aggressive towards both con- and heterospecifics!

Outdoor expanse of water for breeding *Osphronemus goramy*

References:

• RAINBOTH, W. J. 1996: Fishes of the Cambodian Mekong – FAO Species Identification Field Guide for Fishery Purposes. Food and Agriculture Organization of the United Nations: 218.

Outdoor expanse of water for rearing *Osphronemus goramy*

The growing *Osphronemus goramy* find ideal living conditions in the vegetated expanses of water

The fishes being taken about a year to reach saleable size as foodfishes

Nets are suspended in the water of rivers for rearing

The smaller rivers and canals are the best sources of water for rearing the Giant Gourami

Osphronemus laticlavius ROBERTS, 1992

Explanation of the species name:

Latin latus = "broad" + *clavius* = "with a purple stripe" (originally on a senator's toga in ancient Rome).

English name:

Red Giant Gourami

Synonyms:

Osphronemus goramy

Original description:

ROBERTS, T. 1992: Systematic revision of the southeast Asian anabantoid fish genus *Osphronemus*, with descriptions of two new species. Ichthyol. Explor. Freshwaters 2 (4): 358, Fig. 4.

Natural distribution:

Borneo: north-eastern Sabah; eastern Malaysia: Sandakan region.

Biotope data:

Large lakes and rivers with dense vegetation in places.

Reproduction:

No precise details available. Probably a bubblenest spawner like *Osphronemus goramy*, where the male constructs a large bubblenest in a ball of aquatic plants, beneath which the fishes spawn during an embrace; the eggs rise into the nest where the brood develops, with the male guarding the nest.

Total length: (size)

On average 40 to 50 cm (16 to 20 in).

Remarks: (differences from other species of the genus)

Juveniles are grey to orange in colour and exhibit 4 to 5 cross-bands. They are peaceful among themselves and readily maintained in the aquarium. Adult fishes of this species exhibit no cross-bands and develop a splendid coloration in which the body is deep black with coral-red fins. The species is unproblematical in its maintenance, but suitable only for very large aquaria.

A male *Osphronemus laticlavius* around 60 cm long

Osphronemus septemfasciatus ROBERTS, 1992

Explanation of the species name:
Latin *septemfasciatus* = "with seven stripes", referring to the stripe pattern in juveniles of this species.

English name:
Red Giant Gourami

Original description:
ROBERTS, T. 1992: Systematic revision of the southeast Asian anabantoid fish genus *Osphronemus*, withdescriptions of two new species. Ichthyol. Explor. Freshwaters v. 2 (No.4), pp.355, Fig. 2-3.

Natural distribution:
The species has to date been recorded in East Kalimantan (Kalimantan Timur) in the Mahakam River system and in West Kalimantan (Kalimantan Barat), as well as in Sarawak and eastern Malaysia.

Biotope data:
Osphronemus septemfasciatus probably lives in large lakes and quiet rivers like *Osphronemus goramy*. It is probably likewise a vegetarian that feeds on emerse and submerse plants plus fruits, seeds, and algae.

Reproduction:
Bubblenest spawner.

Total length: (size)
Probably around 50 cm (20 in).

Remarks: (differences from other species of the genus)
Males exhibit a red body colour with brown-red fins, longitudinal rows of iridescent green dots on the upper third of the body, and metallic golden yellow on the lower third.

References:
- WITTE, K.-U., 1992: "Who is who" bei Riesenguramis, Der Makropode 5/6: 55.

East-Kalimantan

River Kayan Uli/Sarawak

River Kapuas/East-Kalimantan

River Stunggang/Sarawak

The genus *Parasphaerichthys*

For many years this genus contained just one, virtually unknown species, known as the Burmese Chocolate Gourami. But, as the name *Parasphaerichthys* indicates, it is not a true *Sphaerichthys*, and hence not a true chocolate gourami. Although the species was scientifically described as long ago as 1929, it wasn't until April 1978 that it was first imported alive, by A. WERNER. There were just two specimens, and despite excellent maintenance by W. FÖRSCH it wasn't possible to breed them. Because A. WERNER wasn't able to collect the fishes himself, but obtained them from the aquarium in Rangoon, no biotope data were available to provide important details of water parameters.

In 2002 a second species was assigned to this genus by BRITZ & KOTTELAT. The fishes were collected for the first time in February 2000, in the area west of Rangoon, and became known through K. KUBOTA, Siam Pet Fish Trading Bangkok.

Explanation of the species name:

Latin *lineatus* = "lined", "striped", referring to the longitudinal band on the body.

English name:

None

Original description:

BRITZ, R. & KOTTELAT, M. 2002: *Parasphaerichthys lineatus*, a new species of labyrinth fish from southern Myanmar (Teleostei: Osphronemidae) Ichthyl. Explor. Freshwaters,Vol.13, no.3, pp. 243-250

Natural distribution:

Parasphaerichthys lineatus was found in the area between Rangoon and Pathein, in the south-west of Myanmar (Burma). According to information from K. KUBOTA, Siam Pet Fish Trading Bangkok, the types of this species were collected around 35 km west of Rangoon on the road from Rangoon to Pathein.

Biotope data:

Biotope of *Parasphaerichthys lineatus*, around 30 to 35 km west of Rangoon, road to Pathein near Nyaungdon.

Water parameters:

pH value: 7.5-8.0
Conductivity: 305 µS/cm
Water temperature: 25 °C (77 °F)
General hardness: 4 °dGH
Carbonate hardness: 6 °dKH

The research took place around noon. The water temperature can drop to 20 °C (68 °F) or lower in the rainy season and/or at the coldest time of the year.

Reproduction:

FREYHOF observed these fishes beneath a bubblenest.

The author has seen them spawn on the bottom, similar to spawning in *Sphaerichthys osphromenoides*.

Total length: (size)

Around 3 cm (1.125 in).

Remarks: (differences from other species of the genus)

The species should be kept in mineral-rich, alkaline water (pH around 8). Maintenance in a well-planted species tank is advantageous. The water temperature for maintenance should be around 22 °C (71.5 °F), may rise to 26 to 28 °C (79 to 82.5 °F) for short periods, but should also drop to 18 to 20 °C (64.5 to 68 °F) for a few weeks several times per year.

References:

- LINKE, H. 2003: Das goldene Land Myanmar-Reise in die Vergangenheit, Aquarium live, pp. 20-25 Bede Verlag Ruhmannsfelden, Germany.

- LINKE, H. 2004: Das Goldene Land-Birma, Burma, Myanmar, Zoologischer Zentralaanzeiger, 5.

- LINKE, H. 2005: Fische aus dem Goldenen Land, Neuheiten aus Myanmar..Aquaristik Fachmagazin, 184: 35-40.

- DUNLOP, C. 2011: *Parasphaerichthys lineatus* - Ein Zwerg-Schokoladengurami/ *Parasphaerichthys lineatus* - A dwarf chocolate gourami, Betta News 1: 23-28.

Parasphaerichthys lineatus

Parasphaerichthys ocellatus PRASHAD & MUKERJI, 1929

Explanation of the species name:

Latin *ocellatus* = "bearing an ocellus" (eye-spot), referring to the mood-dependent, light-ringed, flank spots on both sides of the body.

English name:

Burmese Chocolate Gourami

Original description:

PRASHAD, B. & MUKERJI, D. D., 1929: The fish of the Indawgyi Lake and streams of the Myitkyina district (Upper Burma). Rec. Ind. Mus. (Calcutta) 31: 217, pl. 8 (figs. 4-4a), fig. 10.

Natural distribution:

The natural habitat lies in the higher mountainous zones of northern Myanmar (Burma). The scientists who described the species give the location as small muddy streams along the Kamaing Jade Mines Road, a few miles from Kamaing.

We were able to collect the species around 150 metres (500 feet) before the village of Yansyn, north of Lonton (Longton), on the Indawgyi Lake.

Another study site was in the River Nanhkwy, around 20 km before the town of Myitkyina on the road from Myitkyina to Mogaung, in Kachin State in the north of Myanmar (Burma), around 1650 km north of Rangoon.

Biotope data:

Around 150 metres (500 feet) before the village of Yansyn, north of Lonton (Longton), on the Indawgyi Lake. Residual pools in a river bed, both sides of the road, around 100 m (330 feet) before the river enters the lake. The biotope contained more water in the direction of the mountainside.

The following water parameters were recorded:

pH value: 8.5 to 9.0
Conductivity: 54 µS/cm
Water temperature: 25 °C (77 °F)
General hardness: 1 °dGH
Carbonate hardness: 2 °dKH

Parasphaerichthys ocellatus in normal coloration

Parasphaerichthys ocellatus in fright coloration

Monastery in the area of the Nanhkwy Bridge in northern Myanmar

Welcome sign at Lake Indawgyi

Temple precinct on Lake Indawgyi

Area of residual water by the Nanhkwy River where we found *Parasphaerichthys ocellatus*

Dense expanses of aquatic plants along the edges of the Nanhkwy River were preferred habitat of *Parasphaerichthys ocellatus*

The water level was very shallow in the margins, making collecting problem-free

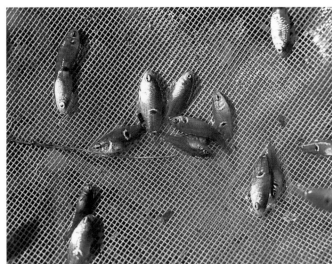

Parasphaerichthys ocellatus shortly after capture in the Nanhkwy River

The research took place at around 15.00 hours in the afternoon. The water temperature was lower in the morning, at around 20 °C (68 °F). It can drop to 15 °C (59 °F) or lower in the rainy season and/or the coldest part of the year. The water was cloudy.

Another study site was in the River Nanhkwy. The collecting sites were the shallow bank zones of the river in the area near the Nanhkwy Bridge. The water parameters there were:

pH value: 7.8 to 8.0
Conductivity: 85 µS/cm
Water temperature (in the afternoon): 20.5 °C (69 °F).
General hardness: 3.5 to 04 °dGH
Carbonate hardness: 2 °dKH
(Research in February 2004: H. LINKE et al.)

Reproduction:
Mouthbrooder (?)

Total length: (size)
Up to 3.5 cm (1.375 in).

Remarks: (differences from other species of the genus)
These fishes are full-grown at a total length of around 3.5 cm (1.375 in) and thus noticeably smaller than the well-known Chocolate Gourami. In addition they are more elongate in body form and the head is more pointed. The caudal fin is rounded. Viewed from above, the body is broader, creating a "beefier" effect. In addition these fishes are remotely similar in appearance only in fright coloration. In normal coloration they look more like a small Blue Gourami, *Trichogaster trichopterus*. In fright coloration they might easily be taken for a totally different species. The spots (ocelli) on the sides of the body, reflected in the scientific name, are visible only in display coloration.

FOERSCH maintained the two *Parasphaerichthys ocellatus* imported in 1978 in a small, well-planted aquarium filled with soft water. The fishes proved secretive and were very

Watercourse at the edge of Lake Indawgyi near Lonton, habitat of *Parasphaerichthys ocellatus*

shy. On the basis of his observations it can be assumed that this species benefits from a low water temperature of around 20 to 22 °C (68 to 71.5 °F) for maintenance. Nowadays we know from our own research that these fishes require mineral-rich water with a pH of around 8.0 to 9.0. The water temperature should rise above 25 °C (77 °F) only for short periods and should vary within the range 20 to 22 °C (68 to 71.5 °F) for long-term maintenance, and can also be allowed to drop to 15 to18 °C (59 to 62.5 °F) for a few weeks.

References:
- Linke, H. 2003: Das goldene Land Myanmar-Reise in die Vergangenheit,
- Aquarium live, pp 20-25 Bede Verlag Ruhmannsfelden, German
- LINKE, H. 2003: Fische aus dem Goldenen Land, Neuheiten.
- LINKE, H. 2004: Das Goldene Land-Birma, Burma, Myanmar, Zoologischer Zentralanzeiger, 5
- DUNLOP, C. 2011: *Parasphaerichthys lineatus*- Ein Zwerg-Schokoladengurami/ *Parasphaerichthys lineatus*- A dwaf chocolate gourami, Betta News, 1, Seite 23-28

The genus *Parosphromenus*

The liquorice gouramies are small group whose maintenance and breeding represents a particular challenge for many aquarists. Their often splendid coloration, especially during courtship, makes them very desirable to keep. Liquorice gouramies are not fishes for the beginner. Their maintenance requires good basic knowledge and a degree of experience. But for optimal maintenance they require not only the requisite environment but also as varied as possible a diet of suitably small foods. Because, with very rare exceptions, *Parosphromenus* species will accept only live food, it is not always easy to provide a balanced diet at all times of the year. But if all these prerequisites can be successfully provided then keeping these fishes can be very enjoyable and it is also possible to breed them.

Liquorice gouramies are found only in South-East Asia. Their natural habitats are often only very small areas or they may be distributed over hundreds of kilometres. They live predominantly in densely vegetated bank zones in small, slow-flowing, so-called blackwater rivers, where they are found mainly at depths of 20 to 100 cm (8 to 40 in), often among plants as well on the bottom among leaf litter (dead leaves) from the trees and bushes along the shoreline.

It is important to set up the aquarium optimally for liquorice gouramies. This will include providing dense groups of plants and bogwood as decor, peat filtration, and plenty of leaf litter (dead oak or beech leaves) on the bottom. This will make the removal of detritus from the bottom somewhat more difficult, but the fishes will have near-natural shelter and hiding-places, as well as spawning sites. Enriching the water with humic substances (leaf litter and peat) is, moreover, an important improvement to the aquatic environment of all blackwater fishes. While soft, slightly acid water is often adequate for maintenance, very soft, very acid water is indispensable for successful breeding; in other words, for maintenance the water parameters should be a pH value between 5 and 6 and a conductivity of 100 to 200 µS/cm, while for successful breeding the pH should be in the range 4.2 to 5 with a conductivity between 20 and 40 µS/cm. Where the source water is hard then this will be possible only via demineralisation.

Liquorice gouramies are small fishes, with the smallest members of the genus attaining a total length of around 2.5 cm (1 in) and the so-called "large" species growing up to 4.5 cm (1¾ in) long. Some species are very attractive with males characterised by especially striking colours. Live food organisms are a very important prerequisite for the optimal maintenance of *Parosphromenus* species.

Liquorice gouramies are concealed brooders that build small bubblenests and spawn in sheltered, cave-like spots among dense vegetation, often close to the bottom. The bubblenest often consists of only a few bubbles of air, or there may even be none at all. The female swims to the spawning site chosen by the male in order to lay her eggs. The male then embraces the female for a few seconds during which the eggs emerge and are simultaneously fertilised. The male then leaves the breeding "cave" briefly to defend the spawning site. The female takes the eggs into her mouth and attaches them to the ceiling of the "cave". Only a few minutes later the male swims back to the female and a further embrace takes place. The total duration of the succession of matings can be up to four hours.

After spawning the brood is tended and guarded exclusively by the male. The larvae hatch from the eggs after three to four days and then hang from the ceiling or walls of the spawning "cave", looking like little dark streaks. At this point in time I have often seen renewed mating and egg-laying. During the intervals between matings the female tends not only the newly laid eggs but also the larvae already hanging in the nest, without problems or losses. The fry swim free after on average seven days; they are very small and require the tiniest possible live food as first food. Usually the freshly-hatched nauplii of a small *Artemia* species are an appropriate size and well suited as the first food.

But brood care can also sometimes take a different course: my *Parosphromenus linkei* had spawned once again. The male had constructed a bubblenest consisting of only a few small air-bubbles in a small cave. After a rather reserved courtship the female indicated her willingness to spawn to the male by waggling her body, waving the posterior part of it to and fro, and developing so-called "sexy eyes", a black, usually vertical bar through the eye. The female next followed the male into the cave, where he embraced her, prompting her to release eggs which he immediately fertilised. This process lasted for around three to four hours. Once the spawning was over the female left the spawning site. Thereafter the male assumed the care of the spawn alone. Normally the female wouldn't go anywhere near the spawning cave thereafter. The spawning took place in the afternoon, but astonishingly next morning it was no longer the male but the female in the breeding cave. She was tending the eggs, from time to time taking one into her mouth and gently "washing" it before returning it to the nest a few seconds later. The male was no longer to be seen in the vicinity and didn't show himself at intervals in front of the spawning cave, either. This brood care by the female alone was also to be seen on the third day, when the larvae hatched and hung from the ceiling of the cave, where they were very inconspicuous. And the brood care by the female alone continued at this stage too. Unfortunately there was an explanation for the absence of the male. It appears he became ill after the mating and was no longer capable of tending the spawn. Four days later I found him dead. It would appear that shortly after spawning the female noticed the inability of the male to look after the brood because of illness and took over the care of the spawn herself. Just eight days after spawning the fry swam free and slowly left the breeding cave. And at this point in time the brood care by the female also ceased.

Parosphromenus species can be distinguished by their different, species-typical total length and by different caudal-fin forms (for example the tail may be lance-shaped or rounded or have just a single prolonged ray in the middle of the fin). They can also be differentiated by the number of fin-rays in the dorsal fin (i.e. the length – shorter or longer – of the dorsal fin) and by species-typical body markings such as dark, often turquoise-coloured spots on the sides of the body or light dot-like spots on the unpaired fins.

Distinguishing the species is often possible only on the basis of display coloration, caudal-fin form, and ventral-fin length in males. Females of the various species can be told apart only with great difficulty, often not at all, on the basis of external appearance. Only during the brief courtship phase do females of the various *Parosphromenus* species sometimes also exhibit a species-typical coloration, but this is usually limited to brown to red-brown colour in the unpaired fins, rarely with additional markings. Hence it is wise to keep all species separate and never keep the species with no clear distinguishing characters together in the same tank. Sometimes adult body size can be used to differentiate the species (for example, *Parosphromenus ornaticauda* and *Parosphromenus parvulus* with a total length around 2.5 to 3.0 cm (1 to 1.25 in) and *Parosphromenus quindecim* around 4.0 to 4.5 cm (1.5 to 2 in), in addition to completely different coloration and markings).

Parosphromenus alfredi KOTTELAT & NG, 2005

Explanation of the species name:

Dedication in honour of Eric ALFRED, former Director of the National Museum of Singapore.

Original description:

KOTTELAT, M. & NG, P. K. L. 2005: Diagnoses of six new species of *Parosphromenus* (Teleostei: Osphronemidae) from Malay Peninsula and Borneo, with Notes on other species. The Raffles Bulletin of Zoology 2005, Supplement 13: 102-103.

Type locality:

Western Malaysia: Johor Province: Kota Tinggi. Road from Mawai to Desaru.

Species-typical characters:

In comparison to all other *Parosphromenus* species, males of *Parosphromenus alfredi* have very long ventral fins, whose anterior ray is prolonged and very light in colour. This can be regarded as species-typical.

Similar species:

Parosphromenus tweediei is one of the most striking species of the *rubrimontis* group. This species exhibits a broad, bright red band in the central part of the unpaired black fins. This central red band may also have a narrow turquoise margin as a transition to the outer black part of the fin. The unpaired fins

have a narrow, light, outer margin. The ventral fins are normal in length and slightly turquoise in colour.

Parosphromenus rubrimontis exhibits a larger dark spot at the base of the caudal fin, with a broader, red, semicircular band adjoining, sometimes with a narrow, turquoise-coloured margin distally, and then a broader outer black band. This coloration and pattern are repeated in the dorsal and anal fins. The unpaired fins have a narrow, light outer margin. The ventral fins are comparatively short and turquoise-coloured. The species has a longer dorsal fin in comparison to the other members of the *rubrimontis* group.

Parosphromenus sp. Tanjung Malim is often incorrectly regarded

Parosphromenus alfredi ♂

Parosphromenus alfredi ♂

Biotope of *Parosphromenus alfredi* in the Sedili area

Parosphromenus alfredi was found mainly in the heavily vegetated bank zones

Small watercourse in the Sedili area, biotope of *Parosphromenus alfredi*

as a local form of *P. rubrimontis*, but has only a narrow red band in the predominantly black caudal fin, and there may be a similar band in the dorsal and anal fins. The ventral fins are normal in length and black in colour, with a delicate, lighter, anterior upper portion to the hard ray. The unpaired fins have a narrow, light outer margin.

Natural distribution:

The species is apparently found only in the area to the east and south-east of Kota Tinggi in the south-east of western Malaysia. These fishes have to date been recorded several times in the swampy areas along the road running in a south-easterly direction from Mawai to Desaru.

We found the species in a small river by the sand and gravel track (temporary road), around 4 km from the turn-off to Sedili from National Highway 212 at Sanjung, close to the kilometre marker showing 26 km to Kota Tinggi. At Kampong Mawai Baru the 212 joins National Highway 99, which in turn feeds into National Highway 3 and from there it is 56 km to Johor Baru, western Malaysia.

Coordinates:
01° 51' 23 N   104° 03' 54 E

Biotope data:

These fishes live in clear, slightly brown water to typical black water which is very acid

Aquatic plants and dead wood on the bottom provide good cover for the fishes

National Highway 212 from Sedili to Sanjung

with a pH value of 4.5 to 5.0. The species was numerous in the area at the time the collections were made, but is now (2012) very rare and possibly endangered as a result of cultivation and the creation of numerous oil-palm plantations.

The small stream that crosses the temporary road (sand and gravel track), around 4 km after it branches off from National Highway 212, contained clear, slightly brown water with a pH of 5.38 and a conductivity of 19 µS/cm. The water temperature measured 28.3 °C (83 °F) and air temperature was 33.4 °C (91.4 °F).

There was a further collecting site only 4.5 km further on, at the edge of an oil-palm plantation. This was a small woodland watercourse with clear, dark brown water with a slight to moderate current. The water parameters were pH 4.76, conductivity 21 µS/cm, and water temperature 29.0 °C (84.2 °F).

Coordinates:
01° 50' 02 N   104° 04' 26 E

(Research in September by YAP, P. & LINKE, H.)

Maintenance:
As described for the genus.

Reproduction:
As described for the genus.

Total length: (size)
Up to 3.5 cm (1.375 in).

The water in the *Parosphromenus alfredi* biotopes are always brownish in colour with a slight current

Parosphromenus allani BROWN, 1987

Explanation of the species name:

Dedication in honour of Allan BROWN, who first discovered these fishes.

Original description:

BROWN, B. 1987: Two new Anabantoid species. Aquarist and Pondkeeper 52 (3): 34.

Type locality:

Eastern Malaysia: Sarawak: area around the town of Sibu.

Species-typical characters:

Parosphromenus allani exhibits a bright red colour on the central parts of the unpaired fins, resembling bars. This is bordered by a broad, dark, often black band, in turn followed by a light, narrow, outer margin. There is a large dark spot at the base of the caudal fin. The rest of the caudal fin is red. The broad posterior, outer margin is dark brown with a light outer edging. The soft portion of the dorsal fin also sometimes exhibits a large dark round spot. The ventral fins are normal in length and have a faint turquoise coloration.

Similar species:

By comparison, *Parosphromenus* sp. Sungai Stunggang has no or only a small, dark spot on the caudal base and lots of red in the caudal fin. The red in the dorsal and anal fins is less extensive and sometimes adorned with turquoise bars. While the red of the caudal fin is bordered by only a faint dark band, the corresponding dorsal- and anal-fin markings are broader and bolder. The unpaired fins have light narrow margins. The ventral fins are somewhat longer and bright solid turquoise in colour. The central stripes on the body are narrower and in different positions.

Parosphromenus opallios has bright red on the central parts of the unpaired fins, in the form of narrow bars. This is followed by a broad dark, usually black band, and in between the two there is a turquoise-coloured band of variable width, especially in the dorsal and anal fins. The unpaired fins have light margins. The base of the caudal fin exhibits a smaller dark spot. The

Parosphromenus allani ♀

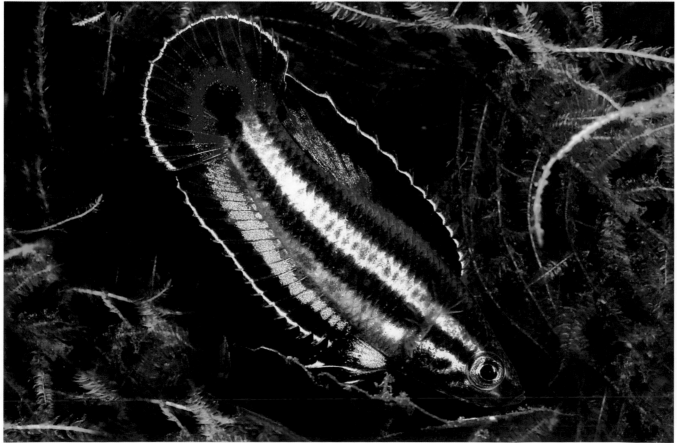

Parosphromenus allani ♂

A small blackwater river that crosses the road at Nibong, near Sibu

The water was dark red-brown in colour with a slight current

First look at *Parosphromenus allani* immediately after capture

ventral fins are somewhat longer and coloured turquoise right to their tips.

Parosphromenus quindecim is one of the largest liquorice gouramies known to date. *Quindecim* (Latin) means fifteen and refers to the 15 hard rays in the dorsal fin. The species exhibits a broad brown (also sometimes red) band in the central part of the unpaired fins, followed by a turquoise-coloured band, and then a dark, narrow band extending to the edge of the fins. The unpaired fins have light margins. The caudal fin has a large, dark, star-shaped spot at its base. The ventral fins are of normal length and exhibit no noteworthy coloration.

Natural distribution:

Eastern Malaysia: Sarawak: The species is found in the area of the town of Sibu. *Parosphromenus allani* was first discovered in 1986, on the eastern side of the road to the swimming bath and in the direction of Jalan Ulu Oya, 10.6 km from the intersection with the Jalan Tang Sang. The swimming bath is next to the school. According to the data of A. and B. Brown, these fishes live in small water-filled ditches and streams, up to a metre (40 in) deep, with clear, brown water and a pH of 5.4. (Research in July 1986: BROWN, A & B)

It is always very important to measure water parameters

Parosphromenus allani in a small blackwater river that crosses the road at Nibong, around 150 metres (500 feet) before the large bridge over the Batang Rajang

A detailed analysis of the water in the biotope provides a lot of information and thereby facilitates maintenance in the aquarium as well

During our own research in August 2008 we found *Parosphromenus allani* in a small blackwater river that crosses the road at Nibong, around 150 metres (500 feet) before the large bridge over the Batang Rajang, around 30 km south-east of Sibu, after the turn-off to the airport. The road there is a dual carriageway (four-lane). (Research in August 2008: YAP, P., LO, M., CHIANG, N , NGAI, K. M., & LINKE, H.)

Coordinates: 02°10′05 N   112°00′55 E (LINKE, H.)

Biotope data:

The blackwater stream was dark brown with a slight current. The pH measured 4.65 and the conductivity 5 µS/cm at a water temperature of 27.5 °C (81.5 °F). The underwater visibility was around 40 cm (16 in). *Parosphromenus allani* was found almost exclusively in the densely vegetated bank zones. In around 90 minutes we managed to catch around 25 *P. allani* (18 females and 7 males).

Maintenance:

Maintenance in the aquarium is problem-free given suitably soft acid water with the addition of humic substances. Good water quality is very important.

Reproduction:

This little bubblenest spawner has already been bred several times. The species like to spawn in small caves. An important factor is very soft and very acid water with the addition of humic substances.

After the fry become free-swimming they can be left with the parents in the breeding aquarium, and will grow very rapidly given appropriate feeding with freshly-hatched *Artemia* nauplii. However, here too very good water quality is important.

Total length: (size)
Up to 3.5 cm (1.375 in).

Remarks: (differences from other species of the genus)
Parosphromenus allani is very uncommon in the aquarium hobby and only very rarely imported.

References:

- BROWN, A. & B. 1987: A Survey of Freshwater Fishes of the Family Belontiidae in Sarawak. The Sarawak Museum Journal, 37: 155-172, 3 pls.
- BROWN, A. & B. 1987: A Summary of the Anabantoids Found in Sarawak in July 1986. AAGB Labyrinth 7(4), 34: 2-6
- GECK, J. 1992: Das Labyrinthfischportrait Nr. 68 *Parosphromenus allani* BROWN, 1987. Der Macropode 14 (9/10): 73-74.
- LINKE, H. 2009: Allani und Co Aquarium life. Bede Verlag, Ruhmannsfelden, Germany.

Parosphromenus anjunganensis KOTTELAT, 1991

Explanation of the species name:
Latin adjective meaning "of Anjungan".

Original description:
KOTTELAT, M. 1991: Notes on the taxonomy and distribution of some Western Indonesian freshwater fishes, with diagnoses of a new genus and six new species (Pisces: Cyprinidae, Belontiidae, and Chaudhuriidae). Ichthyological Exploration of Freshwaters, 2 (3): 281.

Type locality:
Borneo: Kalimantan Barat: Sungai Kepayang, 7 km SE of Anjungan on road to Pontianak.

Coordinates: 0° 20' N 109° 08' E. (KOTTELAT, M. 1990)

Species-typical characters:
Parosphromenus anjunganensis looks somewhat more elongate compared to other *Parosphromenus* species. A species-typical feature is the uniformly wine-red unpaired fins with a very narrow, usually light blue margin and no inclusions of other colours. The ventral fins are comparatively prolonged in males of this species.

Similar species:
Parosphromenus sp. Red remains smaller and exhibits faint areas of dark coloration at the base of the unpaired fins (and sometimes on the membranes), as well as a striking dark spot at the base of the caudal fin.

Natural distribution:
The species first became known from collections by Maurice KOTTELAT et al. in April 1990 and was collected again in July 1990 by BAER, NEUGEBAUER, and the author, and subsequently imported alive. We caught these fishes in a narrow water-filled ditch by the road from Anjungan to Sungai Penjuh, around 8 km south-west of Anjungan, in the Kapuas area in West Kalimantan (Kalimantan Barat) on the island of Borneo.

Biotope data:
The biotope was a typical blackwater with corresponding parameters. The following water parameters were measured in the almost stagnant water (low-water period): pH 4.5; carbonate hardness less than 1 °dKH; electrical conductivity 39 µS/cm at a water temperature of 27.6 °C (81.7 °F). The biotope was home to a second *Parosphromenus* species, which was likewise scientifically described by KOTTELAT, M. in 1991 and given the name *Parosphromenus ornaticauda*. (Research in August 1990: LINKE, H.)

Parosphromenus anjunganensis was not very numerous in this biotope but was more frequent in a larger, less acid, watercourse nearby. Both species were variably numerous depending on the water conditions. *Parosphromenus anjunganensis* apparently

Parosphromenus anjunganensis ♂

Collecting in the blackwater ditches between Anjungan and Sungai Penjuh in Kalimantan Barat

prefers less acid water. The water in the habitat studied was strongly coloured brown and clear, with a slight current, and again very soft and very acid, with a pH of 5.4. The electrical conductivity measured only 22 µS/cm at a water temperature of 28.3 °C (83 °F).

The species was also recorded in a small river around 7 km south-west of Anjungan by the road from Pontianak to Sanggau in West Kalimantan (Kalimantan Barat). These fishes supposedly also live in the Mandor River, around 20 km away to the east, which would mean a somewhat larger habitat.

Maintenance:
As described for the genus.

Reproduction:
As described for the genus.

Total length: (size)
Attains a total length of around 3.5 to 4.0 cm (1.375 to 1.5 in).

Remarks: (differences from other species of the genus)
Liquorice gouramies with striking blue coloration on the unpaired fins are sometimes also labelled as *Parosphromenus anjunganensis*; however they are not *Parosphromenus anjunganensis* but possibly an as yet unknown species.

References:
- LINKE, H., 2006: *Parosphromenus ornaticauda* – Ein Kleinod aus Kalimantan. Aquaristik Fachmagazin 191: 40-43.

Parosphromenus bintan KOTTELAT & NG, 1998

Explanation of the species name:
Named for the type locality, the island of Bintan.

Original description:
KOTTELAT, M & NG, P. K. L. 1998: *Parosphromenus bintan*, a new osphronemid fish from Bintan and Bangka Island, Indonesia; with redescription of *P. deissneri*. Ichthyol. Explor. Freshwaters, 8 (3): 263-272.

Type locality:
Indonesia: Riau archipelago: island of Bintan, just before km 45 on road from Tanjung Uban to Tanjung Pinang.

Species-typical characters:
Parosphromenus bintan has predominantly black fins, with a delicate light blue to turquoise-coloured band in the outer part of the dorsal, caudal, and anal fins. This is bordered on both sides by a broader black margin. The unpaired fins have light margins. The base of the caudal fin is largely black. The dark, delicate blue-coloured ventral fins are moderately long and the lower portion of their hard rays is light blue in colour. They exhibit no red coloration.

Similar species:
Parosphromenus harveyi exhibits turquoise-coloured stripes in the unpaired fins, but, compared to similar species, more in the central part of the fins and not in the edge region. This species again has no red coloration in the fins.

Parosphromenus sp. Pontian Besar exhibits a similar coloration and markings, but, compared to the other species, has a significant amount of red interior to the turquoise-coloured stripes in the unpaired fins, especially the caudal fin.

Parosphromenus cf. *bintan* (blue line) has bolder and narrower turquoise banding and a narrower black margin in the outer part of the unpaired fins. This species exhibits no red in the fins.

Natural distribution:
The island of Bintan, just before km 45 on the road from Tanjung Uban to Tanjung Pinang; (1° 06' 40.1" N   104° 30' 09.8", H. H. TAN et al.)

The island of Bangka, 2 km east of the road to Pangkalpinang; blackwater stream in heath forest.

Parosphromenus bintan has been confirmed as occurring on the Indonesian islands of Bintan and Bangka. The species has so far been recorded almost all over the island of Bangka, where it is frequently found living syntopic with the species *Parosphromenus deissneri*.

Biotope data:
GERSTNER,K., LINKE, M. and the author found this species together with *Parosphromenus deissneri* in the south of the

Parosphromenus bintan ♂ Bangka island

Black water habitat of the Bangka island

island of Bangka, in blackwater streams which in places were heavily vegetated. The water here was mostly flowing, very clean and clear. The water temperature was on average 25.5 °C (78 °F). The water was very mineral-poor and had a pH value of 4.8 to 5.5. (Research in July 1993: LINKE, H.)

A year later FRANK, K. and NEUGEBAUER, N. again recorded these fishes on the island of Bangka, in a blackwater stream near a bridge on the road from Airbara to Pajoeng, close to the kilometre marker Airbara 8 (km) - Pajoeng 23 (km). The watercourse flowed through primary forest and had a width of on average 4 metres (13 feet) and a depth of up to 1.5 metres (5 feet). Even towards the end of the dry season there was still a large amount of water there. (Research in October 1994: NEUGEBAUER, N.)

Maintenance:

As described for the genus.

Reproduction:

As described for the genus.

Total length: (size)

The species grows to around 3.5 cm (1.375 in) long.

References:

- GECK, .J. & KOPIC, G. 2005: Prachtguramis. Aquaristik aktuell, special publication 1/2005: 30-36.

Parosphromenus cf. *bintan*

Explanation of the name:

This possibly good species exhibits numerous parallels to the species *Parosphromenus bintan*.

English name:

Blue-line Liquorice Gourami

Species-typical characters:

Parosphromenus cf. *bintan* exhibits a bolder but narrower turquoise to bright blue banding and a narrower black edging to the outer part of the unpaired fins. The species exhibits no red in the fins.

Similar species:

In comparison to *Parosphromenus* cf. *bintan*, *Parosphromenus bintan* has broader, more weakly coloured stripes in the unpaired fins and no red coloration.

Parosphromenus harveyi likewise exhibits turquoise-coloured stripes in the unpaired fins, but by comparison more in the central part of the fins. This species again has no red coloration in the fins.

Parosphromenus sp. Pontian Besar exhibits a similar coloration and markings, but, compared to the other species, has a significant amount of red interior to the turquoise-coloured banding in the unpaired fins.

Natural distribution:

The species is found in the Sungai Tungkal region in the north of the province of Jambi in Sumatra, Indonesia (details from the trade).

Biotope data:

The species probably lives in blackwater swamp regions like all other *Parosphromenus* species, but no concrete biotope data are available at present (2012).

Maintenance:

As described for the genus.

Reproduction:

As described for the genus.

Total length: (size)

Up to 3.5 cm (1.375 in).

Remarks: (differences from other species of the genus)

Belongs to the *bintan* group.

These fishes have been available in the trade for some time, under the name *Parosphromenus* sp. blue-line.

Parosphromenus cf. *bintan* ♂

Parosphromenus deissneri (BLEEKER, 1859)

Explanation of the name:

Dedication in honour of F.H. DEISSNER, who was a doctor in Java and on Bangka for many years.

Original description:

BLEEKER, P. 1859: *Osphromenus deissneri*. Natuurkd. Tijdschr. Neder. Indie, vol. 18: 376-377.

Type locality:

The Sungai Baturussak area on the Indonesian island of Bangka.

Synonyme:

Osphromenus deissneri
Osphronemus deissneri
Parosphromenus deissneri sumatranus
Parosphronemus deissneri

In addition it has also been confused with many other members of the genus, as it was formerly erroneously assumed that there was just this one species.

Species-typical characters:

The most important characteristic of *Parosphromenus deissneri* is the lanceolate caudal fin with a prolonged central ray, often in both sexes, though in females the lanceolate caudal-fin form is less well-developed and the central ray is usually only slightly prolonged. The species exhibits grey to turquoise-coloured stripes on the unpaired fins and a faint red colour between the central stripes on the body.

Similar species:

Parosphromenus linkei is distinguished by a pronounced lanceolate caudal-fin form in males, also seen to a greater or lesser extent in females. Sometimes, however, females exhibit only a prolonged central ray in the caudal fin. A further characteristic is a central spot on either side of the body, and this can sometimes be doubled or more rarely even tripled. In display coloration these fishes often exhibit a wine-red body coloration, and males may have a striking light dot pattern on the body and fins, as well as light, narrow fin margins.

Female *Parosphromenus deissneri*

Displaying male *Parosphromenus deissneri*

Parosphromenus filamentosus has a lanceolate caudal fin with a much-prolonged central ray in males. The caudal fin is largely red and bordered by a broad, dark stripe. The unpaired fins are bordered by a narrow light stripe.

Parosphromenus paludicola has the largest number of spines in the dorsal fin and hence the longest (base length) dorsal fin of all *Parosphromenus* species. These fishes have a lanceolate caudal fin and sometimes one or even two flank spots. Depending on provenance, however, there are also specimens with no flank spots. Again depending on provenance, the unpaired fins are transparent and colourless to orange in colour.

Natural distribution:

At the beginning of the 1990s KOTTELAT, M. (pers. comm.) recorded *Parosphromenus deissneri* in the central part of the Indonesian island of Bangka, 25 km north of Koba, between the villages of Desa Kurau and Desa Balilik.

Klaus GERSTNER, Mike LINKE, and the author collected the species in 1993 in the south of Bangka and imported it alive for the first time. These fishes were living in the company of several other labyrinthfish species such as *Betta burdigala*, *Betta schalleri*, and *Sphaerichthys osphromenoides*. Our study site lay around 60 km south of Koba on the road to Toboali between the villages of Djeridja and Bikang, around 4 km from Bikang. The habitat was a large area of woodland, swamp, and flood plain in which a stream crossed the road. The water level was very high as the result of previous heavy rainfall (rainy season, end of June) and the stream had a moderate current in most places. At this time the water temperature was 24 to 25 °C (75 to 77 °F). The carbonate and general hardness were both less than 1° (German), the pH was 5.0, and the electrical conductivity 18 μS/cm. The water was clear and dark red-brown in colour. The water depth was on average easily a metre (40 in). The bottom consisted of light, loamy sand. In addition to the dense marginal growth of normally emerse plants, well beneath the water's surface at this time of year, there were also a strikingly large number of long-stemmed and long-leaved *Cryptocoryne* in the biotope. The *Parosphromenus* were living in the dense vegetation and were difficult to catch there because of the high water level. Surprisingly there were also two other species to be found in this biotope.

A year later Klaus FRANK and Norbert NEUGEBAUER found *Parosphromenus deissneri* south-west of Koba in a blackwater stream near a bridge on the road from Airbara to Pajoeng,

Biotope of *Parosphromenus deissneri* at Kura, Bangka island

Parosphromenus deissneri lives predominantly in very dark, clear blackwater

The fish fauna on the Bangka island is very diverse and very interesting for ornamental fish enthusiasts

Parosphromenus deissneri also often lives hidden in dark rainforest streams

close to the kilometre marker Airbara 8 km - Pajoeng 23 km. The watercourse was flowing through primary forest and had an average width of 4 metres (13 feet) and a depth of up to 1.5 metres (5 feet). Even towards the end of the dry season (October) the water level in this biotope was comparatively high.

In 2008 we found *Parosphromenus deissneri* in additional habitats. One of these was a small blackwater river with widenings and linked to a large swamp area on either side of the road from Pangkalpinang to Toboali, 27 km south of Pangkalpinang, 3 km from Kurau. The water parameters were as follows: pH 4.72 and conductivity 4 µS/cm at 27.0 °C (80.6 °F). The water here was dark red-brown in colour with the current varying from place to place. The marginal zones were densely vegetated, and these were the areas preferred by the *Parosphromenus deissneri.*

Coordinates for Kurau:
02°19′30 S   106°13′10 E (LINKE, H.)

A further research site lay only a few kilometres from Sempan on the road from Sungailiat to Puding Besar, where a small stream, around 3 metres (10 feet) wide, crossed the road. It was a cloudy, brownish watercourse with a pH of 5.49 and a conductivity of 19 µS/cm at a water temperature of 27.3 °C (81 °F).

Coordinates for Sempan 1:
01°58′05 S   106°00′16 E (LINKE, H.)

Only around 1 km further on a small watercourse, only around 1 m (40 in) wide, flowed out of the forest and widened next to the road, then only a few metres further on emptied into a somewhat larger watercourse that subsequently crossed the road from Sungailiat to Puding Besar between Sempan and Puding Besar. This was a clear blackwater with a pH of 4.84 and a conductivity of 8 µS/cm at a water temperature of 27.6 °C (81.7 °F).

Coordinates for Sempan 2:
01°58′22 S   106°00′15 E (LINKE, H.)

Maintenance:

Parosphromenus deissneri is predominantly an inhabitant of blackwater biotopes and hence requires soft, acid water for maintenance in the aquarium. The addition of humic substances is also very important. Like all other liquorice gouramies, *Parosphromenus deissneri* normally eats only suitably small live food, and rarely takes frozen or dry foods.

Reproduction:

Bubblenest spawner.

Very soft, acid water is a prerequisite for successful breeding. If possible the carbonate hardness should be less than 1 °dKH, general hardness around 1 to 2 °dGH, conductivity 20 to 30 μS/cm, pH 4.0 to 4.5, and water temperature around 26 to 28 °C (79 to 82.5 °F). The aquarium should be set up as described for the genus.

Total length: (size)

Up to 4.0 cm (1.5 in).

Remarks:

A species-typical character is the prolonged central ray in the caudal fin, accompanied by a lanceolate caudal fin in adult males. Displaying adult males also exhibit red areas on the body between the central stripes.

Numerous other liquorice gouramies have been and will continue to be labelled as *Parosphromenus deissneri*. Not until 1998 did KOTTELAT & NG throw some light on the problem. According to them the species *P. deissneri* lives exclusively on the Indonesian island of Bangka, and, as is so often the case with *Parosphromenus* species, with a second *Parosphromenus* species in the same habitat. It is very easily distinguished from other *Parosphromenus* species by having a prolonged, lanceolate caudal fin, and exhibits no red in the fins. To date *Parosphromenus deissneri* has apparently only very rarely been maintained in the aquarium.

The small number of specimens currently still in the aquarium hobby are thought to have been imported by Alfred WASER and subsequently successfully bred by Günter KOPIC.

References:

• GECK, J. and KOPIC, G. 2005: Prachtguramis, Aquaristik aktuell, special publication 1/2005: 30-36.

• KOTTELAT, M & NG, P. K. L.1998: *Parosphromenus bintan*, a new osphronemid fish from Bintan and Bangka islands, Indonesia, with redescription of *P. deissneri*. Ichthyological Exploration of Freshwaters 8: 263-272

• LINKE, H. 2008: *Parosphromenus deissneri* - wieder erfolgreich eingeführt.

• Betta News, Journal of the European Anabantoid Club/Arbeitskreis Labyrinthfische im VDA, 2008 (4): 16

Parosphromenus deissneri ♂ after capture

It is very important to measure the water parameters in all biotopes studied

Parosphromenus deissneri biotope between Sungailiat and Puding on Bangka

Explanation of the species name:

Latin *filamentosus* = "filamentous", "with filament(s)", referring to the form of the caudal fin

Original description:

VIERKE, J.. 1981: *Parosphromenus filamentosus* sp. n. aus SO-Borneo. Senkenbergiana biol. Frankfurt a.M., 61 (5/6): 363-367.

Type locality:

A swampy region along the road between Banjarmasin and Pleihari, in south-east Kalimantan (Borneo, Indonesia).

Species-typical characters:

Parosphromenus filamentosus has a lanceolate caudal fin with a greatly prolonged central ray in males. The caudal fin is largely red in colour and bordered with a dark stripe. There is no transitional colour in between, or only very rarely a narrow, very faint turquoise area. The unpaired fins have a narrow light marginal band.

Similar species:

Parosphromenus linkei, but exhibits one to two dark, often turquoise-coloured, spots on the flanks and a pattern of small light spots on the unpaired fins.

In *Parosphromenus deissneri* the most important characteristic is the lanceolate caudal fin in both sexes, though in females the lanceolate caudal-fin form is less well-developed and the central ray is usually only slightly prolonged. The species exhibits grey to turquoise-coloured stripes on the unpaired fins and no red.

Parosphromenus paludicola has a much longer (base length) dorsal fin and, depending on natural provenance, exhibits one to two spots on the flanks as in *Parosphromenus linkei*.

Parosphromenus pahuensis likewise exhibits dark spots on the flanks and a pattern of small light spots on the unpaired fins.

Natural distribution:

The natural habitat of this species lies south of the town of Banjarmasin in Kalimantan, in the south of the island of Borneo.

Parosphromenus filamentosus ♂

Parosphromenus filamentosus ♂

Rice field with irrigation ditches south of Banjarmasin in Kalimantan Selantan

The habitats here are mainly slow-flowing watercourses as well as irrigation ditches for the fields and plantations.

In 2009 we also recorded this species in the area of the town of Buntok, around 200 km north of Banjarmasin, in the eastern Barito river system. It appears that all *Parosphromenus* species known to date from southern Kalimantan have a very much larger distribution region than previously assumed.

Biotope data:

The research in the Buntok area involved a blackwater swamp region on the road from Buntok to Ampah, around 9 km from the centre of Buntok. A swamp region with small trees and scrub. The water parameters were pH 4.72, conductivity 24 µS/cm, and water temperature 28.8 °C (84 °F). (Research in June 2009: LINKE, H.)

Coordinates: 01° 41'07 S   114° 52' 55 E (LINKE; H.)

Maintenance:

For optimal maintenance these fishes should be provided with separate, small, shallow aquaria, decorated with areas of dense aquatic plants and small caves. The species is robust in comparison to the majority of *Parosphromenus* species and hence easier to maintain. A regular partial water change every 8 - 10 days is nevertheless advisable. These fishes should be given exclusively live food.

Reproduction:

If the fishes are properly maintained then breeding may take place in the normal aquarium. However, a separate breeding tank for each pair is advisable if you want to rear a worthwhile

Swampy area between plantations south of Banjarmasin in Kalimantan Selantan

Small watercourses with little current are the habitat of *Parosphromenus filamentosus*

The typical locality of *Parosphromenus filamentosus* lies to the south of Banjarmasin in Kalimantan Selatan

There are numerous swampy areas in the area south of Banjarmasin

number of young. These fishes usually spawn beneath a small bubblenest in a cave. The brood care is undertaken by the male. The fry swim free after around 6 days at a water temperature of around 27 °C (80.5 °F) and then need very small live food such as freshly-hatched *Artemia* nauplii.

Total length: (size)
Up to 4 cm (1.5 in).

Remarks: (differences from other species of the genus)
On the basis of the description of the species from the Buntok area, the distribution region of *Parosphromenus filamentosus* appears to be very much larger than previously assumed.

It is possible that *Parosphromenus* cf. *filamentosus* and *Parosphromenus filamentosus* are one and the same species.

References:
- DONOSO-BÜCHNER, R. 2002: Der Fadenschwanzprachtgurami *Parosphromenus filamentosus* Der Macropode – Zeitschrift der IGL 24 (7/8): 184-189.

- SCHMIDT, J. 1989: Der Faden-Prachtgurami - *Parosphromenus filamentosus* VIERKE, 1981. DATZ (Die Aquarien- und Terrarienzeitschrift) 10: 593-595.

- VIERKE, J. 1982: Wieder ein neuer Prachtgurami von Borneo Das Aquarium, 153: 122-126.

- VIERKE, J., 1982: Schön und pflegeleicht: Der Faden-Prachtzwerggurami aus Borneo-Pflege und Zucht von *Parosphromenus filamentosus* Aquarium Magazin 5: 300–305.

Explanation of the name:

This possibly good species exhibits numerous parallels to the species *Parosphromenus filamentosus*.

English name:

Pointed-Tail Liquorice Gourami

Species-typical characters:

Parosphromenus cf. *filamentosus* exhibits numerous parallels to *Parosphromenus filamentosus* in its appearance. *Parosphromenus* cf. *filamentosus* likewise exhibits a lanceolate caudal fin with a greatly prolonged central ray in males. The caudal fin is largely red and bordered with a broad, dark stripe. There is an intermediate bright turquoise-coloured transitional area in all the unpaired fins, which are edged with a thin light stripe. The ventral fins are much prolonged and usually dark, sometimes bright turquoise, in colour. The species is more boldly coloured in comparison to *Parosphromenus filamentosus*.

Similar species:

Parosphromenus filamentosus has a lanceolate caudal fin with a greatly prolonged central ray in males. The caudal fin is largely red in colour and bordered with a broad, dark stripe. There is no noticeable transition area between these colour zones, or only a very faint turquoise-coloured stripe. The unpaired fins have a narrow light marginal band.

Parosphromenus linkei, but exhibits one to two dark, often turquoise-coloured, spots on the flanks and a pattern of small light spots on the unpaired fins.

In *Parosphromenus deissneri* the most important characteristic is the lanceolate caudal fin in both sexes, though in females the lanceolate caudal-fin form is less well-developed and the central ray is usually only slightly prolonged. The species exhibits grey to turquoise-coloured stripes on the unpaired fins and no red.

Natural distribution:

The species is found in the region around Palangkarya. We frequently found these fishes in watercourses on the road from Palangkaraya to Kuala Kurun. These were tributaries of the

Parosphromenus cf. filamentosus ♀

Parosphromenus cf. filamentosus ♂

Habitat of *Parosphromenus* cf. *filamentosus* by the road from Palangkaraya to Kuala Kurun near Bawan in Kalimantan Tengah

Sungai Kahayan, which flows along the south-eastern edge of the town of Palangkaraya. The distribution region lies to the west of the great Sungai Kapuas in Kalimantan Tengah, Borneo/Indonesia.

The species was also caught by Alfred WASER in 1991, in the area between Kasongan and Tangkiling. For many years an illustration on the Internet showed it erroneously as *Parosphromenus* sp. from Pundu.

Biotope data:

A clearwater stream around 5 metres (16.5 feet) wide with a slight brownish coloration, which crosses the road to Kuala Kurun around 74 km after the bridge over the Sungai Kahayan at Palangkaraya, near Bawan. The watercourse flows for around 1.5 km to the settlement of Desa Pahawan.

Coordinates: 01° 40' 06 S   113° 56' 35 E (LINKE, H.)

The water had a slight to moderate current, and was clear and completely shaded by trees. The following water parameters were measured: pH 4.15 and conductivity 10 µS/cm at a water temperature of 26.7 °C (80 °F). The bottom consisted of light sand. The margins were cloaked with aquatic plants and large emerse bushes. The depth of the water was between 50 and 120 cm (20 and 48 in). The bottom was firm and easy to walk on.

Maintenance:

The species is very colourful and unproblematical in its maintenance as long as the water is mineral-poor and slightly acid. The addition of humic substances is also important.

Otherwise as described for the genus.

Reproduction:
As described for the genus.

Total length: (size)
Up to 3.5 cm (1.375 in).

Remarks: (differences from other species of the genus)
The distribution region of *Parosphromenus* cf. *filamentosus* lies around 300 km north-east of the type locality of *Parosphromenus filamentosus* at Banjarmasin, and is separated from it by two major river systems, the Sungai Barito and the Sungai Kapuas. Further research is required to determine whether *Parosphromenus* cf. *filamentosus* is a good species or just a local form of *Parosphromenus filamentosus*.

The bottom consists predominantly of light sand

In places the bottom is covered in large accumulations of dead leaves

The areas beneath the overhanging banks were the preferred habitat of many underwater denizens

Dead wood and leaves litter on the floor of the biotope create good cover for small fishes

The water in the biotope of *Parosphromenus* cf. *filamentosus* is flowing

In places there are dense stands of aquatic plants

Parosphromenus gunawani SCHINDLER & LINKE, 2012

Explanation of the species name:
Dedication in honour of Gunawan "Thomas" Kasim, Sindo Aquarium, Kota Jambi, Insel Sumatra, Indonesia.

English name:
Danau Rasau Liquorice Gourami

Original description:
SCHINDLER, I. & LINKE, H., 2012: Two new species of the genus *Parosphromenus* (Teleostei: Osphronemidae) from Sumatra. Vertebrate Zoology 62 (3) 2012; 399-406

Species-typical characters:
These fishes exhibit a similar colour pattern to *Parosphromenus rubrimontis*, but by comparison *Parosphromenus gunawani* has broader longitudinal bands on the head and body and, as far as is known to date, a different colour scheme – instead of red it exhibits only a brown coloration.

Similar species:
Parosphromenus rubrimontis exhibits a large dark spot at the base of the caudal fin, with a broad, red, semicircular band adjoining, sometimes with a narrow, turquoise-coloured margin distally, and then a broader outer black band. This coloration and pattern are repeated in the dorsal and anal fins. The unpaired fins have a narrow, light outer margin. The ventral fins are comparatively short and turquoise-coloured. The species has a longer dorsal fin in comparison to the other members of the *rubrimontis* group.

Natural distribution:
The species lives in the Danau Rasau (Lake Rasau), a lake-like widening, with affluents, of the Sungai Batang Hari near Rantanpanjang, in the Province of Jambi, island of Sumatra, Indonesia. Rantanpanjang is around 76 km north-east of Kota Jambi and Tanjung, and from there it is around 15 km up the Sungai Batang Hari to the Danau Rasau.

Biotope data:
The Danau Rasau is an extremely blackwater swamp-lake, overgrown in places with trees and scrub, while the water

Parosphromenus gunawani ♀

Parosphromenus gunawani ♂

Parosphromenus gunawani ♂

The water in the lake is coloured very dark red-brown compared to drinking water

The lake is largely covered in floating plants

Parosphromenus gunawani is kept in black water after capture

Families of fishermen looking for good places to fish

surface is predominantly covered by floating plants. The water has a pH of 4.1 and a conductivity of 30 μS/cm at a water temperature of 29.3 °C (84.7 °F). The water is heavily stained dark red-brown and everywhere flowing slightly. (Research in May 2007: SIM, T., CHIANG, N., LINKE, M. & LINKE, H.) The species lives in the company of *Betta coccina*, *Trichogaster leeri*, *Trichogaster trichopterus* (only a very low population density), *Belontia hasselti*, and *Sphaerichthys osphromenoides*.

Coordinates: 01° 22'S   103° 56' E (LINKE,H.)

The Danau Rasau, a large blackwater swamp lake to the north-east of Kota Jambi, Sumatra

Fisherman at work. The lake is mostly only 80 to 120 cm deep

The fishes are carefully sorted into bags after capture

Parosphromenus gunawani being studied after capture

Maintenance:
As described for the genus.

Reproduction:
As described for the genus.

Total length: (size)
Around 3.5 cm (1.375 in).

Remarks: (differences from other species of the genus)
The species was for many years traded as *Parosphromenus* sp.
Jambi and *Parosphromenus* sp. Danau Rasau.

Explanation of the species name:

Dedication in honour of the German-English aquarist Willi HARVEY.

Original description:

BROWN, B. 1987: Two new Anabantoid species. Aquarist and Pondkeeper 52 (3): 34.

Type locality:

Western side of peninsular Malaysia

Synonyms:

Parosphromenus deissneri

Species-typical characters:

Parosphromenus harveyi exhibits turquoise-coloured stripes in the unpaired fins, but, compared to similar species, more in the central part of the fins than the marginal region. This species too exhibits no red coloration in the fins.

Similar species:

Parosphromenus bintan exhibits stripes in the unpaired fins but these are broader and more faintly coloured by comparison, and there is no red coloration.

Parosphromenus sp. Pontian Besar exhibits a similar coloration and markings, but, compared to the other species, has a significant amount of red interior to the turquoise-coloured stripes in the unpaired fins.

Parosphromenus cf. *bintan* exhibits a bolder but narrower turquoise to bright blue banding and a narrower black edging to the outer part of the unpaired fins. The species exhibits no red in the fins.

Natural distribution:

The species is known from Batu Arang, near Rawang, north of Kuala Lumpur. The fishes are, however, also found in the blackwater swamp regions south of the Sungai Bernam, in areas by the road between the villages of Sungai Besar and Kalumpang, in the north of the province of Selangor in western Malaysia.

Biotope data:

These fishes live predominantly in heavily vegetated blackwater streams with a slight current and soft, very acid water (pH around 5.0 and electrical conductivity around 20 µS/cm) and a water temperature around 27 °C (80.5 °F).

Maintenance:

As described for the genus.

Parosphromenus harveyi ♂

Courting male *Parosphromenus harveyi*

Courting male in front of the breeding cave

Reproduction:

As described for the genus.

Total length: (size)

Up to 3.5 cm (1.375 in).

Remarks: (differences from other species of the genus)

Possibly belongs to the *bintan* group.

Explanation of the species name:

Dedication in honour of the author of this book, Horst LINKE, who discovered the species.

Original description:

KOTTELAT, M. 1991: Notes of the taxonomy and distribution of some Western Indonesian freshwater fishes, with diagnoses of a new genus and six new species (Pisces: Cyprinidae, Belontiidae, and Chaudhuriidae). Ichthyological Exploration of Freshwaters, 2 (3): 282.

Type locality:

Creeks in Pudukuali, 2 km north of Sukamara and Tarantang, Kalimantan Tengah, Borneo.

Species-typical characters:

Parosphromenus linkei differs from other liquorice gourami species by having a pronounced lanceolate caudal-fin form in males, also seen to a greater or lesser extent in females. Sometimes, however, females exhibit only a prolonged central ray in the caudal fin. A further characteristic is a central spot on either side of the body, and this can sometimes be doubled or more rarely even tripled.

In display coloration these fishes often exhibit a wine-red body coloration, and males may have a striking light dot pattern on the body and fins, as well as light, narrow fin margins.

Similar species:

Parosphromenus pahuensis likewise exhibits spots on the flanks but has a rounded caudal fin.

In *Parosphromenus deissneri* the most important characteristic is the lanceolate caudal fin in both sexes, though in females the lanceolate caudal-fin form is less well-developed and the central ray is usually only slightly prolonged. The species exhibits grey to turquoise-coloured stripes and no red on the unpaired fins.

Parosphromenus filamentosus has a lanceolate caudal fin with a much-prolonged central ray in males. The caudal fin is largely red and bordered by a broad, dark stripe. These fishes have no pattern of dots. The unpaired fins are bordered by a narrow light stripe.

Parosphromenus linkei ♀

Parosphromenus linkei ♂ without lateral spot

Parosphromenus linkei with double spots

Parosphromenus linkei with green lateral spot

Parosphromenus linkei , reddish variant

Parosphromenus paludicola has the largest number of spines in the dorsal fin and hence the longest (base length) dorsal fin of all *Parosphromenus* species. These fishes have a lanceolate caudal fin and sometimes one or even two flank spots. Depending on provenance, however, there are also specimens with no flank spots. Again depending on provenance, the unpaired fins are transparent and colourless to orange in colour.

Natural distribution:

Parosphromenus linkei was first discovered in 1990 by Norbert NEUGEBAUER, Ingrid. BÄR, and the author, in a swamp region 2 km north of Sukamara in the south-west of Central Kalimantan (Kalimantan Tengah). Research in 2009 demonstrated that the distribution of *Parosphromenus linkei* is much larger and the large Sungai Asambaru (105 km east of Pangkalanbun) may perhaps form the natural boundary to the east. It remains to be seen whether the fishes in the vicinity of Palangkaraya, around 300 km further east, are also *Parosphromenus linkei*, and in this work they will be termed *Parosphromenus* cf. *linkei*.

Biotope data:

The area north of Sukamara is a blackwater swamp region with a water depth of between 20 and 100 cm (8 and 40 in). The water's surface was largely heavily shaded by numerous emerse-growing marsh plants, sometimes up to 2 metres (80 in) tall. Hence the water had a temperature of only around 25 °C (77 °F), despite the bright sunshine. The only route for traffic through the swamp was a wooden causeway on wooden piles (for bicycles and small motorbikes), about 2 km long and 1.50 metres (5 feet) wide, running in a north-easterly direction from Sukamara to the village of Pudukuali, 4 km away.

The wooden path to Sukamara, for many years, this is the only way to get there

At around 09.00 on the day of our research the water had the following parameters: carbonate and general hardness less than 1° (German), pH 3.55, and conductivity 9 µS/cm at a water temperature of 24.5 °C (76 °F). The water was clear, heavily stained dark red-brown, and still with a slight current in places.

We were able to find numerous other fish species in this swamp region, including a second liquorice gourami species, *Parosphromenus opallios* KOTTELAT & NG, 2005 (sometimes known as *Parosphromenus* sp. Sukamara).

At another research site, on the road from Sampit to Pangkalanbun, these fishes were found in a blackwater ditch crossed by three long bridges.

Coordinates: 02° 28' 36 S   112° 06' 43 E (LINKE; H.)

The water was heavily stained dark red-brown and had only a slight current. The following water parameters were measured: pH 4.71 and conductivity 10 µS/cm at a water temperature of 28.7 °C (83.7 °F).

Maintenance:

The maintenance and breeding of this liquorice gourami isn't difficult provided the most important prerequisites are taken into account. Soft water with a general hardness of up to 8 °dGH (lower is better) and a very low carbonate hardness are important for optimal maintenance, and the pH should be

Large area of aquatic plants in the blackwater swamp region in north of Sukamara

between 5 and 6. The aquarium should have a volume of around 50 litres (11 gallons), and be arranged with areas of dense vegetation and small caves on the bottom. The substrate should be a thin layer of fine sand, which can be siphoned off during maintenance and then returned to the tank after rinsing. However, it is also possible to dispense with substrate completely, for reasons of hygiene. A peat filter is very important for adding humic substances to the water, and dead oak or beech leaves as bottom décor will also make a very positive contribution to the aquatic environment.

Reproduction:

An aquarium set up as for maintenance is also very suitable for breeding, only the water parameters need to be altered appreciably, i.e. the general hardness should be around 1 °dGH, but above all the carbonate hardness must be less than 1 °dKH and the pH between 4.3 and 4.8. This equates to an electrical conductivity of around 20 µS/cm at a water temperature of 25 to 27°C (77 to 80.5 °F).

Total length: (size)

The species attains a total length of around 3.5 cm (1.375 in).

Remarks: (differences from other species of the genus)

Parosphromenus linkei is one of the loveliest liquorice gourami species. Its maintenance is easier than for other species, as its tolerance appears to be somewhat greater when it comes to water parameters for optimal maintenance.

References:

- GECK, J. & KOPIC, G. 2005. Prachtguramis. Aquaristik aktuell, special publication 1/2005: 30-36.
- SCHMIDT, J. 1997: Das Labyrinthfischportrait Nr. 74 *Parosphromenus linkei* KOTTELAT, 1991. Der Macropode 19 (11/12): 125-128.

The blackwater swamp region in north of Sukamara, typical locality for *Parosphromenus linkei*

Explanation of the name:

This fish is very similar to the species *Parosphromenus linkei*.

English name:

None

Species-typical characters:

Parosphromenus cf. *linkei* exhibits numerous parallels to *Parosphromenus linkei* in its appearance.

The bolder markings and coloration are striking. *Parosphromenus* cf. *linkei* are a bolder orange to red in colour. Their appearance is more attractive. Here too there are spots on the sides of the body, varying in number between one and three on each side. Males of this species again have a lanceolate caudal fin and light edgings to the unpaired fins.

Similar species:

Parosphromenus pahuensis likewise exhibits spots on the flanks, but has a rounded caudal fin.

In Parosphromenus deissneri the most important characteristic is the lanceolate caudal fin in both sexes, though in females the caudal-fin form is less well-developed and there is usually only a single prolonged central ray. The species exhibits grey to turquoise-coloured stripes and no red on the unpaired fins.

Parosphromenus filamentosus has a lanceolate caudal fin with a much-prolonged central ray in males. There are no dots on the body. The caudal fin is largely red and bordered by a dark stripe. The unpaired fins are bordered by a narrow light stripe.

Parosphromenus paludicola has the largest number of spines in the dorsal fin and hence the longest (base length) dorsal fin of all *Parosphromenus* species. These fishes have a lanceolate caudal fin and sometimes one or even two flank spots. Depending on provenance, however, there are also specimens with no flank spots. Again depending on provenance, the unpaired fins are transparent and colourless to orange in colour.

Parosphromenus cf. linkei ♀

Parosphromenus cf. linkei ♂

Blackwater area in north of Palangkaraya

Bank zone in the blackwater area, habitat of numerous labyrinthfish species

The bridge spanning the large swamp on the road from Palangkaraya to Kuala Kurun

Natural distribution:

These fishes were found in various biotopes in the area between Pundu and Palangkaraya in Central Kalimantan (Kalimantan Tengah), Island of Borneo, Indonesia.

Biotope data:

These fishes are found almost exclusively in blackwater biotopes. We recorded the species in a blackwater swamp region by the road from Palangkaraya to Kuala Kurun, around 15 km after the large bridge over the Sungai Kahayan, immediately after the third steel-girder bridge thereafter. The terrain was open with tall, dense scrub in places. The fishes were living among dense vegetation in the shallow bank zones.

Coordinates: 02° 05'08 S   113° 57' 16 E (LINKE, H.)

The water was very dark red-brown in colour. The following

Small plants of island dotted the blackwater swamp

The vegetated margins are particularly popular habitat for small fishes

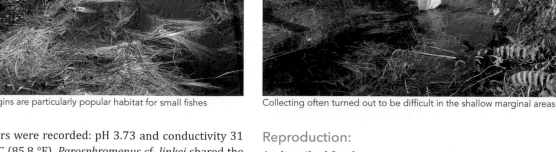
Collecting often turned out to be difficult in the shallow marginal areas

water parameters were recorded: pH 3.73 and conductivity 31 µS/cm at 29.9 °C (85.8 °F). *Parosphromenus* cf. *linkei* shared the biotope with *Parosphromenus parvulus*, *Betta anabantoides*, *Betta edithae*, *Betta* sp., *Sphaerichthys selatanensis*, and *Sphaerichthys acrostoma*. (Research in June 2009: LINKE, H. et al.)

Maintenance:

As described for the genus. But in this case the water should be very low in minerals, acid, and have humic substances added.

Reproduction:

As described for the genus.

Total length: (size)

Up to 3.5 cm (1.375 in).

Explanation of the species name:
Dedication in honour of Peter NAGY de Felsö Gor, Salzburg, Austria.

Original description:
SCHALLER, D. 1985: *Parosphromenus nagyi* spec. nov., ein neuer Prachtgurami aus Malaysia. Aquar. Terrar. Z. 38 (7): 301-303.

Type locality:
Asian Highway Route 18, 16 km south of Kuantan, eastern part of the Malayan peninsula, Malaysia (western Malaysia).

Species-typical characters:
In comparison to other species, *Parosphromenus nagyi* exhibits very short membranes in the dorsal, caudal, and anal fins. The rays protrude noticeably beyond the membranes. This gives the impression of this species having smaller fins and the rays protruding beyond the fin membranes like claws. Males of this species in display coloration exhibit a beige or smoke-grey to black body with no stripes as well as variable markings and coloration on the unpaired fins. The species has a rounded caudal fin and sometimes exhibits a faint dark spot on the dorsal fin.

Similar species:
In *Parosphromenus ornaticauda* the margins of the unpaired fins are lighter, and, above all in display coloration, there is a ruby-red central area on the caudal fin, bounded on either side by thick black bars then light edges. Hence also the name *ornaticauda*, meaning with a decorated tail, and the common name of Ruby Liquorice Gourami. *Parosphromenus ornaticauda* is the smallest *Parosphromenus* species known to date.

Males of *Parosphromenus parvulus* exhibit a series of dot-like spots on the dorsal and anal fins. The unpaired fins are bordered with a narrow light margin. Further species-typical characters of adult males are round red spots on a dark background on the dorsal and anal fins and a transparent, colourless to smoke-grey caudal fin. During courtship and other display the underside of the body becomes deep black. The ventral fins remain predominantly colourless and transparent.

Natural distribution:
The northernmost point in the distribution lies north of Kuantan in the Ceratin area. The main distribution of the species lies in the area around Kuantan. The southernmost habitats lie south of the great Sungai Pahang and may, on the basis of the latest information, extend to north of Kota Tinggi, inevitably with concomitant differences in coloration.

Biotope data:

North of Kuantan:
West of the village of Kampong Cerating, on the right-hand side (inland of) National Highway 3 heading in the direction of

Parosphromenus nagyi ♂ in normal coloration

Kuantan, after several businesses, there is a large, modern railway suspension bridge, and on the left-hand side immediately thereafter (around 100 m (110 yards) after the railway bridge) there is a small blackwater stream that crosses the narrow road beneath a small road bridge. The stream contained numerous plants and was readily negotiable, in most places only 10 to 30 cm (4 to 12 in) deep at the end of August. The water had a slight current and a pH of 3.8.

Coordinates:
04° 07. 58 N   103° 22. 14 E (LINKE, H.)

South of Kuantan:

National Highway 3 from Kuantan to Pekan (N3 Kuantan-Johor Bahru), between kilometre markers 305 and 306 heading for Kota Bahru or kilometre markers 22 and 23 heading for Kuantan, narrow blackwater ditches on both sides of the road, with dense rainforest beyond.

Coordinates:
03° 39. 74 N   03° 17. 77 E (H. LINKE)

Small blackwater river that crosses the road from Pekan to Kampong Terlang (and thence to National Highway 12) beneath a bridge. There is a large inundation region in the forest on either side. The road runs almost parallel to and south of the great Sungai Prahang.

The bridge is about 100 metres (110 yards) before the Kampong Sungai Ganchong Aceh.

Coordinates:
03° 31. 58 N   103° 18. 40 E (LINKE, H.)

Flooded forest region by the old bridge, on the left-hand side coming from Segamat, near to National Highway 12 from Kuantan to Segamat, a few kilometres after the turn-off to Pekan, heading in the direction of Kuantan (around 37.3 km to Kuantan).

Coordinates:
03° 37.30 N   103° 07.47 E (LINKE, H.)

West of Kuantan:

Narrow, often up to 5 metres (16.5 feet) wide, clear, slightly brownish-coloured watercourse in an oil-palm plantation. The water depth was on average 50 cm (20 in). There was a sandy track (a southern turn-off from National

Display coloration of *Parosphromenus nagyi*

Parosphromenus nagyi in normal coloration

Parosphromenus nagyi biotope in an oil-palm plantation

Biotope in the palm woodland west of Kuantan

Sometimes there were still unpolluted watercourses flowing through the oil-palm plantations. Astonishingly *Parosphromenus nagyi* occurred here as well

Highway 2 (Kuala Lumpur-Kuantan) after kilometre marker 23 to Kuantan or shortly before kilometre marker 237 to Kuala Lumpur) leading into this oil-palm plantation near a temple with very broad steps built on the hill opposite. The sandy track led past alternating oil-palm, banana, and rubber plantations and then into a large oil-palm plantation on either side, where the watercourse crossed the sandy track after several hundred metres. (Distance from National Highway 2 around 1 km.)

Coordinates: 03° 42.89 N   103° 09.32 E (LINKE, H.)

Maintenance:

As described for the genus.

Reproduction:

As described for the genus.

Total length: (size)

Up to 3 cm (1.25 in).

There are broad channels on either side of the National Highway from Kuantan to Mersing to drain away rain water, and these too are habitats of *Parosphromenus nagyi*. They are blackwaters and the population density of *Parosphromenus nagyi* is high there

Explanation of the species name:

The name is derived from the Greek *opallios* = "opal", a gemstone which can be of almost any colour.

Original description:

KOTTELAT M. & NG P. K. L. 2005: Diagnoses of six new species of *Parosphromenus* (Teleostei: Ospronemidae) from Malay Peninsula and Borneo, with Notes on other species. The Raffles Bulletin of Zoology, Supplement 13: 101-113.

Type locality:

Kalimati, Pangkalanbun area, Arut River basin: Kalimantan Tengah, Borneo.

Species-typical characters:

Parosphromenus opallios has bright red on the central parts of the unpaired fins, in the form of narrow bars. This is followed by a broad dark, usually black band, and in between the two there is a turquoise-coloured band of variable width, especially in the dorsal and anal fins. The unpaired fins have light margins. The base of the caudal fin exhibits a smaller dark spot. The ventral fins are somewhat longer and coloured turquoise right to their tips.

Similar species:

Parosphromenus sp. Stunggang exhibits a smaller, dark spot on the caudal base and lots of red in the caudal fin. The red in the dorsal and anal fins is less extensive and sometimes adorned with turquoise bars. While the red of the caudal fin is bordered by only a faint dark band, the corresponding dorsal- and anal-fin markings are broader and bolder. The unpaired fins have light narrow margins. The ventral fins are somewhat longer and bright solid turquoise in colour.

Parosphromenus allani has a bright red colour on the central parts of the unpaired fins, resembling bars. This is bordered by a broad, dark, often black band, in turn followed by a light, narrow, outer margin. There is a large dark spot at the base of the caudal fin. The soft portion of the dorsal fin also sometimes exhibits a large dark round spot. The ventral fins are normal in length and have a faint turquoise coloration.

Parosphromenus quindecim is one of the largest liquorice gouramies known to date. *Quindecim* (Latin) means "fifteen" and refers to the 15 hard rays in the dorsal fin. The species exhibits a broad brown (also sometimes red) band in the central part of the unpaired fins, followed by a turquoise-coloured band, and then a dark, narrow band extending to the edge of the fins. The unpaired fins have light margins. The caudal fin has a large, dark, star-shaped spot at its base. The ventral fins are of normal length and exhibit no noteworthy coloration.

Parosphromenus opallios ♂

Natural distribution:

Fishes of this species were first discovered by BAER, I., NEUGEBAUER, N. and LINKE, H. in July 1990 in the south-west of Kalimantan, only a few kilometres north of Sukamara. Here they live, along with numerous other species, in the same biotope as *Parosphromenus linkei* KOTTELAT 1991. Both species live in a large, sometimes heavily vegetated swamp region, around 2 km north of Sukamara in the direction of Pudukuali. This is a blackwater biotope with flowing water in places. A narrow wooden causeway, around 1.5 kilometres long, is the only route to the north. It has been built through the swamp region and hence makes interesting observations possible. The water was heavily stained dark brown. The carbonate and general hardness were both less than 1° (German), pH 3.55, and conductivity 9 µS/cm at a water temperature of 24.5 °C (76 °F). The measurements were made in the morning after extensive rainfall. *Parosphromenus opallios* was very numerous in this area and was found mainly in the shallow, heavily vegetated bank zones and other areas of plants.

The holotype of this species was collected in 2001 by KUBOTA, K. in the Pangkalanbun area, in the drainage of the Arut River.

Coordinates:
02° 45' S   111° 36' E (KUBOTA, K.)

Clumps of plants were to be found mainly in the shallows, habitat of *Parosphromenus opallios*

There are dense clumps of plants growing along the banks. This is the habitat of *Parosphromenus opallios*

We also found the species in the Sungai Benipah (Nippa), a small blackwater river, 14 km south-west of Kubu, 40 km south of Pangkalanbun.

Coordinates: 02° 56'11 S   111° 37' 45 E (LINKE, H.)

The water was dark red-brown. The water temperature measured 28.9 °C (84 °F), the pH 3.91, and the conductivity 12 µS/cm. There was very little water flow. The surrounding area has only scrub, with no tall trees to provide shade. The banks were heavily vegetated. (Research in June 2009: H. LINKE)

Maintenance:

As described for the genus.

Reproduction:

Bubblenest spawner.

Total length: (size)

The species grows to around 3.0 cm (1.25 in) long.

Remarks: (differences from other species of the genus)

The species is a typical blackwater dweller and should be kept only in water with roughly the same parameters.

The species was for many years traded as *Parosphromenus* sp. Sukamara.

Parosphromenus ornaticauda KOTTELAT, 1991

Explanation of the species name:

From Latin *ornatus* = "ornate" and *cauda* = "tail", in allusion to the conspicuous colour pattern in the caudal fin.

Original description:

KOTTELAT, M. 1991: Notes on the taxonomy and distribution of some Western Indonesian freshwater fishes, with diagnoses of a new genus and six new species (Pisces: Cyprinidae, Belontiidae, and Chaudhuriidae). Ichthyological Exploration of Freshwaters 2 (3): 283-284.

Type locality:

Borneo: Kalimantan Barat: Sungai Pinyuh, 8 km SE of Anjungan on the road to Pontianak, 0° 20' N 109° 08' E.

Species-typical characters:

The species-typical characters all relate to males. The margins of the unpaired fins are light, and, above all in display coloration, there is a ruby-red central area on the caudal fin, bounded on either side by thick black bars then light edges. Hence also the name ornaticauda, meaning with a decorated tail, and the common name of Ruby Liquorice Gourami. Along with *Parosphromenus parvulus*, *Parosphromenus ornaticauda* is the smallest *Parosphromenus* species known to date.

Similar species:

Males of *Parosphromenus parvulus* exhibit a series of dot-like spots on the anal fin. The unpaired fins are bordered with a narrow light margin. Further species-typical characters of adult males are several round red spots on a dark background on the dorsal and anal fins and a transparent, colourless to smoke-grey caudal fin. During courtship and other display the underside of the body becomes deep black. The ventral fins remain predominantly colourless and transparent.

Parosphromenus nagyi exhibits very short membranes in the dorsal, caudal, and anal fins in comparison to other species.

Parosphromenus ornaticauda ♂ in normal coloration

Parosphromenus ornaticauda ♂ in nuptial coloration

Mating scenes of *Parosphromenus ornaticauda*

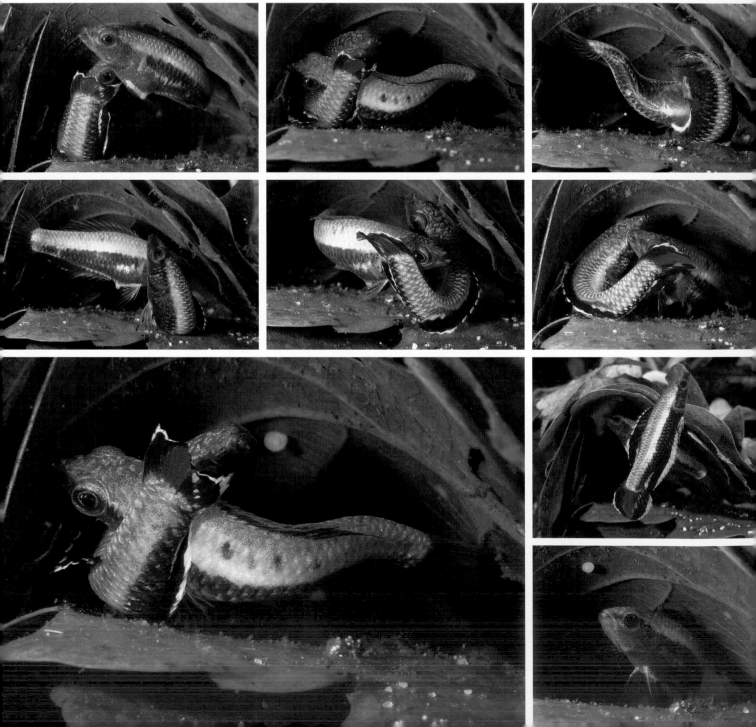

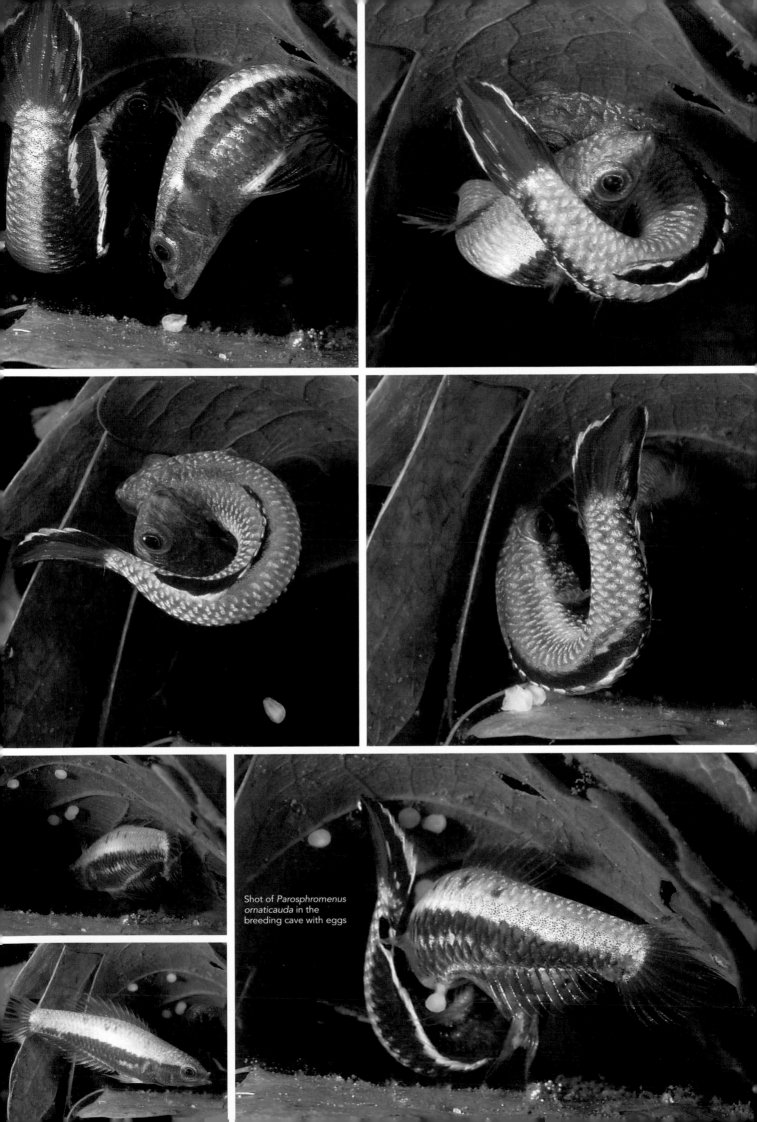

Shot of *Parosphromenus ornaticauda* in the breeding cave with eggs

The rays protrude noticeably beyond the membranes. This gives the impression of this species having smaller fins. The species has a rounded caudal fin and sometimes exhibits a faint dark spot on the dorsal fin. Males of this species in courtship or other display coloration exhibit a very dark to black body as well as variable markings and coloration on the unpaired fins.

Natural distribution:

The species first became known from collections by Maurice KOTTELAT et al. on 21.04.1990, and on the basis of this information was collected again by BAER, I., NEUGEBAUER, N. and the author in July 1990 and subsequently imported alive. We found these fishes in a narrow water-filled ditch by the road from Anjungan to Sungai Penjuh, around 8 km south-west of Anjungan, in the Kapuas area in West Kalimantan (Kalimantan Barat) on the island of Borneo.

Catching *Parosphromenus ornaticauda* in a water-filled ditch by the road from Anjungan to Sungai Penjuh

Biotope data:

The biotope was a typical blackwater with corresponding parameters. The water was almost stagnant and the following parameters were measured (low-water period): pH 4.5; carbonate hardness less than 1 °dKH; electrical conductivity 39 µS/cm at a water temperature of 27.6 °C (81.7 °F). (Research in August 1990: LINKE, H.)

Parosphromenus ornaticauda was comparatively numerous in this biotope. The fishes were to be found mainly in the zones of dense vegetation along the banks. There was a second *Parosphromenus* species living in the same biotope, and this too was taxonomically processed by KOTTELAT in 1991 and described as *Parosphromenus anjunganensis*. Depending on the water chemistry, the two species were variably numerous and were found east as far as the Sungai Mandor. The terrain was predominantly scrub and swamp, but sometimes also interspersed with small wooded areas.

Maintenance:

The water for maintenance should be soft, for breeding very soft. For optimal maintenance it should be acid - ie the pH should lie between 5 and 6 - and for breeding between 4.2 and 4.8.

Reproduction:

The prerequisite for successful breeding is biologically very clean, very soft, acid water. The addition of humic substances is likewise very important. The breeding aquarium should contain a number of areas of dense plants extending to the water's surface. A number of small caves or overhangs near the bottom, as well as slow-turnover filtration of the water, is likewise very important. At the same time the breeding aquarium should not be too deep, ie no more than 25 to 30 cm (10 to 12 in). The male may construct a small bubblenest, but doesn't invariably do so. The male, exhibiting his splendid colours, displays in order to entice the chosen female to the pre-selected spawning site. This may be a cave or maybe only an overhang. If the female is attracted by his display, she swims to the spawning site and remains there for the subsequent matings. During this period the female too assumes a different coloration, with the lower half of her body, from tip of snout to caudal peduncle, appearing dark grey or black.

The male follows the female into the spawning cave and very soon takes her into an embrace, during which he wraps himself round his partner from below in a U-shape and induces her to release eggs by gentle pressure of his head and tail on her ventral region. This is, however, preceded by a very large number of "dummy runs" with no eggs released. Immediately after mating, which lasts for only a few seconds, the male leaves the spawning site and the female tends the eggs. Because the male remains in a slightly looser curve for a few seconds following the embrace, the female is able to release herself from his grip and take the newly expelled eggs, often lying on the anal fin of the male, into her mouth and spit them onto the ceiling of the cave, where, if the water parameters are appropriate, they adhere and remain. The male usually leaves the spawning site immediately after mating to defend the territory, but comes back after a minute or two and then immediately mates with the female again. This "marriage ritual" lasts for around 5 to 7 hours. Thereafter the female leaves the spawning site and the male alone takes on the brood care. The number of eggs (and maybe free-swimming fry) is small. The little *Parosphromenus ornaticauda* can take freshly-hatched *Artemia* nauplii without problem as their first food.

Total length: (size)

Parosphromenus ornaticauda attains a total length of around 2.5 cm (1 inch), sometimes even around 3 cm (2.25 in) under aquarium maintenance.

Remarks: (differences from other species of the genus)

This is a very colourful species, but unfortunately only in males and usually only during courtship and other display.

References:

• LINKE, H. 2006: *Parosphromenus ornaticauda* – Ein Kleinod aus Kalimantan. Aquaristik Fachmagazin 191: 40-43.

Explanation of the species name:

Latin adjective *pahuensis* meaning "of Pahu", derived from the place name Muarapahu, near to the type locality.

Original description:

KOTTELAT, M. and NG, P. K. L. 2005: Diagnoses of six new species of *Parosphromenus* (Teleostei: Osphronemidae) from Malay Peninsula and Borneo, with Notes on other species. The Raffles Bulletin of Zoology, Supplement 13: 104-105.

Type locality:

Island of Borneo: East Kalimantan (Kalimantan Timur), blackwater river in a forest region, which empties into the Sungai Mahakam downstream of the village of Muarapahu.

Parosphromenus pahuensis ♂

Species-typical characters:

Parosphromenus pahuensis exhibits two to three dark spots on the flanks and has a rounded caudal-fin form. In addition the species exhibits a large number of small light spots on the delicately orange-coloured unpaired fins.

Similar species:

Parosphromenus linkei exhibits one or two round flank spots but has a lanceolate caudal-fin form.

Parosphromenus paludicola sometimes exhibits one or two round flank spots but has a lanceolate caudal-fin form and a longer dorsal fin.

Natural distribution:

The species is found in the basin of the Sungai Mahakam. It was discovered in the area around the villages of Melak and Muarapahu. These two settlements lie on the Sungai Mahakam at a distance of around 30 km apart, north-west of the region of swamps and lakes in the Mahakam lowlands in East Kalimantan (Kalimantan Timur, Borneo).

Biotope data:

The species is to date known only from blackwater areas. The water is very soft and very acid.

Muarapahu area:

Blackwater river in a forest region, which empties into the Sungai Mahakam downstream of the village of Muarapahu. (Research by KOTTELAT, M. in August 1991.)

Coordinates: 00° 14' N   116° O7' E (KOTTELAT, M.)

Melak area:

These fishes have been collected in a stream around 2 to 3 metres (6.5 to 9.5 feet) broad and up to 1.5 metres (5 feet) deep, in the area of the village of Melak. The watercourse lies outside the village, around 2 km after the fork to the orchid farm, in an open valley and surrounded by meadowland. The water was clear and slightly brown, with a slight current. There were "*Cabomba*-like" aquatic plants in the stream. The water parameters are given as pH 5, general hardness less than 3 °dGH, and a water temperature of 26.9 °C (80.5 °F). The fishes were found almost exclusively beneath overhangs The stream was partially shaded. (Research in May 1996: DICKMANN, KNORR, and GRAMS.)

Maintenance:

As described for the genus.

Reproduction:

These fishes spawn beneath an overhang or in a cave-like hole in an accumulation of leaves near the bottom. The male tends and guards the spawning site. Compared to other members of the genus, the *Parosphromenus pahuensis* male builds a large bubblenest.

The female enters the spawning cave to mate. The male now exhibits a little-changed coloration while the female is far more intensely coloured than usual, and unlike the male, exhibits a dark vertical stripe through the eye (so-called "sexy eyes"). The fishes spawn at intervals of 3 to 4 days during the breeding period. Hence it is possible that a further mating will result in eggs being deposited in the bubblenest among the developing larvae.

One spawning lasted for around 3 hours. The eggs developed very well at a water temperature of around 27 °C (80.5 °F), a pH of 4.8, and a conductivity of 50 µS/cm. The fry swam free after

Parosphromenus pahuensis tending eggs

Between spawnings the female *Parosphromenus pahuensis* remains in the breeding cave tending the spawn

The female *Parosphromenus pahuensis* also tends larvae already present in the nest without problem

The newly fertilised eggs and freshly-hatched larvae are clearly visible in the bubbleenst

7 days and left the nest. Just three days previously a further spawning took place and the eggs were placed in the nest next to the larvae, where both developed well side-by-side. The male continued to perform the brood care. The female did not attack the larvae during the spawning. As additional observations revealed, the pair spawned at intervals of 3 to 4 days given good feeding with black mosquito larvae and *Artemia* nauplii for the free-swimming fry and juveniles in the tank. The offspring were still invisible to the observer as the fry and juveniles remained in the gaps among the dead leaves lying on the bottom.

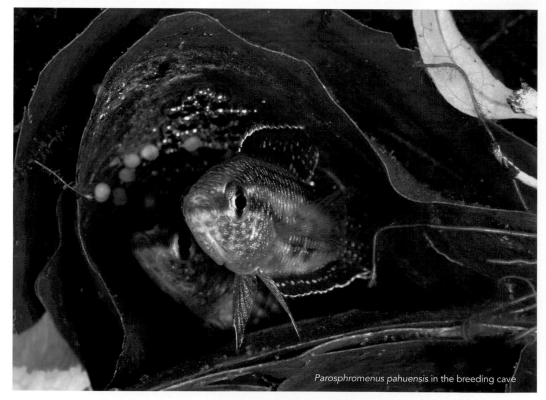

Parosphromenus pahuensis in the breeding cave

Total length: (size)
Probably up to 3.5 cm (1.375 in).

Remarks: (differences from other species of the genus)
The species is to date still very uncommon, and only a very few

specimens have been imported.

The species was for many years traded as *Parosphromenus* sp. Melak.

References:
• LINKE, H. 2008: Sensationell: Der "Honeymoon" Prachtgurami ist wieder da! VDA-aktuell 2008 (4): 29.

Explanation of the species name:

Latin *paludicola* (noun in apposition) = "marsh-dweller" or "swamp-dweller".

Original description:

TWEEDIE, M. W. F. 1952: Notes on Malayan fresh-water fishes. Bulletin of the Raffles Museum 24: 69-71.

Type locality:

North-eastern part of western Malaysia: Terrenganu: Kuala Merchang/Kuala Berang area.

Species-typical characters:

Parosphromenus paludicola has the largest number of spines in the dorsal fin and hence the longest (base length) dorsal fin of all *Parosphromenus* species.

These fishes have a lanceolate caudal fin and sometimes one or even two flank spots. Depending on provenance, however, there are also specimens with no flank spots. Again depending on provenance, the unpaired fins are transparent and colourless to orange in colour.

Similar species:

Parosphromenus linkei is distinguished from this and other dwarf gourami species by a pronounced lanceolate caudal-fin form in males, also seen to a greater or lesser extent in females.

Sometimes, however, females exhibit only a prolonged central ray in the caudal fin. A further characteristic is a central spot on either side of the body, and this can sometimes be doubled or more rarely even tripled. In display coloration these fishes often exhibit a wine-red body coloration, and males may have a striking light dot pattern on the body and fins, as well as light, narrow fin margins.

In *Parosphromenus deissneri* the most important characteristic is the lanceolate caudal fin in both sexes, though in females the lanceolate caudal-fin form is less well-developed and involves only a prolonged central ray. The species exhibits grey to turquoise-coloured stripes on the unpaired fins and no red.

Parosphromenus filamentosus has a lanceolate caudal fin with a much-prolonged central ray in males. The caudal fin is largely red and bordered by a broad, dark stripe. The unpaired fins are bordered by a narrow light stripe.

Parosphromenus pahuensis likewise exhibits dark spots on the flanks and a pattern of small light spots on the unpaired fins.

Natural distribution:

The northernmost distribution regions lie in southern Thailand in the Sungai Kolok region.

Province of Terrenganu: drainage of the Sungai Trengganu, close to the villages of Kuala Brang and Kampong Merchang.

The presumed most southerly occurrence is around 100 km

Parosphromenus paludicola ♂

Parosphromenus paludicola juvenile from Kuala Brang

south of Terrenganu in an area around 6 km west of Paka, via National Highway 122 to around 1 km after Kampong Santong, where there is a right turn after another 500 metres (550 yards) in the direction of Kampong Bungkus. Here a small watercourse crosses the road.

Biotope data:

The species lives predominantly in small flowing watercourses in areas of marshy forest. These are usually small forest streams, sometimes flowing through cultivated plantation regions and thus through dense vegetation (trees and tall brush) that admits hardly any sunlight.

Watercourse near the village of Kuala Brang:

The water had a slight current and a temperature of 26 °C (79 °F). The total and carbonate hardness were both less than 1° (German). The conductivity measured 6 µS/cm at a water temperature of 26 °C (79 °F). The pH was between 5.5 and 6.2. The water was very clear and coloured slightly brownish. Heavy deposits of iron in the form of flakes lay on the plants and banks. The water was 50 cm (20 in) deep. *Parosphromenus paludicola* was mainly to be found in the vegetated, on average

40 cm (16 in) deep, bank zones but sometimes also in only 5 cm (2 in) of depth. The stream had an average width of 1.5 to 2 metres (5 to 6.5 feet). (Research in 1979 and 1985: LINKE, H.& S..)

Watercourse in the area west of Paka, after Kampong Santong in the direction of Kampong Bungkus:

This was a swamp region with wooded areas. The water was heavily stained brown (pH 4.69) and flowing slightly. (Research in August 2006: LINKE, H. et al.)

Coordinates: 04°37.68 N   103°22.48 E (H. LINKE)

I first managed to catch around 15 specimens during a collecting trip in 1979, and thereafter first import live fishes of this species to Germany. At that time we found them near to the villages of Kuala Brang and Kampong Merchang, in small flowing watercourses in areas of marshy forest. These were usually small forest streams, sometimes flowing through cultivated plantation regions and thus through dense vegetation (trees and tall brush) that admitted hardly any sunlight.

The fishes pictured here were caught in a stream close to the village of Kuala Brang. The water had a slight current and a

Two male *Parosphromenus paludicola* from the area north of Terrengganu

Terengganu type

Female *Parosphromenus paludicola* from the area north of Terrengganu

temperature of 26 °C (79 °F). The total and carbonate hardness were both less than 1° (German). The conductivity measured 6 µS/cm at a water temperature of 26 °C (79 °F). The pH was 5.5. The water was very clear and coloured slightly brownish. Heavy deposits of iron in the form of flakes lay on the plants and banks. The water was 50 cm (20 in) deep. *Parosphromenus paludicola* was mainly to be found in the vegetated bank zones, sometimes in only 5 cm (2 in) of depth. The stream had an average width of 2 metres (6.5 feet).

Around 25 years later the landscape in this area had noticeably changed, so the biotopes investigated in 1979 may have partially or completely vanished.

Maintenance:

Aquaria with a water volume of at least 50 litres should be used for maintenance. These fishes will tolerate water parameters ranging from acid to neutral without problem. The species is also tolerant as regards the mineral content of the water. A pH between 4.5 and 7.0 and a general/carbonate hardness of up to 10 degrees (German) are accepted without problem. Good water quality is especially important, as is a wealth of plants providing plenty of hiding-places and protection from tankmates. Optimal maintenance is possible only in so-called species tanks. A water temperature of between 25 and 29 °C (77 and 84 °F) is advised.

South Thailand type

Female *Parosphromenus paludicola* from southern Thailand

Male *Parosphromenus paludicola* from southern Thailand

Watercourse in the area north of Terrengganu, habitat of *Parosphromenus paludicola*

Entrance to the village of Kuala Brang

Water parameter checks in the field are very important for aquarium maintenance

Biotope of *Parosphromenus paludicola* near Kuala Brang

Species diversity isn't very high in *Parosphromenus paludicola* biotopes

Watercourse to the north of Kuala Brang

Reproduction:

For successful breeding the water should be very soft and acid (pH around 4.2 to 4.5 at a carbonate hardness of around (better, less than) 1° and a water temperature around 28 °C (82.4 °F)). The breeding tank should be set up as described for maintenance.

Total length: (size)

Up to 3.5 cm (1.375 in)

Remarks: (differences from other species of the genus)

Various authors speak of "possibly different species" because of the large distribution region extending from southern Thailand to the Kemaman region, north of Cerating/Kuantan in the north to central east of western Malaysia. The supposition that the fishes from the distributional region around Wakaf Tapai (between Kuala Berang and Kuala Terengganu on National Highway 14) are another species is to date pure speculation. It is far more likely that these fishes are the "real" *Parosphromenus paludicola* if they come from Wakaf Tapai, as this is the type locality of the species, the source of the holotype described by TWEEDIE. There is no argument that the extensive distribution of this species at different sites might well also result in forms with a different appearance.

Parosphromenus paludicola after capture in the typical locality

References:

- LINKE, H.1980: Auf Fischsuche in Malaysia - Die ganz normale Geschichte über den Fang eines seltenen Labyrinthfisches. Tatsachen und Informationen aus der Aquaristik (TI) 52: 33-35.

- NAGY, P.1980: Erste Erfolge mit dem Labyrinthfisch *Parosphromenus paludicola.* Das Aquarium 135 (9): 459-463.

- LINKE, H. 1986: Zwei Zwerge aus Malaysia. Tatsachen und Informationen aus der Aquaristik (TI) 78: 11-13.

Parosphromenus parvulus VIERKE, 1979

Explanation of the species name:
Latin *parvulus* = "very small", "tiny".

Original description:
VIERKE, J. 1979: Ein neuer Labyrinthfisch von Borneo – *Parosphromenus parvulus* nov. spec. Das Aquarium 120: 247-251.

Type locality:
Mentaya River system, 250 km north-west of Banjarmasin, South Borneo (Kalimantan).

Species-typical characters:
Males of *Parosphromenus parvulus* exhibit a series of dot-like spots on the anal fin. The unpaired fins are bordered with a narrow light margin. Further species-typical characters of adult males are round red spots on a dark background on the dorsal and anal fins and a transparent, colourless to smoke-grey caudal fin. During courtship and other display the underside of the body becomes deep black. The ventral fins remain predominantly colourless and transparent.

Similar species:
Parosphromenus ornaticauda, but this species has a different coloration, exhibits no red spots in the dorsal and anal fins, and has a red caudal fin bordered above and below by black. The species exhibits light margins to the unpaired fins, especially broad in the anal fin. *Parosphromenus parvulus* remains smaller.

Natural distribution:
In the original description the natural distribution is given as the Mentaya River system, 250 km north-west of Banjarmasin, South Borneo. The area in question lies on the upper Sungai Sampit.

The species was first recorded in the area of the village of Palangan, also shown on maps as Balangan, on the Sungai Kenyota, near to the Sungai Sampit in Kalimantan Tengah.

The habitat in question is a small stream, known as the Planduk (deer stream), behind the rectory at the edge of the village of

Female *Parosphromenus parvulus*

Male *Parosphromenus parvulus*

Parosphromenus parvulus male with the species-typical red spots in the dorsal and anal fins

Palangan. Four specimens of what was then a new species were caught there by E. KORTHAUS and W. FOERSCH (Pater Heinz STROH pers. comm.).

Jürgen KNÜPPEL and the author found this species in 1988 in the region north of Palangkaraya, around 180 km north-west of Banjarmasin in Central Kalimantan (Kalimantan Tengah), and thus around 100 km east of Palangan; this apparently doesn't constitute a major distributional distance for this genus of small labyrinthfishes.

BAER, NEUGEBAUER, and the author were also able to find *Parosphromenus parvulus* in the so-called "deer stream" in Palangan, along with a further *Parosphromenus* species.

Hiroyuki KISHI from Tokyo, a member of the "Borneo Team", has reported its occurrence in the Sungai Sampit area, Katingan, Kahayan basin.

My most recent researches have shown that *Parosphromenus parvulus* has a very much larger distribution region than previously assumed, extending from Palangan in the west to Buntok in the east, i.e. a distance of around 500 km.

In the Babugus area, a blackwater river by the road from Palangkaraya to Buntok, around 17 km in the direction of Buntok after the road from Palangkaraya to Kota Kurun forks off, around 5 km before the village of Barbugus.

Blackwater ditches along the track from Palangkaraya to Buntok in the area between Sungai Kapuas and Sungai Barito, around 32 km before Buntok.

Blackwater biotope with small trees and scrub by the road to Ampah, around 9 km east of Buntok, and hence east of the great Sungai Barito.

Biotope data:

Northern Palankaraya area:
These fishes were collected among thickets of plants in the shallow bank zones of a blackwater river, by the road to Kasungan around 27 km north-west of the town of Palangkaraya. The water was heavily stained brown, very soft, and also very acid with a pH of 4.1. The electrical conductivity measured 24 µS/cm at a water temperature of 28.2 °C (82.8 °F) in the shallow bank zones. (Research in 1988: LINKE, H..)

Palangan area:
Narrow watercourse with little current and brownish water, at the forest margin and the edge of the village of Palangan. These fishes were recorded exclusively in the tangle of marginal plants trailing in the water. The pH was 4.6, the electrical conductivity 18µS/cm at a water temperature of 24.5 °C (76 °F). (Research in July 1990: NEUGEBAUER, N. & LINKE, H..)

East of Palangkaraya:
Blackwater river in the area near Babugus Village, by the road from Palangkaraya to Buntok, around 17 km in the direction of Buntok after the road from Palangkaraya to Kota Kurun forks off. The pH was established at 3.9 and the conductivity at 29 µS/cm at around 28 °C (82.5 °F). The population density of the species was very high here.

Coordinates: 01° 56' 30 S   114° 04' 37 E (LINKE, H., 2007)

Between Kapuas and Berito:
Blackwater ditches along the track from Palangkaraya to Buntok, in the area between Sungai Kapuas and Sungai Barito, around 32 km before Buntok. The water parameters here were

Parosphromenus parvulus biotope near the village of Babugus

measured as pH 3.70 and conductivity 20 µS/cm at 29.3 °C (84.75 °F). The water colour was dark red-brown. The water was almost stagnant and its depth measured 20 to 30 cm (8 to 12 in) on average.

Coordinates:
01° 35' 52 S   114° 45' 42 E (LINKE, H., 2009)

East of the Barito:
Blackwater swamp region with small trees and scrub, by the road to Ampah, around 9 km east of Buntok. The following water parameters were measured: pH 4.72 and conductivity 24 µS/cm at a water temperature of 28.8 °C (83.8 °F). The biotope was largely exposed to the sun.

Coordinates:
01° 41' 07 S   114° 52' 55 E (LINKE, H. 2009)

Parosphromenus parvulus after capture. The species is sometimes very numerous in the biotope

Inundation zone at the edge of the track between Sungai Kapuas and Sungai Barito. A small watercourse at the edge of the woodland permits the fishes to survive during dry periods

Blackwater biotope of *Parosphromenus parvulus* at the edge of the track from Palangkaraya to Buntok

Maintenance:

For the maintenance of this species the pH should be between 5 and 6 and the conductivity up to 100 µS/cm. The water quality should always be very good. Humic substances are important and strongly recommended. Otherwise as described for the genus.

Reproduction:

The species is a bubblenest spawner, requiring small caves for the purpose. For breeding the water should be of very good quality, very soft (no measurable carbonate hardness and conductivity 20 µS/cm), and very acid (pH 4.0 to 4.3).

Total length: (size)

To date total length is known to be barely 3.0 cm (1.25 in).

References:

• FOERSCH, W. & KORTHAUS, E. 1979: Bemerkungen zum Biotop und zur Pflege von *Parosphromenus parvulus* nov. spec. Das Aquarium 120: 250-251.

• LINKE, H. 2007: Ein seltener Gast im Aquarium - *Parosphromenus parvulus*. Aquaristik Fachmagazin Tetra Verlag Germany, 201: 40-43.

Explanation of the species name:

The name is derived from the Greek words *phoinix* (= crimson) and *oura* (= tail), and relates to the coloration of the caudal fin. It is a noun in apposition.

English name:

Langgam Liquorice Gourami

Original description:

SCHINDLER, I. & LINKE, H., 2012: Two new species of the genus *Parosphromenus* (Teleostei: Osphronemidae) from Sumatra. Vertebrate Zoology 62 (3) 2012: 399-406.

Species-typical characters:

Males of the species are characterised by a lanceolate caudal fin and extremely prolonged anterior ventral-fin rays.

Similar species:

Parosphromenus tweediei

Parosphromenus filamentosus has a lanceolate caudal fin with a much-prolonged central ray in males. The caudal fin is largely red and bordered by a broad, dark stripe. The unpaired fins are bordered by a narrow light stripe.

Natural distribution:

A river and swamp region as well as watercourses in a swampy scrub and woodland terrain, along the track at Langgam, broad sand track with ferry, 67 km west of the town of Kerincikiri (paper manufacture) in the central part of Sumatra Riau (Riau Province).

Coordinates: 01°04.35 N   096°07.59 E (LINKE, H.)

The species was found in a swamp region, surrounded by scrub and woodland, associated with the small Sungai Kamparkiri which joins the Sungai Kampa in the vicinity of the settlement of Langgam. The habitat included not only the main river but also numerous watercourses large and small. Except for

Parosphromenus phoenicurus ♀

Parosphromenus phoenicurus ♂

Parosphromenus phoenicurus ♂

this residual area of scrub and trees, the landscape here is characterised by clear-felling and the cultivation of fast-growing plants for paper manufacture. There are only a few remaining undisturbed zones along the Sungai Kamparkiri and the Sungai Kampa.

Biotope data:
These fishes live in blackwaters. The were to be found in shallow bank zones cloaked in marginal plants and scrub and in shallow water with a slight current in the wooded areas. The water parameters were: pH 5.25 and conductivity 7 μS/cm at 26.8 °C (80.25 °F). In January (high-water period) the water was only faintly brown-coloured and had a slight to somewhat stronger current. The visibility was around 100 cm (40 in). Our observations and research took place during the rainy season (high water). (Research in 2008: LINKE, H., CHIANG, N. & SIM, T.)

Maintenance:
As described for the genus.

Reproduction:
As described for the genus.

Total length: (size)
Up to 4 cm (1.5 in).

Remarks: (differences from other species of the genus)
Males of the species are characterised by a lanceolate caudal fin and extremely prolonged anterior ventral-fin rays.

River and swampy area by the Sungai Kamparkiri in the Langgam area

The sand track from Keirincikiri to Langgam

Comparison of water colour from the biotope of *Parosphromenus phoenicurus* in the rainy season

Biotope by the road from Keirincikiri to Langgam

Parosphromenus phoenicurus ♂ after capture

Bank zone in the Sungai Kamparkiri

Biotope of *Parosphromenus phoenicurus*

Explanation of the species name:

Latin *quindecim* = "fifteen", referring to the number of spines in the dorsal fin.

Original description:

KOTTELAT M. and NG P. K. L. 2005: Diagnoses of six new species of *Parosphromenus* (Teleostei: Ospronemidae) from Malay Peninsula and Borneo, with Notes on other species. The Raffles Bulletin of Zoology, Supplement 13: 101-113.

Type locality:

Borneo: Kalimantan Barat: Sungai Pawan basin: Sungai Liong, 4 km north of Nanga Tayap on the road to Sandai; 01° 30' 02" S 110° 34' 19" E, coll. H. KISHI, 11 April 2001.

Species-typical characters:

Parosphromenus quindecim is one of the largest liquorice gouramies known to date. Quindecim (Latin) means "fifteen" and refers to the 15 hard rays in the dorsal fin.

The species exhibits a broad brown (also sometimes red) band in the central part of the unpaired fins, followed by a turquoise-coloured band, and then a dark, narrow band extending to the edge of the fins. The unpaired fins have light margins. The caudal fin has a large, dark, star-shaped spot at its base. The ventral fins are of normal length and exhibit no noteworthy coloration.

Similar species:

By comparison, *Parosphromenus* sp. Stunggang has a small, dark spot on the caudal base and lots of red in the caudal fin. The red in the dorsal and anal fins is less extensive and sometimes adorned with turquoise bars. While the red of the caudal fin is bordered by only a faint dark band, the corresponding dorsal- and anal-fin markings are broader and bolder. The unpaired fins have light narrow margins. The ventral fins are somewhat longer and bright solid turquoise in colour.

Parosphromenus allani has a bright red colour on the central parts of the unpaired fins, resembling bars. This is bordered by a broad, dark, often black band, in turn followed by a light, narrow, outer margin. There is a large dark spot at the base of the caudal fin. The soft portion of the dorsal fin also sometimes exhibits a large dark round spot. The ventral fins are normal in length and have a faint turquoise coloration.

Parosphromenus opallios has bright red on the central parts of

Parosphromenus quindecim in the breeding cave, male below, female above

Mating and brood development stages during

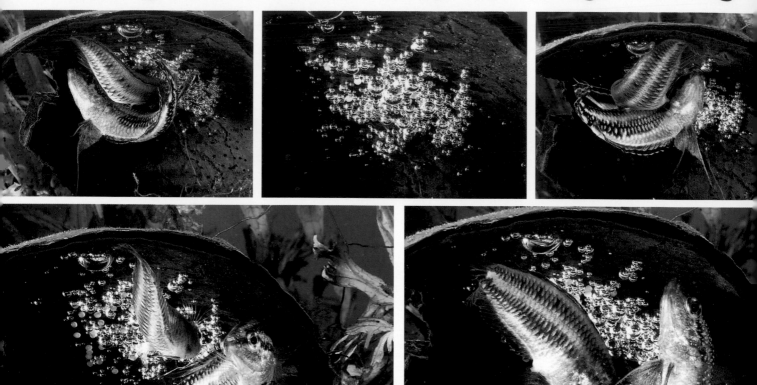

the unpaired fins, in the form of narrow bars. This is followed by a broad dark, usually black band, and in between the two there is a turquoise-coloured band of variable width, especially in the dorsal and anal fins. The unpaired fins have light margins. The base of the caudal fin exhibits a smaller dark spot. The ventral fins are somewhat longer and coloured turquoise right to their tips.

Natural distribution:

The species is found in the rain forest of the island of Borneo. Although prior to its scientific description the species was known as *Parosphromenus* sp. Manis Mata, it does not come from the area of the village of Manis Mata. According to KOTTELAT & NG (2005) the specimen used for the description came from the Sungai Liong, a tributary of the Pawan River, around 4 km north of the village of Nanga Tayap on the road to Sandai in the geographically central region of West Kalimantan (Kalimantan Barat, Borneo). On the basis of my own researches, however, this means not the Pawan but the Kajung River area to the east, separated off by a mountain chain up to 2500 metres (8200 feet) high.

Biotope data:

None available at present. It is unclear from the original description whether the natural habitats of *Parosphromenus quindecim* are blackwaters, but on the basis of experiences with maintenance and breeding in the aquarium, this is a species that lives in very mineral-poor, acid water.

Maintenance:

An aquarium with a water volume of at least 50 litres, well planted in places, should be used for optimal maintenance of these fishes. The substrate can consist of a thin layer of sand. *Parosphromenus* like to live among aquatic plants and dead leaves lying on the bottom, hence not only numerous aquatic plants but also a number of dead oak or beech leaves (or pieces of large Sea Almond leaves) should be introduced as bottom decoration and for a beneficial effect on the water.

However, it is possible to dispense with the sand substrate to make it easier to keep the tank clean.

The water for maintenance should be soft with a pH between 6.0 and 6.5. Peat filtration is highly advisable in this respect.

he reproduction of Parosphromenus quindecim

Reproduction:

The actual spawning procedure begins with the construction of a small bubblenest, with air bubbles, enclosed in a secretion from the mouth, being attached beneath the "roof" of the cave and sometimes also to the sides. The male next entices the ripe female below the bubblenest, where the first attempts at an embrace take place. The male wraps himself around the female, gripping her tightly, and the eggs are then expelled and fertilised by the male. The number of eggs is very small – just

two to (rarely) five. Not every embrace succeeds in producing the release of eggs. During this period the female is noticeably the more active partner. After every successful embrace she collects up the eggs from the body and/or caudal fin of the male or from the bottom, and puts them into the bubblenest. The male often leaves the spawning site after the release of eggs in order to defend the spawning territory against possible rivals or predators, but then very soon returns to the female. The mating takes place in perfect harmony, and often lasts for up to five hours. The female's supply of eggs is finally exhausted and she leaves the spawning site or is driven away by the male. Thereafter the male alone undertakes the care of the clutch, which by now often consists of up to 50 eggs.

Very soft and acid water is now necessary for their successful development. The addition of humic substances (using suitable dead leaves, ie beech leaves or those of the tropical Sea Almond tree, or filtration over peat) is very important for these putative so-called blackwater fishes. The general hardness, and above all the carbonate hardness, should be less than one degree (German). This equates to a conductivity of 20 to 25 μS/cm at a water temperature of 27 to 28 °C (80.5 to 82.5 °F) and a pH between 4.0 and 4.5. After around 3 days at the aforementioned water parameters there will be larvae hanging in the nest. After another 4 days the fry will swim free, though they apparently don't take any food until 24 to 36 hours later. The fry remain hidden most of the time, for example close to the underside of leaves or beneath the roof of the cave. They sometimes also remain for one or two days in the shelter of the cave and under the protection of the still parental male.

The freshly-hatched nauplii of a small *Artemia* species can be taken as first food without problem, but additional initial feeding with rotifers is an advantage.

Total length: (size)
Males attain a total length of around 4.0 to 4.5 cm (1.5 to 2 in). Females remain around 1.0 cm (0.4 in) smaller and are full-grown at around 3.0 to 3.5 cm (1.125 to1.375 in) TL.

Remarks: (differences from other species of the genus)

The species is thought to have first been imported by the trade in 2001 and thereafter was for many years traded as *Parosphromenus* sp. Manis Mata.

The common name "Star-Spot Liquorice Gourami" refers to the half-star-shaped marking on the central part of the caudal fin in males. In the Indonesian language, Manis Mata means roughly "large eye" or "star" and also refers to the pattern on the caudal fin.

References:
• GECK, J. & KOPIC, G. 2005. Prachtguramis. Aquaristik aktuell, special publication 1/2005: 30-36.

Parosphromenus quindecim 8 days after free-swimming

Parosphromenus quindecim 64 days after free-swimming

Parosphromenus quindecim 73 days after free-swimming

Parosphromenus quindecim 81 days after free-swimming

Explanation of the species name:

Latin *ruber* = "red" and *mons* (genitive montis) = "mountain" or "hill", thus "of the red hill". Named for Bukit Merah, a town near the type locality, whose name also means "red hill".

Original description:

KOTTELAT, M. & NG, P. K. L. 2005. Diagnoses of six new species of *Parosphromenus* (Teleostei: Osphronemidae) from Malay Peninsula and Borneo, with Notes on other species. The Raffles Bulletin of Zoology, Supplement 13: 105-106.

Type locality:

Western Malaysia, Perak Province, near kilometre marker 21 on the road from Taiping to Segama, Sungai Bedang area.

Coordinates: 05° 07' 18 N 100° 39' 04 E (TAN, H. H.)

Species-typical characters:

Parosphromenus rubrimontis exhibits a large dark spot at the base of the caudal fin, with a broader, red, semicircular band adjoining, sometimes with a narrow, turquoise-coloured margin distally, and then a broader outer black band. This coloration and pattern are repeated in the dorsal and anal fins. The unpaired fins have a narrow, light outer margin. The ventral fins are normal in length and turquoise-coloured. The species has a longer dorsal fin in comparison to the other members of the *rubrimontis* group.

Similar species:

In comparison to all other *Parosphromenus* species, males of *Parosphromenus alfredi* have very long ventral fins, whose anterior ray is prolonged and very light in colour.

Parosphromenus sp. Tanjung Malim is often incorrectly regarded as a local form of *P. rubrimontis*, but has only a narrow red

Parosphromenus rubrimontis ♂

Parosphromenus rubrimontis ♂

Biotope of *Parosphromenus rubrimontis* in the Pondak Tanjok area

Basil and Lee ChiMing are sorting the fishes after capture

Examining the fishes collected

band in the predominantly black caudal fin, and there may be a similar band in the dorsal and anal fins. The ventral fins are normal in length and black in colour, with a delicate, lighter, anterior upper portion to the hard ray. The unpaired fins have a narrow, light outer margin.

Parosphromenus tweediei is one of the most striking species of the *rubrimontis* group. This species exhibits a broad, bright red band in the central part of the unpaired black fins. This central red band may also have a narrow turquoise margin as a transition to the outer black part of the fin. The unpaired fins have a narrow, light, outer margin. The ventral fins are normal in length and slightly turquoise in colour.

Natural distribution:

Bukit Merah area:
The species lives in a blackwater stream that crosses the road between Kampong Lobak and Bukit Merah. Here the species lives together with, inter alia, *Betta pugnax* and *Sphaerichthys osphromenoides*. Water tests made on several occasions at different times of year produced average pH values of 3.4 to 4.5, general hardness always less than 1° dGH, electrical conductivity up to 20 µS/cm at a water temperature usually in the range 26 to 29 °C (79 to 84 °F). The water depth was mainly around 50 cm (20 in). The banks were densely vegetated with emerse- and submerse-growing plants. In places thick cushions

Blackwater biotope of *Parosphromenus rubrimontis* following rainforest clearance and the planting of oil palms

Biotope of *Parosphromenus rubrimontis* at the edge of the oil-palm plantation at km marker 35/36 on the road from Bukit Merah to Changkat Lobak

Section of biotope at km marker 35/36

of fallen leaves (from the trees on the bank) covered the bottom of the watercourse. The species has been seriously endangered here for a number of years as a result of deforestation and the creation of plantations. (Research in January 1979, 1983, 1986, 2005, and 2006; LINKE, H. et al.)

Around 6 to 7 km north of Kampong Bukit Merah in the direction of Kampong Changkat Lobak - Selama.

A watercourse which flows from a palm plantation and crosses the road from Bukit Merah heading north to Kampong Changkat Lobak, between kilometre markers 35 and 36. Unfortunately the biotope has been much changed and is now too exposed to the sun as a result of deforestation and the planting of palms, including on the other side of the road, and hence overheated. (Research in August 2006; LIM, T. Y. & LINKE, H. et al.)

Coordinates: 05° 05. 09 N 100° 40. 31 E (LINKE, H., 2006)

A watercourse flowing from the rainforest and crossing the road from Kabu Gejah heading south to Pondak Tanjok, north-east of Bukit Merah, only a few kilometres before Pondak Tanjok. On average around a metre (40 in) deep and very difficult of access. Blackwater area (pH 4.63).

(Research in August 2006; H. LINKE)

Coordinates: 05° 02. 00 N 100° 43. 70 E (H. LINKE, 2006)

Biotope data:

The species has to date been recorded only in blackwater biotopes, in very soft, very acid water with pH values between 3.8 and 4.8.

Blackwater biotope in the oil-palm plantation at km marker 35/36

Section of the *Parosphromenus rubrimontis* biotope

Biotope at km marker 35/36 on the road from Bukit Merah to Changkat Lobak

Maintenance:

Parosphromenus rubrimontis should be maintained only in very soft, acid water with areas of dense aquatic plants and hiding-places in the form of dead leaves on the bottom.

Reproduction:

Breeding will succeed only in very soft and very acid water (carbonate hardness less than 1 °dKH and pH 4.2). The newly free-swimming fry are very small, but will take freshly-hatched *Artemia* nauplii. Additional feeding with rotifers during the first days is advantageous.

Total length: (size)

Around 3.5 cm (1.375 in).

Remarks: (differences from other species of the genus)

The species was for many years traded as *Parosphromenus* sp. from Bukit Merah.

References:

- LINKE, H. 1986: Zwei Zwerge aus Malaysia. Tatsachen und Informationen aus der Aquaristik (TI) 78: 11-13.

Parosphromenus sp. Ampah

Explanation of the name:

The epithet refers to the southernmost distribution area known so far, near to the village of Ampah.

English Name:

Ampah Liquorice Gourami

Species-typical characters:

Males of the species exhibit a lanceolate caudal-fin form with the middle ray prolonged.

Natural distribution:

According to Patrick Yap, the exporter in Singapore, this species has so far been found in the affluents of the Sungai Barito. Its distribution extends from Ampah north to the Muarateweh area in the east of the province of Kalimantan Tengah, island of Borneo, Indonesia.

Biotope data:

A blackwater swamp area along the road from Buntok to Ampah, 9 km from the centre of Buntok. The swamp area is dotted with small trees and scrub and large parts of it are exposed to the sun.

The fishes live here in small areas of water with very soft, acid, black water with a pH of 4.7 and an electrical conductivity of 24 µS/cm. The species is syntopic with, among other species, *Parosphromenus parvulus* and *Sphaerichthys selatanensis*.

(Research in June 2009: LINKE, H. et al.).

Coordinates: 01°41.07 S   114°52.55 E (LINKE, H.)

Parosphromenus sp. Ampah ♂

Maintenance:

As described for the genus.

Reproduction:

As described for the genus.

Total length: (size)

Up to 4 cm (1.5 in).

Remarks: (differences from other species of the genus)

The species clearly differs from *Parosphromenus filamentosus* in having a different coloration and different markings. *Parosphromenus filamentosus* is found around 200 km to the south in the area south of Banjarmasin on the Sungai Barito in the south of the province of Kalimantan Selatan, island of Borneo.

Parosphromenus sp. Ampah ♀

Swamp region east of Buntok, biotope of *Parosphromenus sp.* Ampah

Explanation of the unofficial name:

The unofficial name refers to the distribution region, the island of Belitong.

English name:

Belitong Liquorice Gourami

Species-typical characters:

The males of this species are almost black in display coloration, and have light fin margins and only very faint turquoise bars in the dorsal and caudal fins.

On the basis of observations of wild-caught specimens in the aquarium, *Parosphromenus* sp. Belitong may actually be two different species. Only the species apparently most common on the island will be discussed here.

Natural distribution:

A small, narrow forest watercourse between Bantan and Pelulusan, south-east of Tanjungpandan. Scrubland watercourse near Kepayang, on the road from Tanjungpandan to KL. Kampit on the island of Belitong, Indonesia.

Biotope data:

The following water parameters were measured in the biotope between Bantan and Pelulusan: pH 5.3 and a conductivity of 8 µS/cm at a water temperature of 25.5 °C (78 °F). The water showed little movement and was slightly brownish in colour. (Research in September 2008; LINKE, H.)

Coordinates: 02°54′12 S   107°44′29 E (LINKE, H.)

Scrubland watercourse near Kepayang, on the road from Tanjungpandan to KL. Kampit

The following water parameters were measured in the biotope near Kepayang: pH 5.59 and conductivity 9 µS/cm at a water temperature of 25.2 °C (78 °F). The water was flowing and coloured slightly brownish. (Research in September 2008: H. LINKE.)

Coordinates: 02°46′23 S   107°46′52 E (LINKE, H.)

Parosphromenus sp. Belitong ♀

Parosphromenus sp. Belitong ♂ in courtship coloration

Maintenance:

As described for the genus.

Reproduction:

As described for the genus.

Total length: (size)

Up to 4 cm (1.5 in).

Remarks: (differences from other species of the genus)

Possibly belongs to the *Parosphromenus bintan* group.

Parosphromenus sp. Belitong male in normal coloration

Woodland biotope of *Parosphromenus* sp. Belitong at Membalong in the south of the island

Section of biotope in the dark forest region in the south of the island

The species wasn't very numerous in the woodland area near Membalong

There was only a narrow track through the rainforest

Small watercourses with slightly brown water traversed the woodland area. It was difficult to find *Parosphromenus* there

Biotope shots of various habitats of *Parosphromenus* sp. Belitong at Kepayang on the Belitong island

Biotope shots of various habitats of *Parosphromenus* sp. Belitong at Kepayang on the Belitong island

Parosphromenus sp. Calak

Explanation of the unofficial name:
The unofficial name refers to the distribution region, the Danau Calak.

English name:
Calak Liquorice Gourami

Species-typical characters:
Parosphromenus sp. Calak males in display coloration exhibit very dark to black fins with a light margin. The species appears to be one of the smaller liquorice gouramies. There is, however, a second colour form with noticeably paler blue markings on the fins. It may be that there are two species living in the same biotope.

Similar species:
Parosphromenus bintan has predominantly black fins, with a delicate light blue to turquoise-coloured band in the outer part of the dorsal, caudal, and anal fins. This is bordered on both sides by a broader black margin. The unpaired fins have light margins. The base of the caudal fin is largely black. The dark, delicate blue-coloured ventral fins are moderately long and the lower portion of their hard rays is light blue in colour. They exhibit no red coloration.

Natural distribution:
In an affluent of the Danau Calak. The Danau Calak is a large, up to 5 km long, lake-like river widening with several small affluents, and is very rich in fishes. This "lake" lies in the Air Musi river system, near to the village of Bailang, west of Sekayu, north-west of Palembang, province of Sumatra Selatan, Indonesia.

Biotope data:
The water parameters at the "lake" were: water temperature 31.1 °C (88 °F), air temperature 29.1 °C (84.4 °F), pH 6.3, and conductivity 25 µS/cm. In the affluent, the habitat of *Parosphromenus* sp. Calak: pH 5.12 and conductivity 12 µS/cm. The water was slightly turbid and yellowish in colour. (Research in September 2008; LINKE, H.)

Coordinates: 02°57′33 S   103°59′49 E (LINKE, H.)

Maintenance:
As described for the genus.

Reproduction:
As described for the genus.

Total length: (size)
Up to 3 cm (1.25 in).

Parosphromenus sp. Calak ♂ in full colour.

Parosphromenus sp. Calak ♂ in mating coloration, ♀ left, ♂ right

Parosphromenus sp. Calak ♂ in courtship dress

Remarks: (differences from other species of the genus)
Possibly belongs to the *Parosphromenus bintan* group.

References:

• LINKE, H. 2008: Neue *Parosphromenus*.

• Betta News, Journal of the European Anabantoid Club/Arbeitskreis Labyrinthfische im VDA: 2008 (4): 25.

Explanation of the unofficial name:

The unofficial name of this scientifically undescribed species refers to the distribution region near to the town of Dabo (or Dobo).

English name:

Dabo Liquorice Gourami

Species-typical characters:

Males of this species have a rounded caudal-fin form.

Similar species:

Parosphromenus bintan has predominantly black fins, with a delicate light blue to turquoise-coloured band in the outer part of the dorsal, caudal, and anal fins. This is bordered on both sides by a broader black margin. The unpaired fins have light margins. The base of the caudal fin is largely black. The dark, delicate blue-coloured ventral fins are moderately long and the lower portion of their hard rays is light blue in colour. They exhibit no red coloration.

Natural distribution:

Parosphromenus sp. Dabo was found in the area of the town of Dabo, also shown as Dobo on various maps, on Pulau Singkep (the island of Singkep), the southern island of the Kepulauan Lingga (Lingga Archipelago), province of Riau Archipelago, Indonesia.

Biotope data:

Not available at present. The species is thought to live in blackwater rivers, in very mineral-poor, acid water.

Maintenance:

As described for the genus.

Reproduction:

As described for the genus.

Total length: (size)

Around 3.5 cm (1.375 in).

Remarks: (differences from other species of the genus)

Possibly belongs to the *Parosphromenus bintan* group.

Parosphromenus sp. Dabo ♀

Parosphromenus sp. Dabo, ♂ left, ♀ right

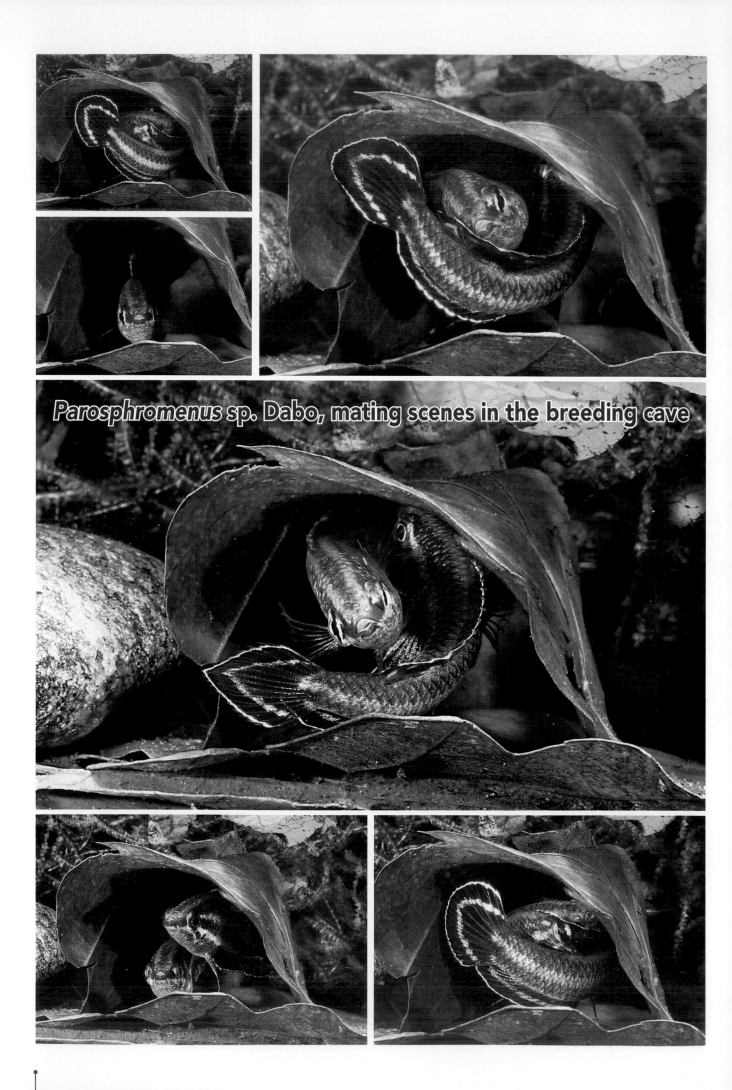

Parosphromenus sp. Dabo, mating scenes in the breeding cave

Parosphromenus sp. Dua

Explanation of the name:

The epithet relates to the natural habitat in the Pari Dua swamp region near the town of Jambi in the east of the province of Jambi, island of Sumatra, Indonesia.

English name:

Dua Liquorice Gourami

Species-typical characters:

Males of the species have a rounded caudal-fin form. These fishes undoubtedly belong to the *P. bintan* group.

Natural distribution:

The species lives in the Pari Dua swamps – areas of blackwater swamp with clumps of trees and scrub – around 15 km east of east of Kota Jambi, province of Jambi, island of Sumatra, Indonesia.

Biotope data:

The Pari Dua swamps are a large "peat marsh" with scrub in places. The soil is deep black and the water coloured dark red-brown. The landscape is partially cultivated bushland. The fishes were comparatively numerous in the natural habitat. The water was typical black water – mineral-poor and very acid – and the biotope very rich in plants.

(Research in May 2012: LINKE, H., BAYER, H., SCHÄFER, R., & SIM, T.).

Maintenance:

As described for the genus.

Reproduction:

Bubblenest spawner.

Total length: (size)

Up to 3.5 cm (1.125 in).

Remarks: (differences from other species of the genus)

Maintenance and breeding in the aquarium are problem-free if the natural water parameters are provided. These fishes are traded as *Parosphromenus* sp. Dua. They undoubtedly belong

Parosphromenus sp. Dua ♀

Parosphromenus sp. Dua ♂

Overgrown groups of plants in the biotope of *Parosphromenus* sp. Dua

Collecting often proved difficult, but the blackwater biotope was easy of access

The habitat at Pari Dua is a large, often densely vegetated, blackwater swamp region

to the *bintan* group. When breeding they construct only a very small bubblenest with only a few bubbles in the nest. In this species the breeding cave is located among dense vegetation, usually well hidden near or right at the bottom. It isn't unusual for spawning to take place at intervals of three to four days, with both eggs and larvae then developing in the nest simultaneously under the attentive care of the male. The larvae swim free after around seven days of development, and can take freshly-hatched *Artemia* nauplii as first food without problem.

Parosphromenus sp. Gawing

Explanation of the name:

The epithet refers to the distribution in the Sungai Gawing, near to the village of Gawing.

English name:

Gawing Liquorice Gourami

Species-typical characters:

Males of this species have an only slightly lanceolate caudal fin with the middle ray prolonged.

Natural distribution:

These fishes were found by the road from Buntok to Palangkaraya at Desa Gawing, around 20 km west of the Sungai Kapuas and thus around 170 km west of the town of Buntok, in the east of the province of Kalimantan Tengah, island of Borneo, Indonesia. The habitat was a blackwater river around 15 to 20 metres (50-65 feet) wide and mostly up to 1.5 m (60 in) deep, with a slight current.

Biotope data:

The species lives in areas of dense vegetation in the bank zone, in very mineral-poor, acid, black water.

(Research in June 2012: LINKE, H., SCHÄFER, R., BAYER, H. & TOMMY, H.).

Maintenance:

As described for the genus.

Reproduction:

As described for the genus.

Total length: (size)

Up to just under 5 cm (2 in).

Remarks: (differences from other species of the genus)

Parosphromenus sp. Gawing exhibits similar coloration and patterning to *Parosphromenus* sp. Ampah, but has a large number of striking, X-like transverse markings on the sides of the body and is a remarkably large size (one female 4.7cm (1.875 in) total length, measured immediately after capture). This is possibly the largest *Parosphromenus* species known to date.

Parosphromenus sp. Gawing, ♂ in front.

Male *Parosphromenus* sp. Gawing. The numerous x-shaped markings on the body are a noteworthy feature of this species

The Sungai Gawing is a large blackwater river

The Sungai Gawing is shaded by dense scrub and trees on its banks

Sometimes the river has shallow, densely vegetated bays at its margins. These too are the habitat of *Parosphromenus* sp. Gawing

Parosphromenus sp. Palangan

Explanation of the unofficial name:

The species was first recorded in the Palangan area.

English name:

Palangan Liquorice Gourami

Species-typical characters:

In males of this species the dorsal and anal fins have a broad area of wine red to red extending outwards from the body, sometimes followed by a narrow, turquoise band, and then a broad, dark margin and a narrow, light outer edging. The caudal fin is delicate red near its base and sometimes exhibits a faint dark spot there. The rest of the caudal fin is colourless and without markings.

Natural distribution:

These fishes live in the same biotope as *Parosphromenus parvulus*, described back in 1979. The habitat is the so-called Planduk (= "deer stream"), at the exit to the village of Palangan (Balangan) on the Sungai Kenyala, near to the Sungai Sampit in Kalimantan Tengha (Central Kalimantan). BAER, NEUGEBAUER, and the author were able to catch not only these fishes but also *Betta foerschi* and *Betta anabatoides* here in 1990.

The "Borneo Team" recorded *Parosphromenus* sp. Palangan in the Parenggean area, around 25 km north-east of Palangan. (Research in February 2000: Hiroyuki KISHI.)

In 1991 the same species was caught in the area of the village of Pundu in the Sungai Cempaga region. Pundu lies on the road from Palangkaraya through Tangkiling to Kasongan on the Sungai Katinganin. From there, after crossing the river, the road runs for around 20 km further in a westerly direction across the Sungai Cempaga in the Pundu area. Downstream the Sungai Cempaga comes to Sampit, as does the road. For many years the species was also known as *Parosphromenus* sp. Pundu, and portrayed as such on the Internet, because it wasn't realised that the fishes with a lanceolate caudal fin caught at Pundu were the same species as that found at Palangan. The error wasn't discovered until 2009.

Biotope data:

The water temperature in the Palangan area was measured at 24 °C (75 °F) at 08.00 in the morning and later at 24.5 °C (76 °F) at

Young male *Parosphromenus* sp. Palangan

Juvenile *Parosphromenus* sp. Palangan, a rarely being imported *Parosphromenus* species nowadays

Palangan, a small village on the Sampit River

11.00 am. The total and carbonate hardness were both less than one degree (German), the pH 4.6, and the water had an electrical conductivity of 18 µS/cm, measured at a water temperature of 24 °C. The fishes lived in the dense, often up to 50 cm (20 in) wide, band of marginal vegetation along the banks and could be caught there only with great difficulty. (Research in July 1990: NEUGEBAUER, N. & LINKE, H.)

In the Pundu area these fishes live in current-free, densely vegetated zones in blackwater streams. Tests on the water revealed a pH of 5.5 and a conductivity of 20 µS/cm at a water temperature of 26.5 °C (80 °F). (Research in 1991: KRUMMENACHER, R. & WASER, A.)

Maintenance:

Maintenance requires small aquaria, up to 50 cm (20 in) long, decorated with plants such as *Microsorum pteropus* (Java Fern), and with no substrate in order to facilitate a high level of cleanliness. The water parameters for optimal maintenance, and above all for breeding, should be general and carbonate hardness no higher than 3 °dGH/dKH and a pH of around 5 to 5.5.

Reproduction:

As described for the genus.

Total length: (size)

Members of this species grow to around 3.0 to 3.5 cm (1.125 to 1.375 in) long.

Toilets by the river are typical sanitary arrangements in Kalimantan

Parosphromenus sp. Palangan is again an inhabitant of blackwaters. These fishes are found predominantly among plants

Parosphromenus sp. Pelantaran

Explanation of the unofficial name:

The unofficial name refers to the distribution region and the village nearest to the collecting site.

English name:

Pelantaran Liquorice Gourami

Species-typical characters:

In display coloration these fishes exhibit a stunning colour pattern of striking light blue stripes in the unpaired fins. Unusually, the hard rays in the unpaired fins are prolonged, as is also known from *Parosphromenus nagyi*, for comparison.

Similar species:

To date (2012) no comparable species (appearance and coloration) are known from Kalimantan.

Natural distribution:

The species was found in a small blackwater river that crosses the road leading from Palangkaraja in the direction of Sampit, around 19 km south of the bridge in Pundu (69 km from the big bridge in Kasungan), and in the Sungai Parit to Rubuk Langgang area near Kruing Pelantaran.

Biotope data:

The biotope investigated at this site was a watercourse on average 8 metres (26 feet) wide and mainly 50 cm (20 in) deep. The water was noticeably brown in colour and had a slight current. The air temperature was measured at 32 °C (89.6 °F) and the water temperature 28.5 °C (83.3 °F). The pH was established as 4.8 and the electrical conductivity as 28 µS/cm. The bank zones were thickly overgrown with scrub and dense stands of trees provided extensive shade. In places there were dense expanses of aquatic plants (*Cryptocoryne*). *Parosphromenus parvulus*, as well as numerous other labyrinthfishes, was also recorded in this biotope.

Maintenance:

As described for the genus.

Reproduction:

As described for the genus.

Total length: (size)

3.5 cm (1.375 in).

Remarks: (differences from other species of the genus)

Possibly belongs to the *Parosphromenus bintan* group.

To date (2011) only two specimens of this species have been imported.

References:

• LINKE, H. 2008: Ein neuer Prachtgurami aus Kalimantan. Aquaristik Fachmagazin, Tetra Verlag,Berlin-Velten, 207: 28-30.

Parosphromenus sp. Pelantaran ♂

Biotope of *Parosphromenus* sp. Pelantaran, 19 km south of Pudu

There are still residues of the rainforest in the vicinity of plantations

Large areas in this region are (still) bushland or plantations

Bank zone with vegetation

Displaying male *Parosphromenus* sp. Pelantaran

Parosphromenus sp. Pelantaran seeking to mate with a female of another species

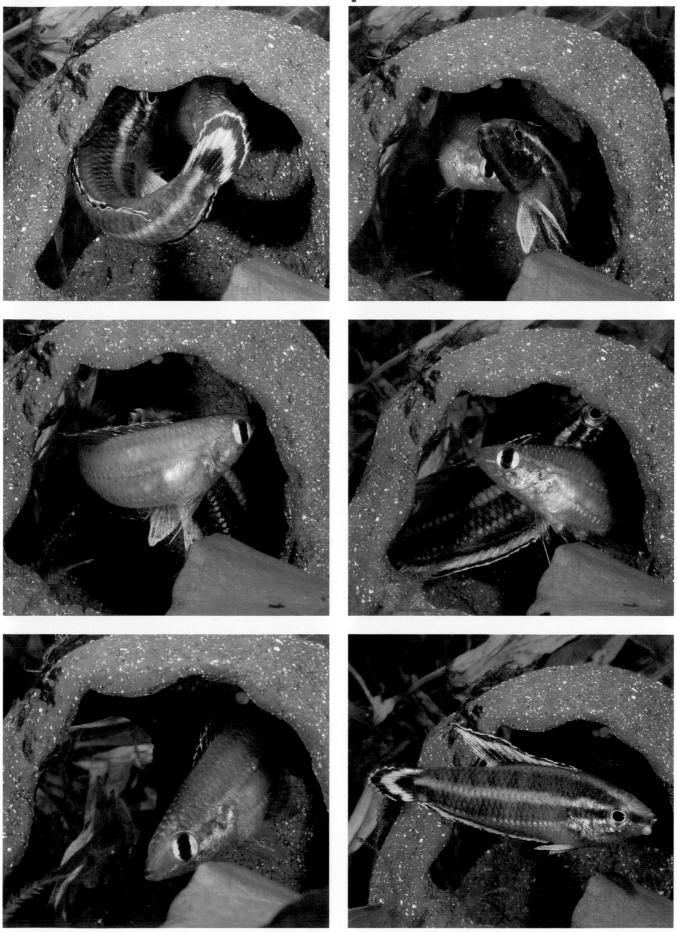

Explanation of the name:

The epithet refers to the distribution of this fish, close to the village of Pematanglumut in the north of the province of Jambi.

English name:

None

Species-typical characters:

Males of this species have a rounded caudal-fin form. The latest information suggests that there are probably two *Parosphromenus* species of different appearance living in the same habitat.

The first of these is *Parosphromenus* sp. Pematanglumut 1, with red-brown coloration in the unpaired fins, and the other is *Parosphromenus* sp. Pematanglumut 2, which has blue-black fins and possibly belongs to the *bintan* group.

Natural distribution:

Blackwater swamp region with dense scrub and trees, east of the road from Kota Jambi to Kualatongkai at Pematanglumut (sandy track following a gas main), around 95 km north of Kota Jambi, province of Jambi, island of Sumatra, Indonesia.

Biotope data:

The habitat of this as yet scientifically undescribed *Parosphromenus* species was among aquatic vegetation and the foliage of scrub growing in the water in gently flowing blackwater areas, in a large blackwater swamp region with a pH of 4.3 and a conductivity of 25 µS/cm at 27.5° C (81.5 °F). The water (low-water period) was strongly coloured dark brown. (Research at the end of May 2007: LINKE, H., CHIANG, N., LINKE, M. & SIM, T.)

Coordinates: 01° 04.84 S   103° 25.34 E (LINKE, H.)

Observations and water parameters from the high-water period (mid January 2008) are also interesting. At this time the water temperature around midday was 27.5°C (81.5 °F; air temperature 29 °C (84 °F) and the pH was measured at 4.48 and conductivity 46 µS/cm. The water was strongly coloured brown, but less so than in May 2007. The visibility was around 60 cm (24 in). The *Parosphromenus* species lived in the company of *Betta simorum*, *Betta coccina*, *Belontia hasselti*, *Sphaerichthys osphromenoides*, *Boraras* sp., inter alia. (Research in January 2008: SIM, T., CHIANG, N. & LINKE, H.).

Maintenance:

As described for the genus.

Reproduction:

As described for the genus.

Total length: (size)

Up to 4 cm (1.5 in).

Male *Parosphromenus* sp. Pematanglumut Form 2 from the *bintan* group

Remarks: (differences from other species of the genus)

Observations of around 35 fishes from this biotope during subsequent aquarium maintenance revealed striking differences in the courtship and threat colours of males. Hence it is presumed that two different species of the genus *Parosphromenus* are represented in the same biotope.

Male *Parosphromenus* sp. Pematanglumut Form 1 with red-brown fin colour

Male *Parosphromenus* sp. Pematanglumut Form 1 with red-brown fin colour

The modest homes of the fishermen in the swamp region

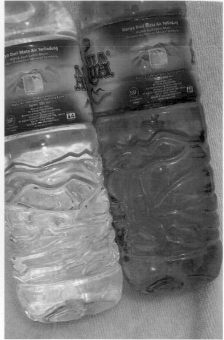

The brown coloration differs from that of drinking water and indicates a possible admixture of humic substances

Unfortunately the biotope has been dramatically altered or even destroyed by industrial usage

The blackwater swamp region at Pematang Lumut

The fish collectors try to provide the fishes with the best possible care until they are bought by exporters

Nets suspended in the water make it possible to keep larger fishes as well

Smaller fishes are placed in plastic containers during collecting

Parosphromenus sp. Pontian Besar

Explanation of the name:

The species was recorded in the Pontian Besar area.

Species-typical characters:

Parosphromenus sp. Pontian Besar exhibits similar coloration and markings to other species, but, compared to them, has a noticeable red interior to the turquoise-coloured stripes in the unpaired fins, especially the caudal fin.

Similar species:

Parosphromenus bintan exhibits broader, less brightly coloured stripes in the unpaired fins and no red coloration.

Parosphromenus harveyi likewise exhibits turquoise-coloured stripes in the unpaired fins, but more in the central part of the fins. This species has no red coloration in the fins.

Parosphromenus cf. *bintan* has bolder and narrower turquoise banding and a narrower black margin in the outer part of the unpaired fins. This species exhibits no red in the fins.

Natural distribution:

These fishes were first recorded by Peter K. L. NG of the National University of Singapore in the area of the village of Pontian Besar, only a few kilometres north of Pontian Kecil, around 50 km west of Johor Bahru, in the south-western coastal region of western Malaysia.

Biotope data:

The natural habitat of *Parosphromenus* sp. Pontian Besar is blackwater swamp regions with very soft, very acid water. The species lives sympatric with *Betta pulchra*, *Betta bellica*, and *Sphaerichthys osphromenoides*.

Maintenance:

As described for the genus.

Reproduction:

As described for the genus.

Total length: (size)

These fishes grow to between 3.0 and 3.5 cm (1.125 and 1.375 in) long.

Remarks: (differences from other species of the genus)

Possibly belongs to the *Parosphromenus bintan* group.

Parosphromenus sp. Pontian Besar ♂

Parosphromenus sp. Red

Explanation of the name:

In reference to the coloration of the fins.

English name:

Red-fin Liquorice Gourami

Species-typical characters:

In *Parosphromenus* sp. Red the lower longitudinal band kinks upwards on the caudal peduncle to the centre of the base of the caudal fin, where it ends in a dark spot. The red to red-brown unpaired fins are dark in the basal region and sometimes exhibit dark striping on the central area. The third, bottom longitudinal band is only faintly coloured.

Similar species:

Parosphromenus anjunganensis, but that species is more elongate in body form, doesn't grow as large, and has larger, uniformly wine-red unpaired fins.

Natural distribution:

The first importation purportedly took place in 1992 via the trade. The natural habitat is at present still (2012) unknown.

Biotope data:

The species is thought to live in blackwater like the other *Parosphromenus* species.

Maintenance:

As described for the genus.

Reproduction:

As described for the genus.

Total length: (size)

Up to 3.5 cm (1.375 in).

Parosphromenus sp. Red ♂

Explanation of the unofficial name:

The unofficial name of this scientifically undescribed species relates to the distribution region near to the village of Sentang in the province of Sumatra Selatan.

English name:

Sentang Liquorice Gourami

Natural distribution:

The village of Sentang lies on the road from Kota Jambi to Kota Palembang in the province of Sumatra Selatan, island of Sumatra, Indonesia. The collecting site is reached via a turn-off to the east in Sentang. After the turn-off in the village of Sentang it is 5 km to a small kampong (at the time of our research the bridge was being rebuilt) where watercourses cross the road. A large part of the terrain is being used for the cultivation of rubber trees.

Coordinates: 01°56.07 S   103°42.53 E (LINKE, H.)

A second collecting site lies 4 km further east, but could not be reached because of the bridge repairs.

The species has also been recorded to the north in the direction of Tebing Merana, via a turn-off to the east from the road from Kota Jambi to Kota Palembang, province of Sumatra Selatan, where the track ends at a river. The species was also collected around 8 km after the turn-off, just after the small settlement of Simpang ti. Here too the landscape was dominated by rubber-tree plantations.

Coordinates: 01°55.62 S   103°41.49 E (LINKE, H.)

Biotope data:

In the Sentang biotopes these fishes were found in dense groups of plants, sometimes in fast-flowing water (high-water period). The water temperature measured 27.2 °C (81 °F) (air temperature around 30 °C (86 °F), at night 25 °C (77 °F)), the pH 5.0, and the conductivity 3 µS/cm. These were blackwater habitats. The water had a slight to strong current and was clear and brown in colour. The underwater visibility was around 1 metre (40 in).

Parosphromenus sp. Sentang ♀

Parosphromenus sp. Sentang ♂

Scenes of the brood care of *Parosphromenus* sp. Sentang showing eggs and larvae

Biotope of *Parosphromenus* sp. Sentang at Simpang ti

In the area to the north at Simpang ti the water temperature was measured at 27.2 °C (81 °F), the pH 5.0, and the conductivity only 3 µS/cm. Here too there were fast-flowing blackwater habitats in places, surrounded by little-frequented rubber-tree plantations and areas of residual woodland and scrub. Here too the water was clear and brown in colour. (Research in January 2008: LINKE, H. CHIANG, N. & SIM, T.)

Inundation zone at Sentang

The collected fishes being examined

It was mainly *Parosphromenus* sp. Sentang and *Boraras* that were found in the biotope at Simpang ti

At the entrance to the village a sign on the bus stop indicates the way to Sentang.

Maintenance:

As described for the genus.

Reproduction:

As described for the genus.

Total length: (size)

Around 3.5 cm (1.375 in).

Remarks: (differences from other species of the genus)

Possibly belongs to the *Parosphromenus bintan* group.

This currently scientifically undescribed species has been and still is traded as *Parosphromenus* singtangnensis as well as *Parosphromenus* sp. Singtang.

Explanation of the unofficial name:

The unofficial name of this scientifically undescribed species refers to its distribution region near to the village of Sungaibertam in the province of Jambi.

English name:

Bertam Liquorice Gourami

Species-typical characters:

Males of this species exhibit a rounded caudal-fin form.

Similar species:

Parosphromenus bintan, but is brighter blue in colour.

Parosphromenus sp. Sentang, but exhibits a bolder stripe pattern.

Parosphromenus sp. Danau Rasau, but exhibits more brown colour in the fins.

Natural distribution:

This scientifically undescribed species comes from the area near Sungaitiga, around 30 kilometres south-west of Kota Jambi, around 10 km from the branch in the road from Kota Jambi (the town of Jambi) to Kota Palembang (the town of Palembang), in Sumatra Selatan, in the province of Jambi, island of Sumatra, Indonesia.

The distribution lies in the area of the village of Sungaibertam (sometimes also written as Sungai Bertam).

Biotope data:

Parosphromenus sp. Sungaibertam lives in dense, broad areas of plants along the banks of small, slow-flowing, shallow watercourses in plantations in the valleys in what is here a very densely vegetated, hilly landscape. These are small, up to 3 metres (10 feet) wide and 60 cm (2 feet) deep watercourses flowing through the swampy lowlands. The water here is very clear with a slight current and a faint brownish colour. The water temperature in January was 25.8 °C (78.5 °F) (air temperature 29 °C (84.2 °F)), the pH 5.01, and the conductivity 2 µS/cm. (Research in 2008; LINKE, H., CHIANG, N. & SIM, T.)

Coordinates: 01° 42.31 S   103° 32.60 E (LINKE, H.)

Parosphromenus sp. Sungaibertam ♀

Male *Parosphromenus* sp. Sungaibertam in full colour

Parosphromenus sp. Sungaibertam mating scenes in the breeding cave

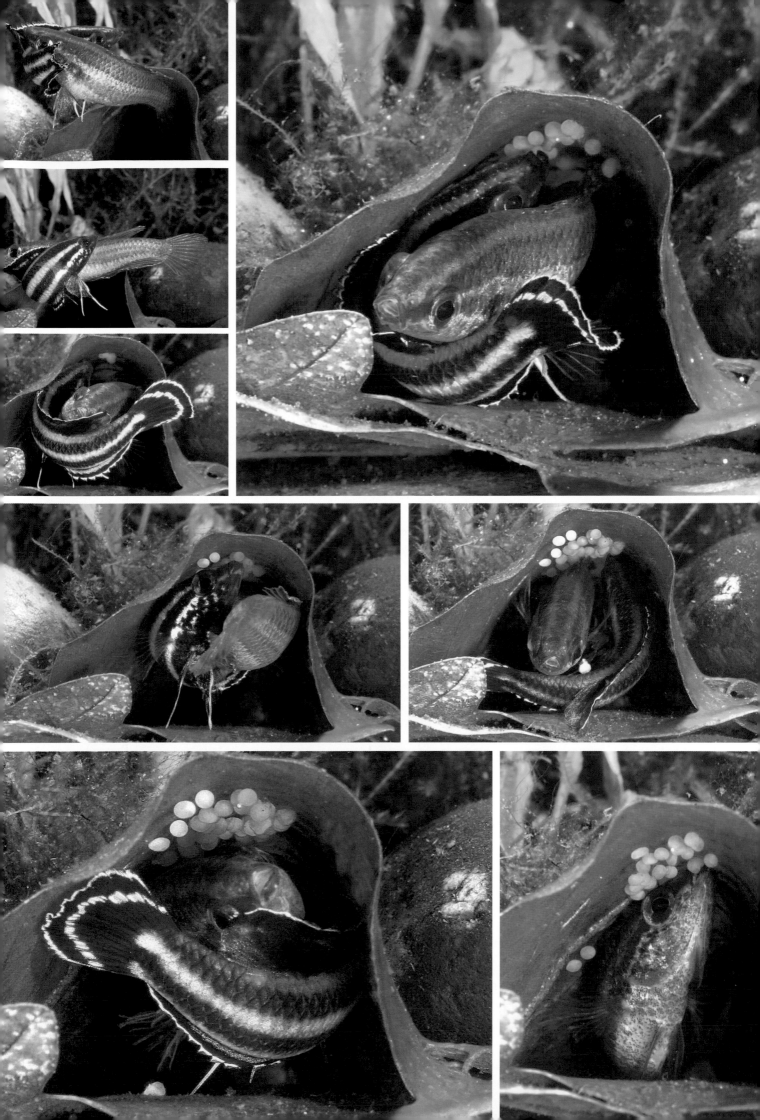

Biotope at Sungaibertam, habitat of *Parosphromenus* sp. Sungaibertam

Vegetated bank zone in the Sungaibertam watercourse

The clear brown watercourse is well suited to collecting. It isn't very deep and easy of access

Parosphromenus sp. Sungaibertam in the inspection container following capture

Maintenance:
As described for the genus

Reproduction:
As described for the genus.

Total length: (size)
Around 3.5 cm (1.375 in).

Remarks: (differences from other species of the genus)
Possibly belongs to the *Parosphromenus bintan* group.

Parosphromenus sp. Sungai Stunggang

Explanation of the unofficial name:
The name relates to the currently known distribution region.

English name:
Parosphromenus allani II

Species-typical characters:
Parosphromenus sp. Sungai Stunggang has a comparatively small, dark spot on the caudal base and lots of red in the caudal fin. The red in the dorsal and anal fins is less extensive and sometimes adorned with turquoise bars. While the red of the caudal fin is bordered by only a faint dark band, the corresponding dorsal- and anal-fin markings are broader and bolder. The unpaired fins have light narrow margins. The ventral fins are comparatively long and bright solid turquoise in colour.

Similar species:
Parosphromenus allani has a bright red colour on the central parts of the unpaired fins, resembling bars. This is bordered by a broad, dark, often black band, in turn followed by a light, narrow, outer margin. There is a large dark spot at the base of the caudal fin. The soft portion of the dorsal fin also sometimes exhibits a large dark round spot. The ventral fins are normal in length and have a faint turquoise coloration. There is a broad, turquoise-green stripe across the anal fin near its base.

Parosphromenus opallios has bright red on the central parts of the unpaired fins, in the form of narrow bars. This is followed by a broad dark, usually black band, and in between the two there is a turquoise-coloured band of variable width, especially in the dorsal and anal fins. The unpaired fins have light margins. The base of the caudal fin exhibits a smaller dark spot. The ventral fins are somewhat longer and coloured turquoise right to their tips.

Parosphromenus quindecim is one of the largest liquorice gouramies known to date. *Quindecim* (Latin) means "fifteen" and refers to the 15 hard rays in the dorsal fin. The species exhibits a broad brown (also sometimes red) band in the central part of the unpaired fins, followed by a turquoise-coloured band,

Female *Parosphromenus* sp. Sungai Stunggang

Male *Parosphromenus* sp. Sungai Stunggang

The Sungai Stunggang, habitat of *Parosphromenus* sp. Sungai Stunggang

The National Highway from Kuching to Lundu in Sarawak

size blackwater river that crosses the road. The fishes were, however, also recorded in watercourses shortly before Lundu. (Research in November 2006: LO, M. & WASER, A.)

Coordinates:
01° 37' 407 N   109° 53' 157 E (WASER, A.)

We found the species in the Sungai Stunggang 73.3 km from Kuching in the direction of Lundu and Semantan, around 16 km after the first bridge over the Batang Kayan (heading away from Kuching), shortly before Lundu. Thereafter there is a second bridge over the Batang Kayan, which flows in a curve and hence crosses the road twice.

Coordinates:
01°37'26 N   109°53'10 E (LINKE, H.)

Biotope data:

The habitat consists of narrow, heavily vegetated watercourses with a gentle current and clear, slightly brownish to heavily brown-stained water; pH around 5.5 and lower.

We found the species in the bank zones where there was less current. The water temperature was 25.5 °C (78 °F), the pH 4.93, and the electrical conductivity 13 µS/cm. The water was cloudy and muddy brown in colour after prolonged heavy rainfall. The visibility measured between 10 and 15 cm (4 and 6 in). The terrain was scrub and rainforest. (Research in January 2008: LO, M. YAP, P. CHIANG, N. NGAI, K. M., & LINKE, H.)

Maintenance:
As described for the genus.

Reproduction:
As described for the genus.

and then a dark, narrow band extending to the edge of the fins. The unpaired fins have light margins. The caudal fin has a large, dark, star-shaped spot at its base. The ventral fins are of normal length and exhibit no noteworthy coloration.

Natural distribution:

Eastern Malaysia: Sarawak: watercourse that crosses the road 49.4 kilometres west of Bau in the direction of Lundu. (Research in July 1986: BROWN, A.& B.).

Recent successful collections were likewise made in the region of the western Batang Kayan basin, on the road from Kuching to Sematan, between the villages of Bau and Lundu, west of the capital city of Sarawak. The habitat was confirmed as the region around and east of the Sungai Stunggang, a medium-

Total length: (size)
Up to 3.5 cm (1.375 in).

Remarks: (differences from other species of the genus).
The species is also known as *Parosphromenus* sp. Biawak and *Parosphromenus* sp. Batang Kayan.

References:
• BROWN, A. & BROWN, B. 1987: A Survey of Freshwater Fishes of the Family Belontiidae in Sarawak

Parosphromenus sp. Tanjung Malim

Explanation of the unofficial name:

The species arrived in the trade with the datum that it was from the Tanjung Malim region.

English name:

Tanjung Malim Liquorice Gourami.

Species-typical characters:

Parosphromenus sp. Tanjung Malim has a narrow red band in the predominantly black caudal fin, and there may be a similar band in the dorsal and anal fins. The ventral fins are normal in length and black in colour, with a delicate, lighter, anterior upper portion to the hard ray. The unpaired fins have a narrow, light outer margin

Similar species:

Parosphromenus rubrimontis exhibits a larger dark spot at the base of the caudal fin, with a broader, red, semicircular band adjoining, sometimes with a narrow, turquoise-coloured margin distally, and then a broader outer black band. This coloration and pattern are repeated in the dorsal and anal fins. The unpaired fins have a narrow, light outer margin. The ventral fins are normal in length and turquoise-coloured. The species has a longer dorsal fin in comparison to the other members of the *rubrimontis* group.

Parosphromenus tweediei is one of the most striking species of the rubrimontis group. This species exhibits a broad, bright red band in the central part of the unpaired black fins. This central red band may also have a narrow turquoise margin as a transition to the outer black part of the fin. The unpaired fins have a narrow, light, outer margin. The ventral fins are normal in length and slightly turquoise in colour.

Parosphromenus alfredi - in comparison to all other *Parosphromenus* species, males of this species have very long ventral fins, whose anterior ray is prolonged and very light in colour.

Natural distribution:

As far as is known at present (details via the trade) the species comes from the Tanjung Malim area in the eastern swamp region of the Sungai Bernam in North Selangor in western Malaysia. The purported distribution region of *Parosphromenus* sp. Tanjung Malim thus lies only a few kilometres east of the natural habitats of *Parosphromenus harveyi*.

Male *Parosphromenus* sp. Tanjung Malim

Parosphromenus sp. Tanjung Malim, ♀ above, ♂ below

Biotope data:

The species probably lives in blackwater swamp regions with very acid, very soft water.

Maintenance:

As described for the genus.

Reproduction:

As described for the genus.

Total length: (size)

Up to 3.5 cm (1.375 in).

Explanation of the species name:

Latin adjective *sumatranus* meaning "of Sumatra", referring to the distribution region, the Indonesian island of Sumatra.

Original description:

KLAUSEWITZ, W. 1955: *Parosphromenus deissneri sumatranus.* Die See- und Süsswasserfische von Sumatra und Java. Senckenbergiana Biologica, 36: 320-323.

Type locality:

Island of Sumatra, watercourse at Jambi (Djambi).

Synonyms:

Parosphromenus deissneri sumatranus

Species-typical characters:

Parosphromenus sumatranus exhibits a striking large black spot in the lower posterior part of the dorsal fin and no pattern in the caudal fin. Males of this species have black tips to the ventral fins as well as red-brown unpaired fins, with the outer half of the dorsal and anal fins black with a light margin. Adult males sometimes also exhibit a slightly prolonged central ray in the caudal fin, giving the latter a slightly lanceolate appearance.

Similar species:

Parosphromenus nagyi - in comparison to other species this species exhibits very short membranes in the dorsal, caudal, and anal fins. The rays protrude noticeably beyond the membranes. This gives the impression of this species having smaller fins. Males of this species in display coloration exhibit a beige or smoke-grey to black body with no stripes as well as variable markings and coloration on the unpaired fins. The species has a rounded caudal fin and sometimes exhibits a faint dark spot on the dorsal fin.

Natural distribution:

In the shallow marginal zones of the Sungai Pijoan, in the Soak Putat, near to the village of Pijoan, on the road from Kota Jambi to Muarabulian in the province of Jambi in the south-east of the island of Sumatra, Indonesia.

Parosphromenus sumatranus, male in normal coloration

Male *Parosphromenus sumatranus* in full colour

The Pijoan River. *Parosphromenus sumatranus* is very numerous in the shallow bank zones.

Biotope data:

These fishes were found in the up to 50 cm (20 in) deep, broad bank zones of the Sungai Pijoan. The water was clear and slightly brownish in colour, with a pH of around 5.0 and a water temperature of 27 °C (80.6 °F). The water had a slight current, even in the sheltered bays, and there were only a few stands of plants in the water, but numerous dead leaves on the bottom instead. Hiding-places and hence shelter were provided by large amounts of scrub growing and foliage trailing in the water. *Parosphromenus sumatranus* were living here in the company of *Sphaerichthys osphromenoides* and *Betta falx*. (Research in May 2007: LINKE, H., CHIANG, N.,LINKE, M., & SIM, T.)

Maintenance:

As described for the genus. The species should be kept in water that is not too acid. A pH of around 5.5 is advisable.

Reproduction:

As described for the genus.

Total length: (size)

Up to 3.5 cm (1.375 in).

References:

- KLAUSEWITZ, W. 1955: *Parosphromenus deissneri* BLEEKER – zum ersten Mal in Deutschland. Die Aquarien- und Terrarien-Zeitschr (DATZ) VIII (10): 257-258.

- LADIGES, W. 1955: *Parosphromenus deissneri*. Die Aquarien- und Terrarien-Zeitschr (DATZ) VIII (12): 333.

- KOTTELAT, M. & NG, P. K. L.1998: *Parosphromenus bintan*, a new osphronemid fish from Bintan and Bangka islands, Indonesia, with redescription of *P. deissneri*. Ichthyological Exploration of Freshwaters, 8: 263-272.

- TAN, H. H. & NG, P. K. L. 2005: The Labyrinth Fishes (Teleostei: Anabantoidei, Channoidei) of Sumatra, Indonesia. The Raffles Bulletin of Zoology Supplement 13: 131.

A collecting station in Pijoan River. The lake-like widening is very rich in fishes.

Many small aquarium fishes that are caught in the shallow bank zones. They are included of *Parosphromenus sumatranus*

Parosphromenus tweediei KOTTELAT & NG, 2005

Explanation of the species name:
Dedication in honour of M. W. F. TWEEDIE, former director of the Raffles Museum in Singapore.

Original description:
KOTTELAT M. & NG, P. K. L. 2005: Diagnoses of six new species of *Parosphromenus* (Teleostei: Osphronemidae) from Malay Peninsula and Borneo, with Notes on other species. The Raffles Bulletin of Zoology Supplement 13: 106-107.

Type locality:
Peninsular Malaysia: Johor: Pontian, Sri Bunian, Kampong Pt.Tekong.1° 27' 59.2"N 103° 26' 07.7"E; NG, P. K. L., 1992

Species-typical characters:
Parosphromenus tweediei is a very striking species of the *rubrimontis* group. This species exhibits a broad, bright red band in the central part of the black unpaired fins. This central red band may also have a narrow turquoise margin as a transition to the outer black part of the fin. The unpaired fins have a narrow, light, outer margin. The ventral fins are normal in length and slightly turquoise in colour. Sometimes, depending on the distribution region, there may also be light blue to turquoise zones in the unpaired fins, especially in the area behind the ventral fins.

Similar species:
Parosphromenus alfredi - in comparison to all other *Parosphromenus* species, males of this species have very long ventral fins, whose anterior ray is prolonged and very light in colour.

Parosphromenus phoenicurus has almost the same coloration as *Parosphromenus tweediei*, but a lanceolate as opposed to a rounded caudal fin.

Parosphromenus rubrimontis exhibits a larger dark spot at the base of the caudal fin, with a broader, red, semicircular band adjoining, sometimes with a narrow, turquoise-coloured margin distally, and then a broader outer black band. This coloration and pattern are repeated in the dorsal and anal fins. The

Parosphromenus tweediei ♀

Parosphromenus tweediei ♂ in full colour

Larvae in the nest of *Parosphromenus tweediei*

unpaired fins have a narrow, light outer margin. The ventral fins are of normal length and turquoise-coloured. The species has a longer dorsal fin in comparison to the other members of the *rubrimontis* group.

Parosphromenus sp. Tanjung Malim has a narrow red band in the predominantly black caudal fin, and there may be a similar band in the dorsal and anal fins. The ventral fins are normal in length and black in colour, with a delicate, lighter, anterior upper portion to the hard ray. The unpaired fins have a narrow, light outer margin.

Natural distribution:

The natural distribution region of these fishes lies in the state of Johor, in the south-west of western Malaysia. There are documented habitats in regions west of Melaka, between Batu Pahat and Keluang, between Keluang and Kulai, as well as near Layang Layang and between Ayer Hitam and Kluang (Keluang).

Biotope data:

These fishes live in sheltering, vegetated, marginal zones in depths of around 1.0 to 1.5 metres (40 to 60 in) in very acid blackwaters with a pH of around 4. (KOTTELAT & NG, 2005)

Maintenance:

Small aquaria with a volume of around 50 litres (11 gallons), with dense plant growth in places and small overhangs and caves, are advisable for successful maintenance. Although these are so-called blackwater fishes in terms of their natural biotope, the water parameters for maintenance can also be somewhat higher, i.e. general and carbonate hardness up to 8° dGH/ dKH (electrical conductivity up to 250 µS/cm) and the pH from 6.0 to 6.5. Peat filtration is highly advisable.

The water temperature can be between 27 and 29 °C (80.5 and 84 °F) and should fluctuate downwards by 2 to 3 °C (3.5 to 5.5 °F) from time to time through water changes.

Reproduction:

Parosphromenus tweediei spawns beneath small overhangs or in caves near the bottom. The male constructs a small bubblenest for the purpose, using air bubbles, enclosed in a secretion from the mouth, which are attached beneath the "roof" of the cave or to its sides. After a long courtship the female follows the male into the spawning cave to lay her eggs. The eggs are released and simultaneously fertilised during an embrace by her partner, and are usually collected by the more active female and deposited in the nest and/or on the ceiling of the cave, where they remain attached for subsequent development. The spawning process can last up to 4 hours. Thereafter the male alone undertakes the care of the brood. The number of eggs is small, often only 12 to 20.

For successful development of the spawn the water must now be very soft, and should, if possible, have an electrical conductivity between 10 and 20 µS/cm and a pH of 4.0 to 4.2. The addition of humic substances using peat filtration or beech/oak leaves is highly advisable. The larvae rupture their egg-shells around 3 days after fertilisation and the almost colourless, slightly grey-coloured fry swim free after around 5 days. The transition to free-swimming can take many hours or sometimes even days. During their first days the fry remain largely concealed among dead leaves or plants. The food offered during these first days should be rotifers used in parallel with very tiny, freshly-hatched, *Artemia* nauplii.

Even with very good water quality and a varied diet the little *Parosphromenus tweediei* grow very slowly.

Total length: (size)
The species attains a total length of around 3.5 cm (1.375 in).

Remarks: (differences from other species of the genus)
Because of their unusual coloration these fishes are without doubt the most attractive *Parosphromenus* species.

The species is one of the round-tailed species, i.e. they have a rounded caudal-fin form. *Parosphromenus tweediei* exhibits a black band at the base of the three unpaired fins, followed by a striking red zone and then another black band distally. The very narrow outer edging to the fins consists of a metallic turquoise to light blue margin. A narrow stripe of the same colour sometimes (apparently depending on the natural provenance) separates the red zone from the outer black band, within the fin. The ventral fins are dark, mainly black, but with turquoise to light blue on the upper anterior part. The body exhibits the *Parosphromenus*-typical two dark longitudinal bands on a beige background, but in display, courtship, and dominant coloration is overall very dark, usually deep black, including the entire breast and belly region.

References:
- GECK, J. & KOPIC, G. 2005. Prachtguramis. Aquaristik aktuell, special publication 1/2005: 30-36.
- LINKE, H. 2006: Ein Juwel unter den Prachtguramis – *Parosphromenus tweediei*. Aquarium live, Gong Verlag GMBH & CO. KG Ismaning. 5: 20-27.

Shots of
Parosphromenus tweediei
in the breeding cave
with eggs and larvae

The genus *Pseudosphromenus*

The genus *Pseudosphromenus* was erected in 1879 by the Dutch doctor and naturalist Pieter BLEEKER, and at present contains two very different species. These are very peaceful fishes that can well be maintained together with other small species. Their appearance is very different. Maintenance in the aquarium is problem-free and extremely interesting. Both are bubblenest spawners and aren't difficult to breed in the aquarium. If kept with inappropriate tankmates they will be very timid and hence hide for most of the time, but in the absence of excessively active and aggressive tankmates they are very interesting fishes to maintain. While *Pseudosphromenus cupanus* was scientifically described as long ago as 1831, there are problems in determining the valid scientific description for the species *Pseudosphromenus dayi*. There is no evidence for the purported original description in 1909 by W. KÖHLER. But the fish had, in fact, been described earlier by KÖHLER, in 1908.

References:

HIERONIMUS, H. & SCHMIDT, J. 1990. Zu Namensgebung und Erstbeschreibungsjahr von *Pseudosphromenus dayi* (KÖHLER, 1908). Der Makropode 12 (11/12): 222-223.

Explanation of the species name:

Latin adjective *cupanus* = "of Caupang", referring to the type locality, a river in southern India.

English name:

Black Spiketail Paradisefish

Names in Sri Lanka:

Tal kossá, Tal kadayá, Ta-but-ti, Heb-bu-ti.

Synonyms:

Polyacanthus cupanus
Macropodus cupanus
Macropodus cupanus cupanus
Polyacanthus cupanus cupanus

Original description:

CUVIER, G, 1831: Hist. Nat. des poissons, vol. 7, Paris: 357.

Natural distribution:

The species was described in 1831 from the Arian-Caupang River, close to the village of Pondicherry in south-eastern India. The natural habitat is given as south-eastern India, the coastal lowlands along the Coromandel coast, sporadically along the western parts of the Malabar coast, and in the north-western parts of the island of Sri Lanka (formerly Ceylon).

Biotope data:

The species lives in narrow watercourses with a slight current, sometimes in irrigation ditches associated with rice fields, and in collections of water and small rivers in the rainforest regions.

Reproduction:

Bubblenest spawner.

The maintenance of the Spiketail Paradisefish presents no problems as they are very undemanding. A small aquarium will suffice, but the bottom should without fail be decorated with small caves and/or half coconut shells and dense planting in places. Fine gravel as substrate and some floating plants will complete the picture. Gentle water movement is also advised. The water temperature can be around 25 °C (77 °F).

Breeding this species is again not difficult. An aquarium of length 60 cm (24 in) or more, 30 cm (12 in) wide and 30 cm deep, will be adequate. The water should be no more than 20 cm (8 in) deep Overhanging pieces of bogwood, as well as rocky caves or inverted, whole or broken flowerpots, should be added to permit the construction of a bubblenest. Groups of plants here and there, plus fine substrate and a water temperature around 27 °C (80.5 °F), are likewise recommended to ensure successful breeding. Long-established aquaria, which are biologically mature and have a good growth of green algae, are preferred by this species for breeding.

Pseudosphromenus cupanus male in courtship dress

The male constructs a small, compact bubblenest in a cave or beneath an overhang. During the period prior to mating there may be "dummy runs" among the plants or near the bottom, independent of the bubblenest and hence the actual spawning site. The pair perform a loose embrace or the activity may also be very intense. The embrace differs from that observed in *Betta* species. The *Pseudosphromenus cupanus* pair swim almost parallel for a short time, "wriggling" so that repeated body contact stimulates them to embrace. But most of the time the male swims very close to his partner and thereby seeks to entice or drive her towards the nest. During this period the male assumes a light coloration, while the female becomes very dark, almost black, in colour.

During the actual spawning both fishes are beneath the bubblenest and now release eggs and sperm during the embrace. Each spawning pass lasts around 12 seconds. Both partners go into a spawning paralysis, then the female is the first to break free and begin to collect up the eggs as they sink downwards. A few seconds later the male also "awakens" and likewise participates in collecting the eggs and putting them into the bubblenest. Sometimes both fishes emerge from the spawning paralysis almost simultaneously. The first spawning passes produce only around 2 to 4 eggs, but the number subsequently increases to 12 to 20. The eggs have a diameter of around 1.2 mm, are oval in shape, and exhibit a light yellow coloration. The spawning act is at an end after around four hours, and the male now assumes the care of the brood. This involves no noticeable aggression towards the female. She is in fact immediately chased away from the bubblenest but is tolerated in its vicinity – i.e. within sight of it.

The larvae hatch after around 48 hours at a water temperature of 27 °C (80.5 °F) and look like white dots with tiny transparent tails hanging in the nest. After a further 24 hours the patterning becomes bolder and clearer and it is possible to detect the form of each tiny fish. The yolk sac is already very small. After a further 24 hours, i.e. four days after spawning, the larvae are almost capable of swimming. The eyes are clearly visible, as is a slightly spotted protective pattern on the body. The yolk sac has by now disappeared, and a few hours later the fry swim free. The parents do not predate on their own fry.

Total length: (size)
Up to 5.5 to 6 cm (2.2 to 2.4 in).

Remarks: (differences from other species of the genus)
Pseudosphromenus cupanus are very peaceful and a little shy. Looking for the somewhat prolonged dorsal fin and lanceolate caudal fin in males is a good method of differentiating the sexes, though sexing is often difficult as these characteristics are seen only in full-grown specimens. Females do not always exhibit a rounded caudal fin but can, like males, develop a pointed, slightly lanceolate caudal-fin form. Sexing is easy during courtship coloration, when the male becomes more colourful and his entire body turns lighter, while the female, by contrast, becomes very dark, almost black.

References:
• DAY, F. 1878: The fishes of India; being a natural history of the fishes known to inhabit theseas and fresh water of India, Burma and Ceylon. vol 1, p. 371. 1967 reprint, Today and Tomorrow's Book Agency, New Delhi, Vol.1
• DERANIYAGALA, P. E. P., 1952: A Colored Atlas of Some Vertebrates from Ceylon. Volume 1, Ceylon National Museums Publication, p 115-116
• ROSSMANN; K.-H. 2009: Die Randgruppe-Spitzschwanzmakropoden und ihre Lebensräume. Der Makropode 1/2009: 13-16.

Pseudosphromenus cupanus,
mating and brood-care scenes beneath the bubblenest

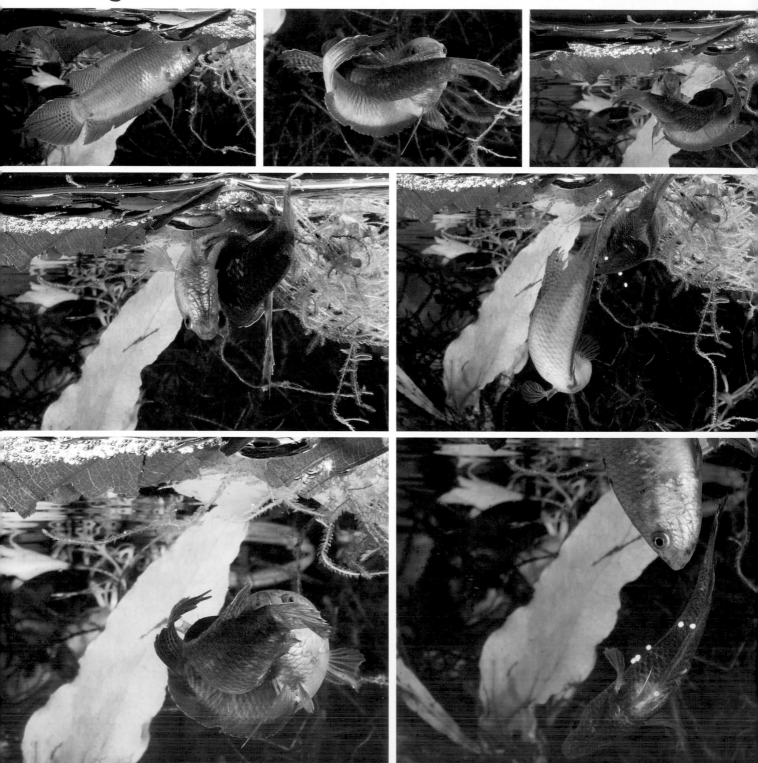

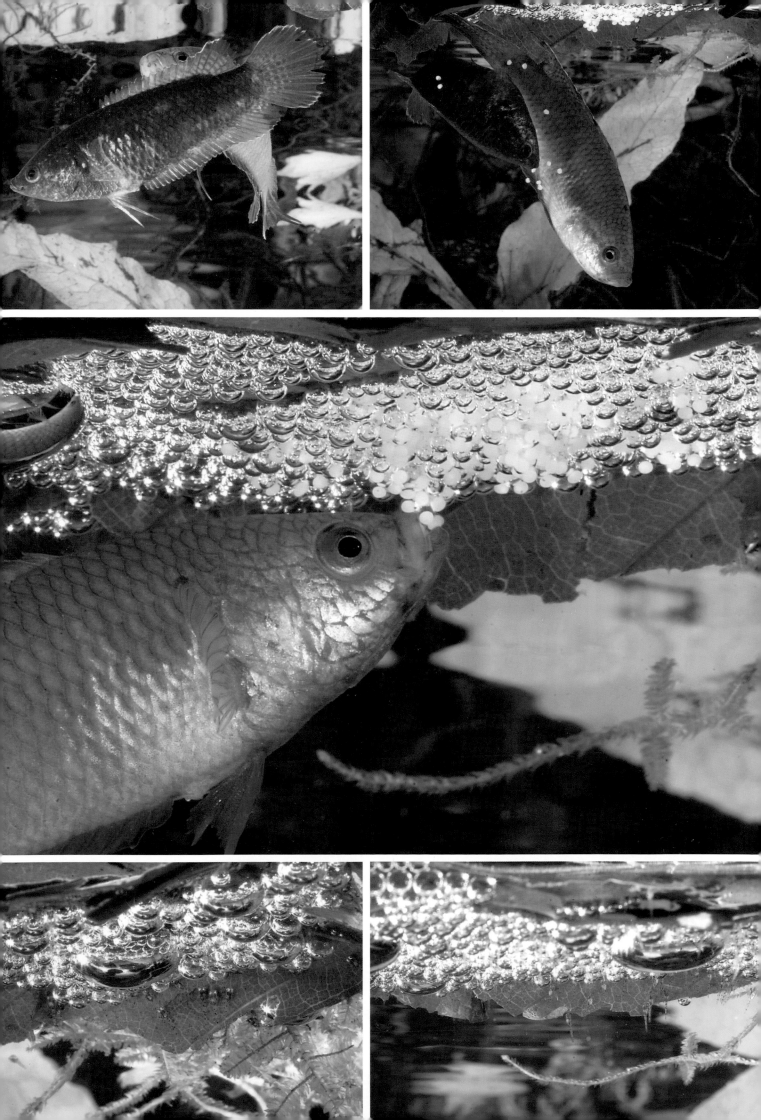

Explanation of the species name:

Dedication in honour of Francis Day

English name:

Red Spiketailed Paradisefish

Synonym:

Polyacanthus cupanus var. *dayi*
Macropodus cupanus var. *dayi*
Macropodus dayi
Parosphromenus dayi
Macropodus cupanus dayi
Parosphromenus deissneri
Polyacanthus dayi
Pseudosphromenus cupanus dayi

Original description:

KÖHLER, W. 1908: Untersuchungen über das Schaumnest und den Schaumnestbau der Osphromeniden. Blätter für Aquarien- und Terrarienkunde 19: 392-396

Natural distribution:

The natural habitat is given as the coastal lowlands of the Coromandel coast in south-eastern India and the coastal regions of western Malaysia. The latter is, however, currently disputed as no recent collections are known from the area. Hence, despite what is stated in numerous other publications, the natural distribution is apparently restricted to south-eastern India and the north-west of Sri Lanka.

Biotope data:

These fishes live in heavily vegetated bank zones in a wide variety of watercourses and swamp regions. Water parameters are apparently of secondary importance.

Reproduction:

Bubblenest spawner.

For breeding the aquarium can be set up as for normal maintenance. The temperature should be raised to 30 °C (86 °F).

Pair of *Pseudosphromenus dayi* beneath the bubblenest

Male *Pseudosphromenus dayi* in full colour

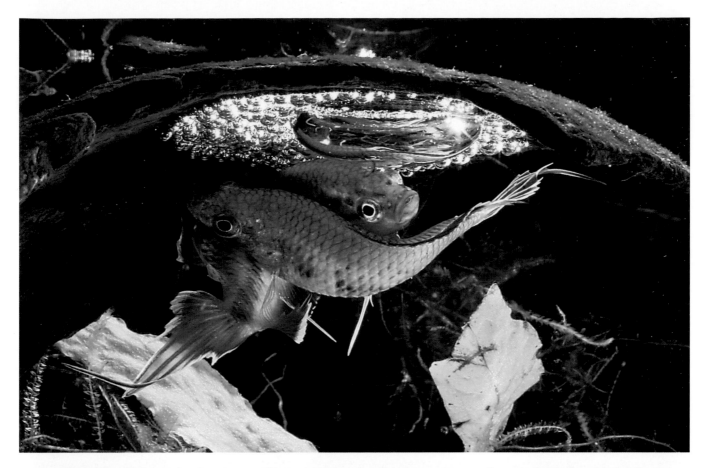

Pseudosphromenus dayi, mating and brood-care scenes beneath the bubblenest

This species often constructs its small, compact bubblenest in caves, but also at the water surface beneath floating plants or in a corner of the aquarium with no additional substrate. Mating is usually harmonious and injury to the female is very rare. The female is often even allowed to take part in the brood care.

Total length: (size)
Up to 7 cm (2.75 in).

Remarks: (differences from other species of the genus)
These fishes look very attractive even as juveniles and are full grown at around 7 cm (2.75 in) (including fins), with the females remaining smaller. The males exhibit a prolonged, pointed dorsal and anal fin, and the central rays of the caudal fin are particularly greatly prolonged. The fins are red with white or light margins.

The maintenance of this species is problem-free. These fishes are peaceful and can readily be housed with other small fishes. However, the aquarium used for their maintenance should not be too small. Tanks 60 cm (24 in) upwards in length are recommended, with a width and depth of around 30 cm (12 in). The tank should contain various groups of aquatic plants as well as caves and some bogwood. The water temperature can be around 26 °C (79 °F). Flake food of various types is greatly enjoyed.

References:
- KÖHLER, W. 1908: Untersuchungen über das Schaumnest und den Schaumnestbau der Osphromeniden. Blätter für Aquarien- und Terrarienkunde 19: 392-396

- ENGMANN, P. 1909: *Polyacanthus* var. Blätter für Aquarien- und Terrarienkunde 20:473-476

- STECHE, O. 1914: Die Fische (von Alfred Brehm, neubearbeitet von Otto STECHE) in: Brehms Tierleben- Allgemeine Kunde des Tierreichs, 4th Edition, Leipzig, p. 590.

- KOTTELAT, M. 1994: Authorship and date of publication of *Pseudosphrome* (Pisces: Belontiidae) J.South Asian nat. Hist., 1 (1): 31-33.

The genus *Sphaerichthys*

The genus *Sphaerichthys* currently (2012) contains four species, all lumped together under the blanket term of "Chocolate Gouramies". All four species are regarded as blackwater fishes that for optimal maintenance require top-quality, very soft, very acid water containing humic substances. Their natural distribution is very diverse and restricted to Malaysia, Sumatra, and Kalimantan (island of Borneo), where they almost always inhabit slow-flowing, clear waters stained dark red-brown and containing large amounts of humic substances. As a result of the latter, as well as the usually very low pH value, the water is very low in micro-organisms. An important prerequisite for optimal maintenance in the aquarium. All four species employ mouthbrooding as their reproductive strategy, though astonishingly observations to date indicate that one species is a maternal mouthbrooder and three paternal. Breeding these fishes isn't always problem-free, but if attention is paid to approximating the water parameters known from the wild then successful breeding is possible.

All four species prefer live food and can be fed to only a limited extent on frozen and artificial foods, especially in the long term. These fishes require an aquarium that if possible is densely planted in places, and provided with small caves and other hiding-places. It is also important to provided good filtration with a gentle current. Filtration over peat is to be recommended. The introduction of dead beech or oak leaves will improve the water quality. The bottom can be left free of substrate or covered with a thin layer of light-coloured sand.

Sphaerichthys acrostoma VIERKE, 1979

Explanation of the species name:
Latin *acer* = "sharp" plus Greek *stoma* = "mouth", referring to the long pointed snout.

English name:
Pointed-head Chocolate Gourami

Original description:
VIERKE, J. 1979: Beschreibung einer neuen Art und einer neuen Unterart aus der Gattung *Sphaerichthys* aus Borneo. Das Aquarium 122: 339-343.

Natural distribution:
The natural habitat of this species is southern Borneo: Mentaya River system, 250 km north-west of Banjarmasin.

The fishes were caught during a collecting trip 1978 made by Mrs. E. KORTHAUS and Dr. W. FOERSCH and his wife. Caught among grass in the inundation zone of a river. Water almost still.

In 1988 Jürgen KNÜPPEL and the current author were able to investigate the natural habitat of this to date rare and still almost unknown species. Our collecting area was, by contrast, only around 150 km north-west of the town of Banjarmasin in the Palangkaraya area in Central Kalimantan.

In 2009 we (expedition members of the EAC/AKL) found the species in a blackwater swamp region at Sengalang, around 15 km north-east of Palangkaraya on the road from Palangkaraya to Kuala Kurun. Here too the population density was not high.

Biotope data:

Biotope north of Palangkaraya:
The collecting area for *Sphaerichthys acrostoma* lay around 150 km north-west of the town of Banjarmasin in the Palangkaraya area in Central Kalimantan. From here a road leads north in the direction of Kasungan. Large areas of the rainforest surrounding the town had been felled and converted into a landscape of swamp and scrub. Watercourses large and small crossed the road. Around 27 kilometres north of Palangkaraya, circa 3 kilometres before the village of Tangkiling, water flowing from a swampy area aggregated to the west of a road bridge to form a clear, dark brown river with a noticeable current. The pH value was 4.2 and the electrical conductivity was recorded as 20 µS/cm at a water temperature of 28 °C (82.5 °F). The bottom consisted of fine white sand with large rocks in places. No aquatic or marsh plants were discernible except that a small number of shallow lagoons, almost cut off by sandbanks, were densely overgrown with grass-like plants, providing shelter and habitat for numerous small fishes. In this dense confusion of vegetation we found three very rare labyrinthfish species as well as numerous other fishes. Specifically the Chocolate Gourami *Sphaerichthys selatanensis* (rather numerous in this area) and *Parosphromenus parvulus*, while by contrast

Male *Sphaerichthys acrostoma*

Juvenile *Sphaerichthys acrostoma*, one day old

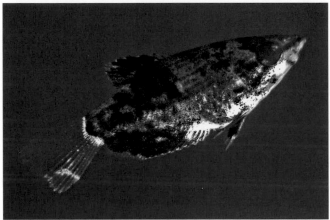
Juvenile *Sphaerichthys acrostoma*, 18 days old

Sphaerichthys acrostoma were uncommon. Despite intensive fishing at the time (1988) we were able to capture only seven specimens about 3 cm (1.125 in) long.

Biotope east of Palangkaraya:

These fishes are found almost exclusively in blackwater biotopes. We found these fishes in a blackwater swamp area on the road from Palangkaraya to Kuala Kurun, around 15 km after the large bridge over the Sungai Kahayan, immediately past the third steel girder bridge thereafter. The landscape was open with dense, tall scrub in places. The fishes were living among areas of dense vegetation in the shallow bank region.

Coordinates: 02° 05'08 S   113° 57' 16 E (LINKE, H.)

The water was very dark red brown in colour. The following water parameters were recorded: pH 3.73 and conductivity 31 µS/cm at 29.9 °C (85.8 °F). *Sphaerichthys acrostoma* lived here along with *Parosphromenus parvulus*, *Parosphromenus* cf. *linkei*, *Betta midas*, *Betta edithae*, *Betta obscura*, and *Betta* sp. in the same biotope. (Research in May 2011: LINKE, H. et al.)

Reproduction:

Male mouthbrooder

The sexes are readily differentiated in *S. acrostoma* of adult size. The male is clearly distinguished from the female by the reddish longitudinal stripe behind the eye and a somewhat more pointed and prolonged dorsal fin. Mating is preceded by a courtship ritual in which the male is very active. The fishes often swim together days before spawning.

Unlike in *S. osphromenoides*, where indications to date are that the female broods the fry in her mouth, in *S. acrostoma* it is evidently the male that performs this task. After spawning he exhibits a greatly enlarged throat sac and usually hides among plants, moving the brood around with gentle chewing motions at long intervals. After 15 days at a water temperature of on average 27 °C (80.5 °F) the male then releases the around 7 mm (0.25 in) long fry from his mouth in the course of the day. There can be up to 40 small, dark coloured *S. acrostoma*, and despite having a very small mouth opening they can immediately take freshly hatched nauplii of *Artemia salina*. Just three weeks later they can already measure around 13-15 mm (0.5-0.6 in) long and already clearly exhibit the species-typical deep body form. Their coloration also changes. The anterior half of the fishes then becomes light brown, while the posterior part of the body turns very dark, almost black. The caudal fin continues to remain without coloration and markings. At the age of two months the young are already on average 3.5 cm (1.375 in) long and it is difficult to satisfy their appetites. After around a year the fishes are full grown at 6 cm (2.375 in) total length. The females remain somewhat smaller. Both sexes have very pointed fins. In addition males exhibit a dark longitudinal band with light margins along the centre of the body.

At the time of writing the F1 generation of these fishes have themselves bred successfully, with some 35 youngsters resulting.

Total length: (size)
Up to 6.5 cm (2.5 in) .

Remarks: (differences from other species of the genus)
Maintenance of these fishes in the domestic aquarium is almost problem-free. If the water parameters measured in the natural habitat are taken into account (soft water with a pH of between 4.5 and 5) then maintenance should prove uncomplicated, although constant good water quality is critical. Regular partial water changes and suitably proportioned filtration are further

Sphaerichthys acrostoma after capture

Blackwater stream at the edge of woodland in South Kalimantan, habitat of *Sphaerichthys acrostoma*

The so-called black water is evidenced by its red-brown colour in the shallows. The bottom consists mainly of light sand

The densely vegetated bank zones are the preferred habitat of *Sphaerichthys acrostoma*

prerequisites. It is very important to enrich the aquarium water with a good dose of humic substances. A number of large clumps of Java Fern and some bogwood, as well as a number of dead beech leaves on the bottom, will provide suitable decor. The addition of humic substances is absolutely essential and will be in part achieved via the layer of dead beech leaves lying on the bottom, and these will also serve as hiding-places and thereby help cure the fishes of shyness. All *Sphaerichthys* species are readily maintained under such conditions and will also breed virtually without problem.

Explanation of the species name:

The species name means "*Osphronemus*-like", referring to the resemblance of the species to fishes of that genus.

English name:

Chocolate Gourami

Original description:

CANESTRINI, J. 1860: Zur Systematik und Charakteristik der Anabantinen. Verh. K. K. Zool.-Bot. Ges. Wien, 10: 697-712.

Synonyme:

Osphromenus malayanus
Osphronemus malayanus
Sphaerichthys osphromenoides osphromenoides
Sphaerichthys osphronemoidesz

Natural distribution:

The natural habitat is given as Malaysia, Kalimantan, Singapore, and Sumatra, although the island state of Singapore has suffered major changes/losses in its original flora and fauna through industrial development.

Biotope in the north-west of western Malaysia:

A watercourse flowing from the rainforest and crossing the road south from Kabu Gejah to Pondak Tanjok only a few kilometres before Pondak Tanjok. Blackwater stream, on average up to a metre (40 in) deep with a pH of 4.63. (Research in August 2006: LIM, T.Y., ZAHAR, Z., & LINKE, H.)

Coordinates: 05° 02.00 N   100° 43.70 E (LINKE, H.)

Biotope in the east of western Malaysia:

South of Kuantan; narrow blackwater ditch at the edge of the road near dense rainforest, both sides of National Highway 3 from Kuantan to Pekan (N3 Kuantan to Johor Bahru).

Between kilometre markers 305 and 306 heading for Kota Bahru or kilometre markers 22 and 23 heading for Kuantan.

Biotope south-west of Kuantan:

Area of flooded forest by the old bridge, left-hand side coming from Segamat, close to National Highway 12 from Kuantan to Segamat, a few kilometres after the fork to Pekan heading for Kuantan (around 37.3 km to Kuantan).

At the time of the research this was a mixed-water biotope following heavy rainfall, but it is normally a blackwater habitat.

Coordinates: 03° 37.30 N   103° 07.47 E (LINKE, H.)

Biotope on the island of Bangka:

River and inundation region in the Sungai Kampa river system on the road to Jebus, around 5 km after the fork in the road leading from Pangkalpinang to Muntok Peltim. The pH

Male *Sphaerichthys osphromenoides*

Male *Sphaerichthys osphromenoides* from Sumatra

Male *Sphaerichthys osphromenoides* from western Kalimantan Barat

value here was 4.37 and the conductivity 4 μS/cm at a water temperature of 25.4 °C (77.7 °F).

Coordinates: 01° 50'20 S   105° 28' 32 E (LINKE, H.)

The species is probably also distributed in the area of the Kapuas River in the north-west of Kalimantan (Kalimantan Barath), only here it has bolder red coloration in the fins. But this population may instead be an as yet undescribed new species or a variant of the species *Sphaerichthys selatanensis*, which is widely distributed in the south of the island of Borneo.

Biotope data:

Chocolate Gouramis live almost exclusively in slow-flowing, heavily vegetated, blackwater streams, with very clean, very soft, mineral-poor and very acid water. The pH usually lies between 4.0 and 4.8, with a conductivity of between 10 and 30 μS/cm.

Reproduction:

Female mouthbrooder

Distinguishing the sexes is not a problem. Males exhibit a bold light margin to the dorsal and anal fins. In addition the upper and lower margins of the caudal fin likewise have light edgings. The dorsal fin is pointed in male fishes, slightly rounded in females.

Spawning follows a brief embrace by the partners; a large number of eggs are released and then taken into the female's mouth. As luck would have it I have been able to observe this procedure and subsequent successful mouthbrooding on several occasions. I have thus established that the female releases the fry after around 19 days at a water temperature of around 27 °C (80.5 °F). The little *Sphaerichthys osphromenoides* have a total length of on average 6.7 mm (0.25 in) at this stage. Measurements were taken from several youngsters and size was found to vary between 6.5 and 6.9 mm. There was no longer any yolk sac visible. The fry exhibited very similar coloration to their parents. They were chocolate brown patterned with dark red spots and a light, almost transparent to faint yellow ring around the centre of the body. The dorsal and anal fins were

likewise dark brown in colour. The fry had small mouths in relation to their size, but were nevertheless immediately able to take freshly-hatched *Artemia* nauplii as their first food without difficulty.

The fry exhibit good growth given frequent feeding and good water quality and can measure around 15 mm (0.625 in) in length after just three weeks. Their body form is by then already noticeably deeper and they already have close to the typical appearance of the parents. The successful development of the

Juvenile *Sphaerichthys osphromenoides*, one day old

Blackwater biotope in the Kuantan area in the east of western Malaysia, home to *Sphaerichthys osphromenoides* as well as numerous other labyrinthfishes such as *Parosphromenus nagyi*, *Betta tussyae*, and *Trichogaster leerii*

Blackwater biotope in western Malaysia, habitat of *Sphaerichthys osphromenoides*

Blackwater biotope in Sumatra, habitat of *Sphaerichthys osphromenoides*

brood took place at average water parameters of conductivity 50 to 95 µS/cm and a pH between 5.0 and 6.3. Very clean water containing humic substances is important. The water temperature was mostly between 27 and 28 °C (80.5 and 82.5 °F). In my opinion warmer water is not essential. For good growth the brood should be fed small amounts several times per day.

Total length: (size)

The species is full-grown at 5.5 to 6 cm (2.125 to 2.375 in) total length.

Remarks: (differences from other species of the genus)

The species differs from *Sphaerichthys selatanensis* by lacking a golden yellow longitudinal band along the centre of the body, an additional, central cross-band, and a silver band at the base of the anal fin. In addition the head doesn't look as pointed. *Sphaerichthys osphromenoides* is somewhat larger when full-grown.

The important requirements for the optimal maintenance of this species can be derived from the biotope descriptions. The aquarium should not be too small in order to guarantee a degree of "water stability". The tank should be 80 cm (31 in) or more in length and 30 cm (12 in) deep, and well planted. These fishes should be kept only in a species tank or with other small, very peaceful species. The water should be as mineral-poor as possible: a general hardness of up to 8 °dGH and a very low carbonate hardness, plus a pH of 4.5 to 5.5, are strongly advised. Filtration over peat is also highly recommended. The resulting enrichment with humic substances is a very important prerequisite for optimal maintenance. The addition of dead beech leaves is also very beneficial for the fishes and for water quality.

The water temperature should be around 26 °C (79 °F) and may periodically rise to 30 °C (86 °F).

The well-known National Highway 3 from Kuantan to Mersing in western Malaysia

Explanation of the species name:

Latin adjective *selatanensis* = "of Selatan", referring to the province of Selatan in south-east Kalimantan, island of Borneo.

English name:

Striped Chocolate Gourami

Synonym:

Sphaerichthys osphromenoides selatanensis

Original description:

VIERKE, J. 1979: Beschreibung einer neuen Art und einer neuen Unterart aus der Gattung *Sphaerichthys* aus Borneo. Das Aquarium 122: 339-343.

Natural distribution:

The species is found in large parts of southern Kalimantan (Borneo).

The type locality is the immediate vicinity of Banjamasin (Kalimantan Selatan).

Biotope data:

The species lives exclusively in very soft, very acid blackwaters (pH around 4.0 to 5.0).

Reproduction:

Female mouthbrooder

Total length: (size)

Around 5.0 cm (2 in).

Remarks: (differences from other species of the genus)

The species has so far been imported only rarely. *Sphaerichthys selatanensis* differs from *S. osphromenoides* in having fewer dorsal spines, a smaller head, and different coloration. By comparison the species exhibits a yellow longitudinal band, sometimes interrupted, along the centre of the body from

Displaying *Sphaerichthys selatanensis*

Sphaerichthys selatanensis ♂

Juvenile *Sphaerichthys selatanensis* at 1 day old

Juvenile *Sphaerichthys selatanensis* at 7 days old

Juvenile *Sphaerichthys selatanensis* at 15 days old

Juvenile *Sphaerichthys selatanensis* at 30 days old

Juvenile *Sphaerichthys selatanensis* at 23 days old

Juvenile *Sphaerichthys selatanensis* at 30 days old

Sphaerichthys selatanensis after capture

the posterior edge of the eye to the caudal peduncle. A further band, strikingly yellow with metallic gold scales, runs along the base of the anal fin. Compared to *Sphaerichthys selatanensis* from the southern parts of Kalimantan, the *Sphaerichthys* that occurs in the Kapus area in north-west Kalimantan (Kalimantan Barat) has a striking red coloration in the fins and lacks the markings typical of *Sphaerichthys selatanensis*. Perhaps this population is a colour variant of *Sphaerichthys osphromenoides* or an additional species.

Biotope of *Sphaerichthys selatanensis* south Pundu, Kalimantan Tengah

Colour comparison with drinking water. The water from the biotope here is particularly dark brown in colour

The population density of *Sphaerichthys selatanensis* is astonishingly high in this biotope

Watercourse on the road from Palangkaraya to Kuala Kurun

The water is strongly coloured red-brown

The bottom consists mainly of light sand and is easy to negotiate

Sphaerichthys selatanensis after capture. The species often attains a total length of 50 mm in the natural habitat

Blackwater biotope in Kalimantan Tengah

The colour intensity can be very well seen using the measuring disc

Measuring equipment is indispensible for the precise measurement of water parameters

The river bottom consists mainly of light sand, its upper layers often stained by the dark red-brown water

Sphaerichthys vaillanti (PELLEGRIN, 1930)

Explanation of the species name:
Dedication in honour of M. L. VAILLANT.

English name:
Vaillant's Chocolate Gourami.

Original description:
PELLEGRIN 1930: Description d'un Anabanttide nouveau de Borneo appartenant au genre *Sphaerichthys*. Bull. Soc. Zool. France Paris, v.55, pp. 242-244

Natural distribution:
Type locality: Borneo: Kalimantan Barat, Kapuas, Sebroeang.

The species is found in the large Kapuas lakes and their affluents as well as in the upper Sungai Kapuas, and in the Sungai Tawang in the centre of north Kalimantan in the province of Kalimantan Barat. The lakes lie in the Kapuas Hulu area, west of the town of Putussibau, south of the border with Sarawak/eastern Malaysia.

Biotope data:
In 1995, after an adventure-filled journey, Olivier PERRIN first found these fishes in Kalimantan Barat, in the upper Kapuas region, in the area of the town of Semitau, east of where the Sungai Tawang flows into the Sungai Kapuas. The location was a broad river up to 3 meters (10 feet) deep. The water was clear and with a slight current. There were no aquatic plants present. In the absence of any measuring equipment no details of water parameters were recorded. After a long (back-packing) journey Olivier PERRIN managed to bring two specimens back alive to Europe, and as they luckily turned out to be a pair he also bred them. This demonstrated that *Sphaerichthys vaillanti* is also a mouthbrooder, and, as in *Sphaerichthys acrostoma*, it is the male that broods the young.

Reproduction:
Male mouthbrooder.

Total length: (size)
Males to 6.5 cm (2.5 in); females remain somewhat smaller.

Remarks: (differences from other species of the genus)
Females are much more colourful compared to males.

Sphaerichthys vaillanti ♂

Sphaerichthys vaillanti ♀

Female *Sphaerichthys vaillanti* is much more strikingly coloured than male

Sphaerichthys vaillanti pair, male in front, female behind

Juvenile *Sphaerichthys vaillanti*, 40 days old

The genus *Trichogaster*

There are four species in this genus, including *Trichogaster trichopterus*, the Three-Spot or Blue Gourami, one of the best-known aquarium fishes. There can be hardly an aquarium shop anywhere in the world that doesn't have these fishes for sale. They are available in a wide variety of cultivated colour forms such as Gold, Silver, Cosby, and Opaline, and in addition these also exist in a wide variety of patterns. Depending on their provenance, the wild forms exhibit variable grey to blue coloration and patterning. They are also prized as food fishes. They are caught with hand nets in irrigation channels and with large sinking nets in lakes and smaller rivers, or by rod and line from the bank.

The large *Trichogaster pectoralis* is particularly popular as a food fish. Its original distribution was purportedly Thailand, Cambodia, and perhaps South Vietnam, but it has now been introduced as a food fish in almost every country in South-East Asia, where it is bred in large numbers for the purpose. Because cultivated specimens can sometimes attain a length of around 25 cm (10 in), they are also good for selling as dried fish. Just as in many latitudes the guests in restaurants are offered a selection of live marine animals, in Malay and Thai restaurants the diners can choose their own *Trichogaster pectoralis* and *Trichogaster trichopterus*. They are displayed swimming in large jars or other containers, to be served a while later on a plate, boiled or grilled.

The farmers in Thailand, for example, like to erect their premises on an island in the sea of rice fields, so to speak. The house and all that goes with it is sited on somewhat higher ground and surrounded by a water-filled ditch, partly as protection against "uninvited visitors", but also in order to have a food supply "right on the doorstep". Because these klongs can be separated off, the fishes placed in them cannot escape into the much-branched system of irrigation channels in the rice fields. A wide variety of fish species are kept in these domestic klongs. As well as *Trichogaster trichopterus*, *Trichogaster pectoralis*, *Trichogaster microlepis*, and *Anabas testudineus*, fightingfishes of the wild form are also found there. They are quickly and skilfully caught out of the water when the locals hold a competition in the evening, then placed in water-filled Thai whisky bottles to await their turn to fight.

At various intervals the water from the klongs is drained off and the large fishes are collected up. Any that aren't used immediately are then cleaned and sorted, and left, without their heads, to dry side by side on a bark mat in the sun.

Three species have no special requirements as regards water parameters when maintained in the aquarium. However, *Trichogaster leerii* should be kept in soft, acid water if possible, because they are fishes that live exclusively in blackwaters in the wild. And the aquarium should always be appropriately spacious relative to the size of the fishes.

Explanation of the species name:

Dedication in honour of J. M. VAN LEER, a colleague of BLEEKER in Indonesia.

English name:

Leeri Gourami
Pearl Gourami

Synonyms:

Osphromenus leerii
Trichopodus leerii
Trichopodus trichopterus

Original description:

BLEEKER, P. 1852: Diagnostische beschrijvingen van nieuwe of weinig bekende vischsoorten van Sumatra. Tiental I-IV. Natuurkd. Tijdsch. Neder. Indië 1-41.

Natural distribution:

The natural habitat of *Trichogaster leerii* is given as Thailand, Malaysia, Sumatra, and Borneo, with Thailand appearing very questionable. Despite trying several times, I have been unable to find these fishes there. *Trichogaster leerii* lives in the blackwater regions of south-eastern Sumatra and in the south of western Malaysia. Its natural occurrence in Thailand and Kalimantan is highly questionable.

Biotope data:

The species lives in watercourses with slow-flowing areas and clear, soft, acid water. Because these are often found in wooded areas they sometimes lie in the half-dark.

Trichogaster leerii live exclusively in very soft, very acid water with a high content of humic substances, predominantly or exclusively in blackwaters. Their occurrence in rice fields and swampy lakes with neutral, mineral-rich water has not been confirmed and would be the exception.

We found this species in the Danau Rasau, a lake-like widening with an outflow to the Sungai Batang Hari at Rantanpanjang, 76 km north-east of Kota Jambi, in the province of Jambi in Sumatra, Indonesia. The site was a large blackwater lake with

Trichogaster leerii ♀

Trichogaster leerii, ♂ above, ♀ below

Male *Trichogaster leerii* beneath the bubblenest

a dense covering of floating plants and overgrown with scrub in places. The water was very strongly stained dark red-brown, with little current anywhere.

Water parameters:
pH value: 4.1
Conductivity: 30 µS/cm
Water temperature: 29.3 °C (84.7 °F)
(Research in May 2007: SIM,T. & LINKE, H.)

Reproduction:
Bubblenest spawner.

A large aquarium should be used for breeding. For successful breeding the total and carbonate hardness should not exceed 4 °dGH and °dKH respectively. The water temperature should be raised to 28 to 30 °C (82.5 to 86 °F) for breeding. The water should be 25 cm (10 in) deep and there

Male *Trichogaster leerii* in full colour

should be several groups of plants extending up to the water's surface. The open areas in between should be partially filled with floating plants. The aquarium should be dimly illuminated until mating takes place, and thereafter normally. An internal filter will suffice to produce clear water and gentle water movement in places.

The male *Trichogaster leerii* builds his bubblenest among or beneath floating plants. The female will take refuge in the area with slight water movement in order to escape the male if his courtship becomes too passionate. Note that clean water is important for rearing. Spawns of up to 700 eggs are not uncommon for adult fishes.

Biotope of *Trichogaster leerii* in Kalimantan

Trichogaster leerii is often to be found near the bank

Trichogaster leerii after capture

Total length: (size)
Up to 12 cm (5 in).

Remarks: (differences from other species of the genus)
The males develop brilliant red gill, breast, and belly areas. Plus the threadlike ventral fins and the anterior half of the anal fin are also red in colour. Moreover males also possess a pointed dorsal fin, while in females this fin is shorter and rounded. Moreover, the unpaired fins in adult males have filamentous extensions up to 1 cm (0.375 in) long. In addition females lack the bright red coloration. Unfortunately the characters that distinguish the sexes are first visible at a total length of around 7 cm (2.75 in) upwards, which means that young fishes exhibit clear sexual differences only at around seven months of age.

Village on the Batang Hari near Danau Rasau

Groups of plants in the blackwater of the Danau Rasau

The fishermen use wire baskets to catch fishes for food in the blackwater swamp lake Rasau

Explanation of the species name:
Latinised Greek microlepis = "small scale".

English name:
Moonlight Gourami

Synonyms:
Deschauenseea chryseus
Osphromenus microlepis
Trichopdus microlepis
Trichopdus parvipinnis

Original description:
GÜNTHER, A. 1861: Catalogue of the fishes in the British Museum, vol. 3, Brit. Mus., 385pp.

Natural distribution:
To date the natural occurrence of this species has been confirmed in Thailand and Cambodia. Because this is again a food fish, it is not uncommon in other South-East Asian countries as well.

Biotope data:
The species lives by preference in well-vegetated water-filled ditches and swampy areas with slight water movement. The water is usually mineral-rich and with a neutral to alkaline pH. Population density isn't high in the natural habitat.

Reproduction:
Bubblenest spawner.

A large, well-planted aquarium should be used for breeding. Males can become very aggressive towards females that are unready or unwilling to spawn.

Total length: (size)
Up to 12 cm (4.75 in).

Remarks: (differences from other species of the genus)
In addition to their basic silver coloration, adult males of this species exhibit an orange breast region and ventral fins of the same colour.

Male *Trichogaster microlepis*

Female *Trichogaster microlepis*

Rice fields are likewise habitats for *Trichogaster microlepis*

Catching *Trichogaster microlepis* in a rice field

Trichogaster microlepis usually build their bubblenests in tangles of vegetation near the bank.

Trichogaster microlepis after capture in Cambodia

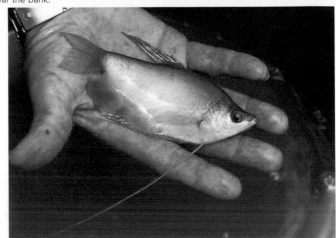

A splendidly coloured male *Trichogaster microlepis* after capture

Explanation of the species name:

Latin *pectoralis* = "having a pectoral fin" (by implication special or notable), referring to the larger pectoral fins in this species.

English name:

Snakeskin Gourami

Synonyms:

Trichopodus pectoralis
Osphromenus cantoris

Original description:

REGAN, C. T., 1910: The Asiatic Fishes of the Family Anabantoidae. Proceedings of the Zoological Society of London 53(4), 767-787.

Natural distribution:

Nowadays the original, natural distribution of this species can be ascertained only with difficulty. In 1910 REGAN cited Siam (Thailand) and Malay Peninsula (western Malaysia) as well as Singapore (?). Nowadays the species is found in Myanmar, Thailand, Malaysia, Indonesia, Cambodia, and Vietnam. It is often farmed as a food fish in these countries.

Biotope data:

The species lives in irrigation ditches associated with rice fields, in ponds and lakes, and in rivers large and small. The water is usually mineral-rich and slightly alkaline.

Reproduction:

Bubblenest spawner.

Because of their size these fishes require spacious aquaria for breeding. There should be dense clumps of plants to provide the female with hiding places, as the male is often aggressive at breeding time. The species is very productive.

Total length: (size)

Up to 20 cm (8 in) in the wild.

Remarks: (differences from other species of the genus)

The sexes can be clearly distinguished by the form of the dorsal fin, which is rounded in females and prolonged into a point in males. Despite mood-related variations, the cross-banding on a grey background is always clearly expressed.

Trichogaster pectoralis eats plants. The species is peaceful and can readily be kept with smaller fishes.

Trichogaster pectoralis

Trichogaster pectoralis

Market *Trichogaster pectoralis* in Thailand

Market *Trichogaster pectoralis* Lake Inle Myanmar

Field of *Trichogaster pectoralis* from east Yangoon in Myanmar

Market *Trichogaster pectoralis* in Palangkaraya Kalimantan

Trichogaster trichopterus (PALLAS, 1770)

Explanation of the species name:

Latinised Greek *trichopterus* = "hair-fin", referring to the long, threadlike ventral fins.

English name:

Three-Spot Gourami

Blue Gourami

Synonyms:

Labrus trichopterus

Nemaphoerus maculosus

Osphromenus insulatus

Osphromenus trichopterus var. *cantoris*

Osphromenus trichopterus var. *koelreuteri*

Osphromenus siamensis

Stethochaetus biguttatus

Trichogaster trichopterus siamensis

Trichogaster trichopterus sumatranus

Trichogaster trichopterus trichopterus

Trichopodus siamensis

Trichopodus trichopterus

Trichopus sepat

Trichopus trichopterus

Original description:

PALLAS, P.S. 1770: Spicilegia Zoologica quibus novae imprimis et obscurae animalium species iconibus, descriptionibus atque commentariis illustrantur. Berolini, 1 (8): 45-46

Natural distribution:

Nowadays it is no longer possible to determine the original, natural distribution. Because this fish is also used for food, it has been introduced into numerous countries and farmed there. Today the species is found – including in various wild colour varieties – in Bangladesh, Myanmar, Thailand, Cambodia, Vietnam, Malaysia, Sumatra, Singapore, and Borneo, and on many of the surrounding islands.

Biotope data:

Three-Spot Gouramis live almost everywhere. They are found in flowing watercourses, but more frequently in bodies of standing water large and small, preferring abundant submerse vegetation as habitat. The water temperature lies predominantly between 25 and 30 °C (77 and 86 °F), often even higher. The species lives predominantly in neutral to alkaline water, and water parameters are apparently of secondary importance. But the species is also sometimes found in blackwaters with very soft, acid water, though the population density is very low in such places. They are thus fishes with a very high degree of adaptability.

Trichogaster trichopterus ♂ , blue variant

Trichogaster trichopterus ♂ , blue-marbled cultivated form

Trichogaster trichopterus ♂ , grey-marbled cultivated form

Trichogaster trichopterus ♂ , silver-marbled cultivated form

Reproduction:

Bubblenest spawner.

The sexes are very easy to distinguish. At spawning time the females develop a very rounded belly; in addition they have a shorter, rounded, dorsal fin, while that of males is longer and prolonged into a point. In the natural habitat, for example in rice-field irrigation ditches or the roadside klongs, *Trichogaster trichopterus* males often construct their bubblenests only a few centimetres apart from one another. Because the water is usually muddy and clouded, the fishes can't see each other even at such short distances. Their bubblenests are up to 25 cm (10 in) in diameter and sometimes extend up to 3 cm (1.25 in)

above the water's surface. They are often exposed to sunshine for long periods. The water temperature can thus rise to more than 33 °C (91.5 °F). Perhaps the bubbles of the nest insulate the brood against overheating, and the oral secretions of the brooding male protect them against excessive bacterial attack.

The container for breeding shouldn't be too small. I have in fact seen Three-Spot Gouramis spawn in aquaria 40 cm (16 in) long and 30 cm (12 in) deep, but this can cause suffering for both the breeding fishes and the brood. The male constructs his bubblenest among plants floating at the water's surface, and it can attain up to 25 cm (10 in) in diameter. Over-passionate males sometimes bite off the anal fin of their females, and

Trichogaster trichopterus ♂, wild-caught from Vietnam

unless there are sufficient hiding-places the passionate attacks of the male can have painful consequences for the female. A larger amount of water is also more "stable" for rearing the fry (which often number more than 1000) and also provides them with more swimming space.

Total length: (size)

Up to 13 cm (5 in), but usually remains smaller.

Remarks: (differences from other species of the genus)

Trichogaster trichopterus is known in various colour forms. Depending on geographical occurrence, there are grey-striped to blue-patterned fishes of this species in the natural habitats. The two dark spots, one on the gill cover, one on the centre of the body, and the third on the caudal peduncle, gave this fish

its popular name. These are very peaceful and warmth-loving fishes. They are cultivated as food fishes.

In 1933 Werner LADIGES described a subspecies, *Trichogaster trichopterus* sumatranus, from the Lake Toba area in Sumatra. These fishes attain a total length of only 10 cm (4 in) and exhibit a predominantly light to dark blue coloration. This subspecies was subsequently invalidated by the author himself and the name is now regarded as a synonym. The same applies to the subspecies *T. t. siamensis*, described by Professor KLAUSEWITZ.

References:

- PAEPKE, H.-J. 2005. Über den Punktierten Fadenfisch *Trichogaster trichopterus* (Belontiidae) und seine Entdecker Joseph Gottlieb KOELREUTER und Peter Simon PALLAS. Der Makropode 27 (09/10):162-169.

Trichogaster trichopterus ♂, wild-caught from Myanmar

Trichogaster trichopterus ♀, wild-caught from Myanmar

Trichogaster trichopterus ♂, silver cultivated form

Trichogaster trichopterus ♀, Cosby cultivated form

Trichogaster trichopterus ♂, gold Prante cultivated form

Trichogaster trichopterus ♂, normal gold cultivated form

Trichogaster trichopterus ♂, gold-marbled cultivated form

TRICHOGASTER TRICHOPTERUS (PALLAS, 1770)   533

Trichogaster trichopterus Breeding Farm

Field of *Trichogaster trichopterus*

Cambodia

Kalimantan

Lake Lamchae Thailand

West Kuantan Malaysia

North-Thailand

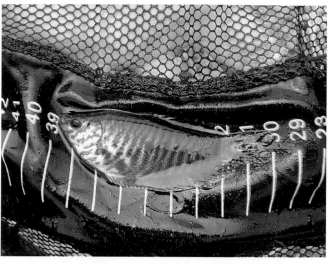

Near Hue in North Vietnam

Market with *Trichogaster trichopterus*

Market *Trichogaster trichopterus* in INDONESIA

Road Shop near Ayutaya Thailand with *Trichogaster trichopterus*

Restaurant *Trichogaster trichopterus* in MALAYSIA

The genus *Trichopsis*

At present there are four species in this genus, erected in 1860 by CANESTRINI in Vienna. The members of this genus are almost exclusively small species, which often produce clearly audible sounds and hence are known as croaking gouramis. These fishes live predominantly in standing waters such as rice fields and associated irrigation ditches, as well as in swamp regions and residual waters, and rarely in flowing watercourses as well. The water parameters of these waters usually lie in the neutral to slightly alkaline range. The water is predominantly mineral-rich and often clouded with suspended material. Almost all the species are found in Thailand. While three of the species have only comparatively small distribution regions, *Trichopsis vittata* is found well outside the borders of Thailand throughout South-East Asia, albeit with slight variations in colour and patterning. The sexes are easy to distinguish by placing the fishes in a transparent container in front of a strong light source, when the elongate egg chamber will be visible in the central part of the body in female fishes, but is absent from males.

Small aquaria with a length of around 60 to 80 cm (24 to 32 in) and densely planted in places should be used for maintenance. The pH can be around neutral (ie 70) or higher, and the water can be mineral-rich.

All the species prefer small live foods. All other sorts of foods are unsuitable for long-term use and render the fishes susceptible to disease.

Trichopsis were assigned to the genus *Ctenops* for a long time. According to LADIGES (1962) *Trichopsis pumila* exhibits a single longitudinal band on the body, *Trichopsis schalleri* two such bands, and *Trichopsis vittata* has two to four longitudinal bands and, depending on mood, a dark spot on the shoulder. Other characters exist for distinguishing the species: thus according to LADIGES *Trichopsis pumila* has 5 hard rays and 20-25 soft rays in the anal fin, while in *Trichopsis schalleri* there are 8-9 hard rays and 19-22 soft rays, and in *Trichopsis vittata* 6-8 hard rays and 24-28 soft rays.

Explanation of the species name:

Latin adjective *pumilus* (feminine *pumila*) = "small", "dwarf".

English name:

Dwarf Croaking Gourami

Synonyms:

Ctenops pumilus
Ctenops pumilus var. *siamensis*
Trichopsis pumilus var. *siamensis*
Trichopsis siamensis

Original description:

ARNOLD, J. P., 1936: *Ctenops pumilis*. Wochenschrift für Aquarien und Terrarienkunde, 33 (11) III: 116.

(However the reference cited, Wochenschrift für Aquarien und Terrarienkunde 1936, volume 33, page 116 doesn't in fact have an original description of *Trichopsis pumilis / Ctenops pumilis* by ARNOLD. In this case just an illustration with caption has been regarded as a description.)

Systematics:

Trichopsis vittata group

Natural distribution:

The natural habitat is given in the literature as Vietnam, Cambodia, and Thailand. No precise locality details are given for the *Trichopsis pumila* mentioned as coming from Cambodia in the work of Walter RAINBOTH. Perhaps they are distributed close to the border with Vietnam. The specimens used for the description were collected in the area north of Saigon in South Vietnam. H. RÖSE reported in 1936 that he had obtained fishes from a sailor who had captured them himself in Saigon.

According to our own researches these fishes are still caught for export in the north of Saigon. At the same time the area around Bangkok in Thailand is also known as a habitat of this species, though it remains unclear whether the fish in question really is *Trichopsis pumila*. That would mean a distribution around 700 km from the type locality of the species in Vietnam.

Biotope data:

It remains to be seen whether or not the *Trichopsis* species found in the vicinity of Bangkok and further to the north actually is *T. pumila*. HERMS reported on this little croaking gourami back in 1953 and described it as *Trichopsis pumilus* var. *siamensis*. We frequently found these fishes in heavily vegetated pool-like accumulations of water and in narrow, shallow ditches in the immediate vicinity of Bangkok. A dense covering of floating plants often left only small areas of the water's surface exposed. The temperature in this biotope reaches around 33 °C (91.5 °F) in areas of shallow water. The oxygen content

Trichopsis pumila ♂ from Thailand

was very low. The fishes were usually found together with *Trichogaster trichopterus*, *Trichogaster pectoralis*, *Trichopsis vittata*, and *Anabas testudineus*, but mainly served the latter as food. The Dwarf Croaking Gourami is usually to be seen in small groups of six or seven individuals, often in the company of small (up to 5 cm (2 in)) long *Trichogaster trichopterus*.

Trichopsis pumila from Thailand

Reproduction:

Bubblenest spawner.

The breeding of this little labyrinthfish species is not difficult. For breeding the temperature should be raised to 28 to 30 °C (82.5 to 86 °F) in an aquarium set up as for maintenance. The male usually constructs his little bubblenest beneath leaves or under an overhanging rock, and a cave with a "roof" is also liked. A male can be put together with two or more females and will often spawn with them in succession in the course of a number of days. During courtship, as well as when displaying, the male may frequently be heard to make croaking or growling noises. Despite its small size *Trichopsis pumila* is quite capable of defending its nest against larger fishes, and at such times its rows of brilliant blue dots make it gleam like a little jewel. In this species too the male embraces the female beneath the bubblenest during the spawning act. Both partners collect up the eggs as they sink towards the bottom and place them in the nest. The female shouldn't be removed from the breeding aquarium after the spawning, as her presence will often result in the male performing intensive brood care. After around four days the fry swim free and the parent fishes can then be removed from the breeding aquarium.

Trichopsis pumila from Vietnam

The sexes are not always easy to differentiate. The females are less colourful and usually have a rounded anal fin. The most reliable way of distinguishing the sexes is, however, to place the fishes in a small glass container in front of a strong light source so that the light shines through. The ovary will then be visible in females as a dark shadow in the rear part of the body cavity, extending in a triangular shape towards the tail.

Total length: (size)
Around 3.5 cm (1.375 in).

Remarks: (differences from other species of the genus)

Small tanks up to 60 cm (24 in) long are adequate for the maintenance of this little croaking gourami. The aquarium should be well planted. Because *Trichopsis pumila* likes to construct its little bubblenest beneath leaves, plants are particularly recommended for breeding tanks. The water parameters are less critical but shouldn't be too hard and alkaline. The water temperature can be around 25 °C (77 °F). For this species the recommended substrate is once again sand or fine gravel. Small caves constructed from rocks, or inverted, diagonally positioned flowerpots, should be added as decoration. The aquarium can have normal lighting and should be sited in a quiet spot. *Trichopsis pumila* is a very peaceful aquarium occupant. Because of its size it is best maintained in a species aquarium, or in a community of its own species or with peaceful fishes of the same size.

Trichopsis pumila from Vietnam are apparently more demanding in their maintenance than other species of the genus.

According to LADIGES (1962) *Trichopsis pumila* exhibits a single longitudinal band on the body, *Trichopsis schalleri* two such bands, and *Trichopsis vittata* has two to four longitudinal bands and, depending on mood, a dark spot on the shoulder.

The scientific names *T. pumilus* and *T. vittatus* were altered in an appendix to the description of *Betta strohi* by SCHALLER and KOTTELAT (1990: 36). The relevant text reads as follows (translated):

"Because *Trichopsis* is a genus name with a feminine ending, then in accordance with the rules of nomenclature the species name of the Croaking Gourami should be *T. vittata*, and that of the Dwarf Croaking Gourami *T. pumila*".

References:

- RÖSE, H. 1936: Zwerg-*Ctenops* von Cichinchina, Wochenschrift für Aquarien und Terrarienkunde, 33 (11): 626-627.
- HERMS, E. 1953: Fische im Gartenteich. Die Aquarien- und Terrarienzeitschrift (DATZ) VI.1 : 279.
- SCHALLER, D. & KOTTELAT, M., 1990: *Betta strohi* sp. n., ein neuer Kampffisch aus Südborneo, (Osteichthyes: Belontiidae), Die Aquarien- und Terrarienzeitschrift (DATZ), 43 (1): 31-37.
- RAINBOTH, W. J. 1996: Fishes of the Cambodian Mekong. FAO Species Identification Field Guide For Fishery Purposes. Food and Agriculture Organization of the United Nations. p.217.

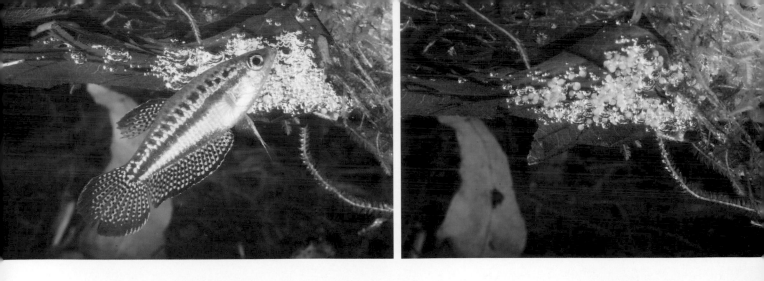

Scenes of brood care of *Trichopsis pumila*

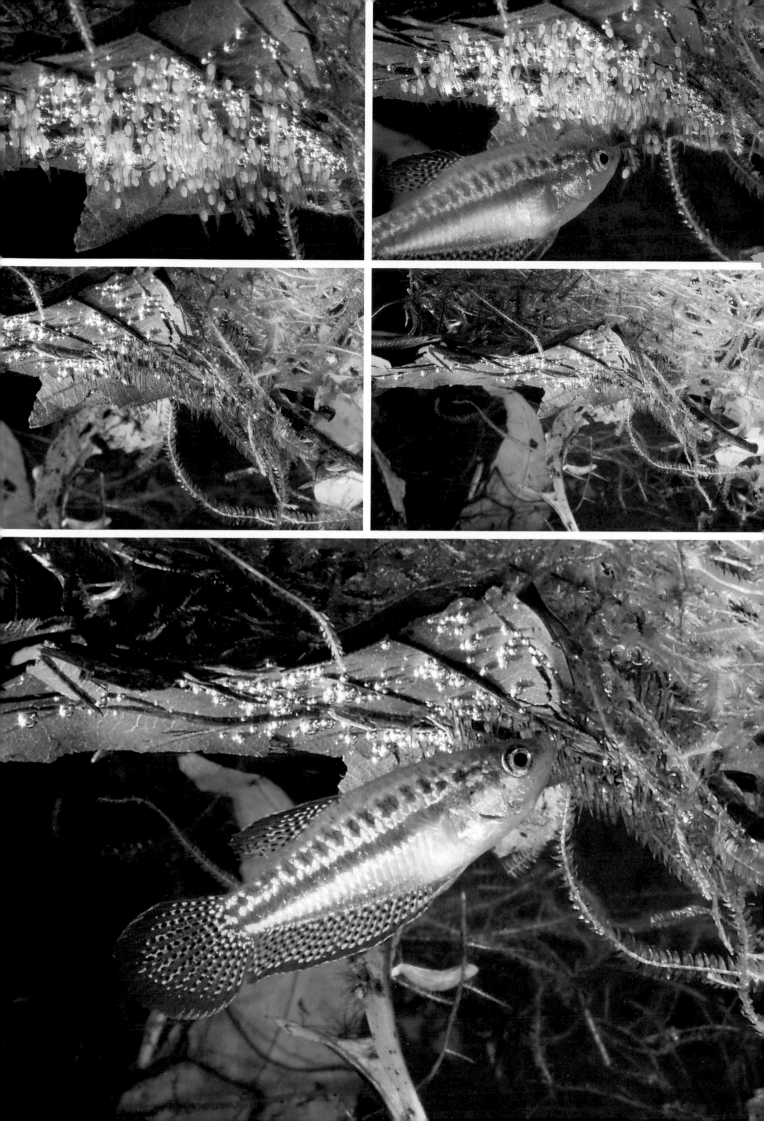

Trichopsis schalleri LADIGES, 1962

Explanation of the species name:
Dedication in honour of Dietrich SCHALLER, who collected the type material.

English name:
Schaller's Croaking Gourami

Synonym:
Trichopsis harisi.

Original description:
LADIGES, W. 1962: *Trichopsis schalleri* spec. nov., ein neuer Gurami aus Thailand. Die Aquarien- und Terrarien Zeitschrift (DATZ), 15(3): 101-103.

Systematics:
Trichopsis vittata group

Natural distribution:
The type specimens were collected in 1961 in the Nam-Mun river in the area of the city of Korat (nowadays called Nakhon Ratchasima), on the southern Korat plateau, around 220 km north-east of Bangkok in Thailand.

Biotope data:
Unfortunately I haven't had the opportunity to collect this species in the Nakhon Ratchasima area. But a large number were netted, along with *Trichopsis vittatus*, *Betta smaragdina*, and other species, in the northern region from Khon Kaen to Nong Khai on the River Mekong. In particular, all the species listed were recorded in the area around the town of Nong Khai, in a wide variety of habitats in rice fields, marsh meadows, klongs, and watercourses within a radius of 100 km. The water temperature was between 28 and 32 °C (82.5 and 89.5 °F). There were invariably dense areas of plants along the banks even if the entire body of water wasn't heavily vegetated, and the water depth was between 5 and 30 cm (2 and 12 in).

Reproduction:
Bubblenest spawner.

Total length: (size)
Around 5 cm (2 in).

Remarks: (differences from other species of the genus)
According to LADIGES (1962) *Trichopsis pumila* exhibits a single longitudinal band on the body, *Trichopsis schalleri* two such bands, and *Trichopsis vittata* has two to four longitudinal bands and, depending on mood, a dark spot on the shoulder. This species also lacks the slight saddle-like indentation in the head typical for *Trichopsis* species and differs in the number of spines in the anal fin and in its coloration.

Displaying male *Trichopsis schalleri*

Fishing in Lake Lamchae in eastern Thailand

Trichopsis schalleri after capture

Lake Lamchae is very rich in fishes, including food fishes

LADIGES writes of the coloration (translated): "In live fishes the longitudinal bands are chestnut brown, and accompanied by rows of intense green-blue iridescent scales that may interrupt the upper band in several places beneath the dorsal. The dorsal fin is bordered by a reddish brown margin, the anal fin in all specimens by a narrow dark edging." On this basis these fishes are similar in appearance to the species *Trichopsis pumila*.

The opinion is sometimes expressed that *Trichopsis schalleri* is only a subspecies of *Trichopsis vittata*. I would like to state my opposition to this viewpoint. In my view *Trichopsis schalleri* is a good species, as were it a subspecies it would not occur together with the species *Trichopsis vittata* in the wild. I have been able to observe and collect *Trichopsis schalleri* and *Trichopsis vittata* together in the same, shared habitat at around eight collecting sites lying more than 40 km apart geographically. There was no separation on either a geographical or an ecological basis.

As with *Trichopsis pumila* and *Trichopsis vittata*, the use of a strong light is recommended for differentiating the sexes.

Trichopsis sp. Hua Hin

Explanation of the name:
The species was first discovered in 1993 by Alfred WASER, in the Hua Hin area.

English name:
Hua Hin Dwarf Croaking Gourami

Species-typical characters:
The species is possibly very closely related to *Trichopsis schalleri*, but by comparison has fewer rows of iridescent light blue dots on the body and fins, exhibits a very attractive red edging to the eye, and purportedly grows somewhat larger.

Similar species:
Trichopsis schalleri
Trichopsis pumila

Natural distribution:
The area investigated was some large pools of water, partially vegetated with reeds and waterlilies, around 16 km south of Hua Hin on the old National Highway 4 which runs through the centre of the town. The collecting site lies around 400 m before the National Highway 4 kilometre marker inscribed 250 km to Bangkok on the road leading into the town (on the same side as the biotope). The distribution region still extant is apparently only small and threatened by ongoing industrialisation and new construction linked to increasing tourism. The region is part of the Praburi National Park.

Coordinates: 12° 25' 35 N   099° 55' 43 E (H. LINKE)

Biotope data:
The study site was a large, flattened, fenced area which at the time of our research was being used as grazing for cattle. The area had only small amounts of scrub and trees. The three large pools of water were only partially vegetated at their margins, with a reed-like plant and waterlilies. The bottom consisted of a mixture of clay and loam and it wasn't possible to wade in the water. The water was clouded and light grey in colour. This biotope extended along the road and inland to the west. It was almost without shade, ie exposed to the sun all day. The air temperature at midday measured around 33 °C (91.5 °F), the water temperature 31.3 °C (88.4 °F), the pH 8.04, and the conductivity 1284 μS/cm. The visibility was barely 5 cm (2 in). (Research in February 2009: TUDSANAVISUT, A., PHOOMCHOOSRI, A. and LINKE, H.)

Maintenance:
As described for the genus.

Reproduction:
Bubblenest spawner.

Trichopsis sp. Hua Hin ♂

Biotope on the old National Highway 4 in Hua Hin

Trichopsis sp. Hua Hin lives in very murky, warm water here

The bottom was very muddy and the shore served as a drinking-place for cattle

Total length: (size)
Around 5 cm (2 in).

Remarks: (differences from other species of the genus)
According to LADIGES (1962) *Trichopsis pumila* exhibits a single longitudinal band on the body, *Trichopsis schalleri* two such bands, and *Trichopsis vittata* has two to four longitudinal bands and, depending on mood, a dark spot on the shoulder.

References:
- LINKE, H. 2009: Neuer *Trichopsis* aus Thailand? Betta News 17 (2): 11.

Explanation of the species name:

Latin adjective *vittatus* (feminine *vittata*) = "banded" or "striped".

English name:

Croaking Gourami

Synonyms:

Ctenops vittatus
Osphromenus vittatus
Osphronemus striatus
Osphronemus vittatus
Trichopodus striatus
Trichopsis striata
Trichopsis vittatus

Original description:

CUVIER, G. 1831: Des poissons á pharyngiens labyrinthiformes. Hist. Nat. Poiss. Paris, 7: 387.

Systematics:

Trichopsis vittata group

Natural distribution:

The natural habitat consists of pond-like accumulations of water, small rivers, canals, ditches, and rice fields. This species has a large distribution region encompassing Sumatra, Malaysia, Laos, Cambodia, and Vietnam.

Biotope data:

The species inhabits a wide variety of habitats including both blackwater and normal biotopes. The species is very variable and adaptable. These fishes often live in the very murky, dirty water of the town. They are often to be found in small shoals in open water, among large groups of plants, or beneath the floating leaves of large aquatic plants.

Habitat in Vietnam:

Rice fields and inundation zones along National Highway 1 from Hue to Da Nang in Vietnam.

Coordinates: 16° 16' 19 N   107° 58' 00 E (LINKE, H.)

The water temperature in the shallow residual water zones along the National Highway measured around 30 °C (86 °F), the water was brownish in colour and clear.

Habitat in western Malaysia:

Swamp region with wooded zones close to National Highway 122, around 1 km after Kampong Santong in the east of western Malaysia. A blackwater stream crosses the road at this point. The pH value was measured at 4.69.

Coordinates: 04° 37' 68 N   103° 22' 48 E (LINKE, H.)

Trichopsis vittata ♂

The following water parameters were recorded: pH 7.5, conductivity 88 μS/cm, water temperature 29.5 °C (85 °F), and hardness 3 °dGH / 2 °dKH.

Reproduction:

Bubblenest spawner.

The aquarium should be arranged in the same way for both maintenance and breeding, except that for breeding the fishes should be kept in pairs and the water temperature raised to 28 to 30 °C (82.5 to 86 °F). The Croaking Gourami likes to construct its bubblenest at the water's surface beneath or among the leaves of floating plants. The nest can attain a diameter of 10 cm (4 in) during brood care. During the courtship period loud croaking noises may be clearly audible, emitted by the male. Mating involves the partners embracing beneath the bubblenest. The subsequent breeding procedure is similar to that for the other species of this genus.

Total length: (size)

Around 7 cm (1.375 in).

Remarks: (differences from other species of the genus)

The maintenance of *Trichopsis vittata* is simple in aquaria 80 cm (31 in) upwards in length, which should be well planted. Floating plants are also highly recommended. Sand should be used as substrate, and some bogwood introduced as décor – the fishes will like to hide beneath this. The water parameters are of lesser importance. The temperature should average 25 °C (77 °F). The tank should be sited in a quiet place. Because this species is rather timid it should be kept only with quiet, peaceful fishes.

Displaying male Trichopsis vittata

Habitat in Kalimantan:

Blackwater swamp region along the road from Buntok to Ampah, 9 km east of Buntok in Kalimantan Selantan, island of Borneo.

Coordinates: 01° 41' 07 S   114° 52' 55 E (LINKE, H.)

The water parameters were pH 4.72, conductivity 24 μS/cm, water temperature 28.8 °C (83.8 °F), and water colour dark red brown.

Habitat in Thailand:

Small river that crosses National Highway 4191 beneath a bridge. The water had a moderate current and a depth of 40 to 90 cm (16 to 35 in) on average. The banks were densely vegetated.

Coordinates:
09° 23' 93 N   099° 12' 93 E (LINKE, H.)

Differentiating the sexes isn't easy. Adult *Trichopsis vittata* often exhibit a distinct shoulder spot and this can be regarded as a reliable character for distinguishing them from the other *Trichopsis* species. Unfortunately females also exhibit this spot and hence it is no use as a method of distinguishing the sexes. Large males usually have prolonged fin edges. But because the geographical forms exhibit different coloration, in this species too the most reliable method of sexing is again using a bright light source, as explained in the introductory section on the genus.

According to LADIGES (1962) *Trichopsis pumila* exhibits a single longitudinal band on the body, *Trichopsis schalleri* two such bands, and *Trichopsis vittata* has two to four longitudinal bands and, depending on mood, a dark spot on the shoulder.

The scientific names *T. pumilus* and *T. vittatus* were altered in an appendix to the description of *Betta strohi* by SCHALLER & KOTTELAT (1990: 36). The relevant text reads as follows (translated):

Field of *Trichopsis vittata* in south Thailand

Field of *Trichopsis vittata* in Cambodia

"Because *Trichopsis* is a genus name with a feminine ending, then in accordance with the rules of nomenclature the species name of the Croaking Gourami should be *T. vittata*, and that of the Dwarf Croaking Gourami *T. pumila*".

References:

- SCHALLER, D. & KOTTELAT, M., 1990: *Betta strohi* sp. n., ein neuer Kampffisch aus Südborneo,
- (Osteichthyes: Belontiidae),
- Die Aquarien- und Terrarienzeitschrift (DATZ), 43 (1): 31-37.

Field of *Trichopsis vittata* near Hue in north Vietnam

Field of *Trichopsis vittata* in East Thailand

Field of *Trichopsis vittata* in West-Malaysia

The African Labyrinthfishes

The final part of this book will deal with the tropical African climbing fishes that have been given the name "bushfishes" by Professor Werner LADIGES. As can be read in numerous publications, they are mainly timid and secretive predators. But if they are maintained properly they will rapidly contradict this statement. Bushfishes are very interesting study subjects and can be maintained without risk with other fishes in the aquarium as long as these tankmates aren't too small. During altercations with other genera their behaviour is mostly defensive, making them attractive as aquarium occupants. With a few exceptions, bushfishes are not among the more colourful members of the labyrinthfishes; but they are very appealing by virtue of their sometimes unusual body form.

These fishes were assigned to the genus *Ctenopoma* by PETERS in 1844. *Ctenopoma* means "with a comb on or by the cover", ie a comb-like or toothed gill-cover. In 1995 the genus was split by NORRIS into a new genus with the name *Microctenopoma* for the bushfishes that practise brood care, and the old genus *Ctenopoma* containing the species that don't, that is the open-spawning bushfishes. The species that practise brood care construct a bubblenest, spawn beneath it, and guard and tend the brood until the fry are independent and free-swimming. In this respect they are similar to the majority of Asian labyrinthfishes. As far as is known they comprise the species *Microctenopoma ansorgii*, *M. damasi*, *M. fasciolatum*, *M. nanum*, and *M. sp.*.

The open-spawning species release their eggs and sperm during a brief embrace in the open water, often only a few centimetres above the bottom, without previously constructing any nest and without any subsequent brood care. On the basis of morphology they comprise – again with the reservation "as far as is known to date" – the species *Ct. acutirostre*, *Ct. kingsleyae*, *Ct. maculatum*, *Ct. multispinis*, *Ct. muriei*, *Ct. ocellatum*, *Ct. pellegrinii*, and *Ct. weeksii*.

The sexes are in general not difficult to distinguish in the species that practise brood care. The different colours and form of the fins make it easy to recognise males and females. But it is a lot more difficult in the open-spawning species. Professor Hans M. PETERS made remarkable discoveries in this regard, for example in *Ct. kingsleyae*. Unlike in the species that practise brood care, the species so far observed and studied in life have small areas of unusual scales on both sides of the body, on the one hand near the posterior edge of the eye and on the other on both sides of the caudal peduncle. These scales have relatively large spines on their posterior edge instead of the normal small teeth. These areas of scales are characteristic of males, and although they are also seen in (almost) fully-grown females, they are smaller. As far as is known at present, these areas of scales serve to enable the male better to grip the female during the very short mating embrace.

There are exceptions, namely the elongate, fast-swimming *Ctenopoma* species such as *Ct. multispinis*, *Ct. machadoi*, *Ct. nigropannosum*, and *Ct. pellegrini*. These species exhibit only the anterior area of scales with spines. It is to be assumed that they too are open-spawners.

The above-mentioned areas of scales are not to be confused with the normal ctenoid (comb-like) scales characteristic of these bushfish species. And the robust teeth on the posterior edge of the operculum are something quite different again; it is precisely these robust spines that often cause serious injuries when the fish is netted, by getting caught up in the material of the net. A problem that should be borne in mind when maintaining all *Ctenopoma* species.

Bushfishes are heavy feeders. The larger species in particular should be given robust foods, though they also enjoy flake.

In the natural habitat the majority of species live in clear, clean, often flowing waters. During my various travels in West Africa I have repeatedly established that these fishes live predominantly in mineral-poor, oxygen-rich, acid water, at temperatures between 24 and 28 °C (75 and 82.5 °F). More rarely they inhabit the hot, oxygen-depleted waters of the open bush country.

Unfortunately these *Microctenopoma* and *Ctenopoma* species are only very rarely offered for sale in the pet trade. It is to be hoped that the situation will change and that aquarists will be able to obtain these interesting objects of study in order to try and answer the numerous still-unanswered questions about them.

References:

NORRIS, S. M., 1995: *Microctenopoma uelense* and *M. nigricans,* a new genus and two new species of anabantid fishes from Africa. Ichthyol. Explor. Freshwaters 6 (4)

The genus *Ctenopoma*

Ctenopoma acutirostre

Ctenopoma kingsleyae

Ctenopoma maculatum

Ctenopoma multispinis

Ctenopoma muriei

Ctenopoma nebulosum

Ctenopoma ocellatum

Ctenopoma pellegrini

Ctenopoma petherici

Ctenopoma weeksii

Ctenopoma acutirostre PELLEGRIN, 1899

Leopard Bushfish.

The popular name refers not to the behaviour of this species, but to its appearance. It exhibits irregular round, dark brown to black spots of various sizes on a muddy yellow to light brown coloured body. Only the pectoral fins and the posterior edges of the caudal, dorsal, and anal fins are transparent and colourless. These fishes grow very slowly and attain a length of up to 15 cm (6 in). In general they attain this size only in very spacious aquaria and not until the age of two to three years at the earliest.

Ctenopoma acutirostre are usually available as juveniles. They are often seen in the trade at a size of around 4 cm (1.5 in) and are readily acclimatised to the aquarium. Because they are usually wild-caught this is a rewarding experience. Acclimatisation to different water parameters presents no problems.

A striking feature of these fishes is the large eyes, pointing to crepuscular or nocturnal activity. In fact they are very lively during the evening hours, unlike during the day when they often remain hidden or floating almost motionless in the water. The deep cleft of a mouth becomes a large oval sucking tube for seizing prey, into which the victim disappears lightning-fast. For this reason the Leopard Bushfish should not be kept with too small tankmates. If these fishes are maintained in richly planted aquaria decorated with bogwood and caves then they rapidly lose their shyness. And they don't come across as brutal predators. Quite the contrary – they are actually rather retiring. If properly maintained in the company of open-water fishes then they often become very confident and are usually to be found at the front of the tank, giving the impression that they are begging for food. Their appearance is impressive by virtue of the enormous body depth and the large head. Sometimes the decorative spot pattern gives way to a dirty brown base colour such that the fish is only with difficulty distinguishable from a dead plant leaf. This camouflage facilitates the hunt for something to eat. If a Leopard Bushfish, thus disguised, drifts into the territory of other fishes then it is only latterly recognised as a fish and chased away. But they never become aggressive as a result, instead skilfully avoiding the attacks of the other fish and drifting away from the danger zone like a leaf.

Their diet should consist of more robust foods. Mosquito larvae, flies, woodlice, and small or chopped earthworms should all be on the menu.

The natural habitat of these fishes is the basin of the Congo in DR Congo including the area around Malebo (formerly Stanley) Pool near to the cities of Kinshasa and Brazzaville, the Stanley Falls, and various tributary rivers. They sometimes live in fast-flowing reaches but also in calm areas of water, usually residing in caves, beneath overhangs, or among dense clumps of plants. So far these fishes have been caught only at night when they have left their protective abodes. The water parameters in the region are known from various reports, and on average the total hardness is 2 to 3° dGH with a pH of 7.0 to 7.5. The water temperature is given as 25 to 29 °C (77 to 84 °F). On this basis these fishes should be kept in water that is as low in minerals as possible.

Head close-up of *Ctenopoma acutirostre*

Ctenopoma acutirostre

Tailspot Bushfish

These are large bushfishes that, according to the original description, are distributed all over West Africa. They grow to barely 20 cm (8 in) long and are very peaceful fishes despite their size. Naturally, small tankmates are not a good idea, as sooner or later they will be regarded as food. But fishes of 6 cm (2.375 in) upwards in length can safely be kept with adult specimens as long as they aren't too slender.

These bushfishes are not particularly colourful: dark grey on the upper half of the body is complemented by silvery green on the lower flanks. A round black spot on the caudal peduncle is boldly expressed only in younger specimens. The pectoral fins have a slight yellow tinge.

Ct. kingsleyae isn't one of the species that practise brood care; it spawns in the open water. The sexes can be distinguished from around 12 cm (4.75 in) upwards, but only in the way described earlier in the introduction to the bushfishes. According to the original description the natural habitat is coastal regions in West Africa, extending from The Gambia across the Niger Delta in Nigeria to Gabon. However, during my trips to West Africa I have so far been able to catch these fishes only in Sierra Leone, where they were particularly numerous in the clear, flowing watercourses of the Kasewe Forest region. The largest specimens were only just 15 cm (6 in) long. Large numbers of them were readily visible in a lake-like body of water there. The water was very mineral-poor and acid. The temperature measured 25 °C (77 °F). The situation was quite different in the north of Sierra Leone, where I caught this species in rice fields and associated channels near to the town of Kambia. The water temperature was 30 °C (86 °F). The fishes here exhibited a slightly different coloration to those at the Kasewe collecting site. While specimens from the forest region exhibited more grey and less green colour, the reverse was true of the northerly specimens from open countryside, which mainly exhibited a very attractive green on the almost black body, so that at first glance I thought I was dealing with another species. Most of the time these fishes were to be seen in groups of several individuals, actively swimming in open areas of water, usually poor in vegetation. When danger threatened they disappeared lightning-fast into the plants in the bank region. Only large aquaria should be used for the maintenance of this species. A length of 130 cm (51 in) upwards, with a breadth and depth of 50 cm (20 in) should be regarded as the minimum dimensions. The tank for maintenance of this species should be densely planted in places but with as much continuous, open swimming space as possible. The temperature can be between 24 and 26 °C (75 and 79 °F). Water chemistry is of lesser importance for maintenance, and, as far as is known to date, also for breeding. A large aquarium is essential for breeding. In addition to dense planting in places there should also be a number of caves or piles of rockwork to provide hiding-places for the females, as males can become rather aggressive towards their partners at mating time. The surface of the water should have a dense covering of floating plants and the temperature should be set to 26 °C (79 °F). *Ct. kingsleyae* release their spawn after a short embrace in the open water. The eggs rise to the surface of the water and develop among the floating plants. The species is very productive. Adult specimens can produce several thousand eggs. The parents should be removed from the breeding tank after spawning as otherwise they will eat their eggs. The larvae hatch after 36 hours and the fry swim free after a further two days. After only around four weeks the black spot develops on the caudal peduncle and the young fishes begin to take in atmospheric air at the water's surface.

Ctenopoma kingsleyae

Single-Spotted Bushfish.

Unfortunately this up to 20 cm (8 in) long bushfish is only rarely seen in the aquarium. The small number of specimens I collected myself proved very retiring, often somewhat timid, in their behaviour when kept with other fishes. Most of them had a total length of only around 6 cm (2.5 in). Only larger specimens exhibit the characteristics that distinguish the sexes. The areas of spine-fringed scales behind the eye and in the vicinity of the base of the tail are visible only at a total length of 10 cm (4 in) upwards. At present it isn't known what size the species attains under aquarium maintenance.

In juvenile dress *Ct. maculatum* exhibits a very appealing coloration. The fishes are dark brown with a light yellow ring running around the body below the dorsal fin and ending between the ventral and anal fins. This pattern is complemented by another light yellow crossband on the head, sometimes suggesting an affinity with the Asian Chocolate Gourami, *Sphaerichthys osphromenoides*. On attaining a size of 4 to 5 cm (1.5 to 2.0 in) this coloration becomes paler, changing to a faint grey-brown body colour with a bold dark spot on the centre of the body. The natural habitat of this species is southern Cameroon, Gabon, and Rio Muni. The watercourses of the Dja, Ntem, and Ogowe river systems are known locations, although I caught this species only very occasionally in these regions. These very secretive fishes can be found in small watercourses with flowing, very soft, pure water, usually hidden in the densely vegetated bank zones. The water depth is often only 20 cm (8 cm). I have been unable to confirm the location given in the original description, the River Krimi (now and for a long time known as the River Kienke), as a habitat of this species.

Ct. maculatum is undemanding when it comes to maintenance in the aquarium. The tank shouldn't be too small. The water temperature should be around 25 to 27 °C (77 to 80.5 °F). It is important to have areas of dense planting and concealed hiding-places in the aquarium, as these fishes are very timid in the absence of such cover. This species is not aggressive and can be kept with any large species without problem.

Ctenopoma maculatum

Many-Spined Bushfish.

This is again one of the large species, and can grow to around 15 cm (6 in) long. These fishes are good and rapid swimmers. While other *Ctenopoma* species often remain motionless for a lot of the time, *Ct. multispinis* are always "on the go". This species should be housed only with large, robust fishes, as it can be termed gluttonous and sometimes aggressive. It can be regarded as ideal for maintenance with the Asian Climbing Perch *Anabas testudineus*, as the two species exhibit similar behaviour. During wet weather *Ct. multispinis* are likewise able to leave the areas of water in which they live and migrate or clamber across land to other areas of water.

The sexes can apparently be distinguished only by the areas of spinous scales. These fishes belong to the open-spawners that release their eggs and sperm in areas of open water after a short embrace; the eggs then develop without any care from the parents.

The natural habitat is a large area in south-eastern Africa, above all tributary streams and swamps in the drainages of large rivers in southern DR Congo, Mozambique, Zambia, Zimbabwe, and Botswana. The specimens pictured here were caught by D. SCHALLER in Mozambique and kindly made available to me for study.

Ct. multispinis is best maintained in spacious aquaria. This isn't absolutely essential but will lead to better development. In addition, if possible several specimens of this species should be maintained together with large, robust fishes, so that any outbreaks of "temperament" that may occur are shared among different tankmates.

Foods such as earthworms, mosquito and other insect larvae, young mealworms, shrimps, fish fry up to 1.5 cm (0.5 in) in length, and other robust fare should be offered. The aquarium must be tightly covered as this species can jump well and accurately. If maintained properly these fishes can become very tame, and it isn't uncommon for them to jump out of the water towards their owner when the aquarium lid is opened for feeding. The water temperature should be around 25 °C (77 °F). The only plants recommended are hardy, robust types such as *Anubias* species from West Africa, for example. The aquarium décor should include several caves as hiding-places for these fishes when resting. A partial covering of floating plants is also recommended to give the fishes a greater sense of security. Water parameters appear to be of secondary importance. Powerful filtration of the water is beneficial for the well-being of the fishes, and the need for a partial water change (a quarter to a third of the aquarium volume) every 8 to 14 days should be self-evident.

Ct. multispinis is a very interesting bushfish species that represents an attractive study subject for the enquiring aquarist and dedicated naturalist.

Ctenopoma multispinis

Ocellated Bushfish

This species can attain up to 8 cm (3 in) in length. The sexes cannot be distinguished on the basis of differences in coloration. This is possible only by virtue of the more prominent areas of spinous scales behind the eyes and on the caudal peduncle in males. Adult females grow somewhat longer than males.

The natural habitat of this species is north-eastern to eastern Africa, the White Nile from Khartoum to the area of Lake Albert, Lake Kyoga, Lake Victoria, and Lake Edward. The southern limit of its distribution lies in the north of Lake Tanganyika and the western in the Chad basin and the adjacent Mayo Kebi basin. Because of this extensive distribution there is a possibility of morphological variation. The habitat of this species consists of small, shallow bodies of water as well as swamps and inundation zones of rivers, plus small pools that are heavily overgrown, ie densely vegetated with plants. These waters are often very oxygen-depleted and attain temperatures of up to 34 °C (93 °F). Research has shown that these usually shallow, warm zones are populated only for short periods; the fishes generally prefer deeper, cooler areas of water.

Ct. muriei are active-swimming fishes. They typically swim rapidly through the middle layers of water in the aquarium, and this should be taken into consideration when setting up. As well as dense planting at the rear of the tank and a number of caves and decorative pieces of bogwood, there should be adequate swimming space available. The water temperature can be around 25 °C (77 °F), and a powerful motorised filter should be used to create a slight current. A covering of floating plants in places will provide this species with more security and permit better observations.

On the basis of available reports the breeding of this species isn't difficult. The breeding tank should not be too small. A length of 100 cm (40 in) upwards is advisable. The water temperature should be raised to 28 °C (82.5 °F). It is advisable to use two to three males to one female.

Ct. muriei is one of the open spawners. Mating usually takes place in the evening twilight. After rapid chasing, during which the males drive the female, one of the males tries to embrace the female during a pause in the chase. The embrace is often over after only three seconds and the eggs expelled rise to the surface of the water. The species doesn't practise any brood care. More than a thousand eggs may be laid. The larvae hatch after around 25 hours.

Ctenopoma muriei female

Ctenopoma muriei male

This is one of the high-backed species that is similar to *Ctenopoma ocellatum* in appearance. The species was described from south-eastern Nigeria, where it lives in larger forest streams with a slight or stronger current, to the east of the River Niger. Predominantly in forested regions along the River Sombreiro and River Imo in the area from the town of Owerri to Port Harcourt and further east. I have found the species in a small forest stream between Umuahia and Aba. The water there was soft, slightly acid, brownish, and clear with a slight current. *Ctenopoma nebulosum* inhabited the deeper, more heavily overgrown and vegetated bank zones.

Fishes of this species can attain a length of up to 15 cm (6 inches). As described in detail in the introduction to this section on African labyrinthfishes, the areas of scales behind the eye and on the caudal peduncle are important characters for differentiating the sexes.

These fishes should be maintained only in larger aquaria. The water temperature should be between 23 and 26 °C (73.5 and 79 °F) and the aquarium can have normal lighting. Floating plants are highly advisable, along with areas of dense planting and other décor to provide shelter. The tank should be tightly covered, as these fishes are accomplished jumpers. These fishes like strong water movement and a varied diet of robust foods is important for good maintenance.

It seems that to date their breeding has been observed only very rarely. They are open spawners that release their eggs into the open water after a short embrace; the eggs then rise to the water's surface, and develop among floating plants. The larvae hatch after 48 hours, and the fry swim free two days later.

Ctenopoma nebulosum should, if possible, be housed only with fishes of approximately the same size.

Ctenopoma nebulosum

Ctenopoma nebulosum

Ctenopoma nebulosum male displaying

Bullseye Bushfish.

This species has been sold under a wide variety of common names such as Bullseye Bushfish, Zulu Bushfish, Leopard Bushfish, and Chocolate Bushfish, for example. The name Leopard Bushfish is the result of frequent confusion of this species with *Ct. acutirostre*, which is very similar in body form.

In general *Ct. ocellatum* grows to a length of 12 to 15 cm (4.75 to 6.0 in) under aquarium maintenance. However, the hobby literature also reports lengths of up to 30 cm (12 in), though these are probably to be regarded as exceptional. Unfortunately this species has only rarely been imported and so hardly any practical details have been reported. These fishes are usually offered at a length of around 5 cm (2 in) in the trade. The sexes can only be determined without problem in large specimens. Because this species too is purportedly an open spawner, whose eggs develop without benefit of brood care, the areas of spinous scales behind the eyes and on the caudal peduncle are the only reliable distinction. At present no differences in coloration or fin form between the sexes are known.

The natural habitat is the Congo river system in DR Congo including the Malebo (Stanley) Pool area near the cities of Kinshasa and Brazzaville, the area of the Stanley Falls, and other parts of the fast-flowing and calm zones of the Congo and its numerous tributaries. It is often found syntopic with *Ct. acutirostre*, and further biotope details are given in the text for that species.

Ct. ocellatum are crepuscular fishes. They often hover almost motionless in the water with only the pectoral fins working like little oars to drive the fish through the water like a leaf, always on the look-out for something to eat. The species is very peaceful and not aggressive towards other species. However under-sized fishes shouldn't be housed together with *Ct. ocellatum*. It has a large appetite and it may happen that an under-sized fish may disappear lightning-fast into the suddenly gigantic, open mouth.

The maintenance of this species is problem-free. The tank for maintenance should have zones of dense planting and additional hiding-places such as rocky caves and bogwood. Maintenance with very fast-swimming species of other genera is not advisable, as *Ct. ocellatum* will then prefer to remain in the quiet, planted areas and there will be few opportunities for observation. At the same time the species is best not kept solitary, as it requires companion fishes. If correctly maintained it will become very tame. Water parameters are apparently of secondary importance, but should, if possible, lie in the soft, slightly acid range. The water temperature should be 25 to 27 °C (77 to 80.5 °F). Slight water movement is advisable, as these fishes like to hover or drift in the current. *Ct. ocellatum* will reward a regular water change (a quarter to a third of the tank volume) every week or two with vitality and health growth.

Ctenopoma ocellatum

Pellegrin's Bushfish.

These slender African labyrinthfishes, which apparently grow to around 12 cm (5 in) long, are purportedly open spawners. Characteristic features of the species include 13 to 15 chain-like crossbands and a usually longitudinally-oriented, dark spot on the caudal peduncle and extending onto the caudal fin itself. These fishes can be regarded as peaceful and can readily be housed with other fishes as long as they aren't too small. They are lively swimmers that require robust morsels of food.

The locality for the specimens used in the original description lies 10 km south of Kindu in the DR Congo. BLEHER also collected the specimens imported in 1986 in the vicinity of Kindu in Congo-Kinshasa. The fishes were living in a small tributary stream of the Lualaba, which flows from the south into the Congo. The water parameters were general and carbonate hardness only slightly above 1° (German), electrical conductivity 65 - 75 μS/cm at a water temperature of 29 °C (84 °F), and a pH of 4.8. On that basis these fishes originate from very soft, acid waters. Nevertheless they have proved to be readily maintained without problem in medium-hard to hard, alkaline water.

Ctenopoma pellegrinii

This fish, which is named in honour of its discoverer and collector, J. PETHERIC, remains very rare in the aquarium to the present day. Its natural distribution region lies in the upper Nile, in the area of Khartoum and Gondokoro in East Africa. As far as is known at present, these fishes live in shallow marginal zones of the Nile as well as associated swamps and the adjacent savannah regions. Adults attain a total length of about 15 cm (6 in). The species may be identical with *Ctenopoma kingsleyae*.

Differentiating the sexes is often problematical. Females are usually plumper and somewhat larger, while large areas of "thorns" are a good indication of males. Like many of the other high-backed bushfishes in the genus, these fishes are open-spawners that lay their eggs in the open water.

Ctenopoma petherici

photo / Jürgen Schmidt-Weisswasser

Mottled Bushfish.

Following a decision by the International Commission for Zoological Nomenclature in 1989, *Ctenopoma oxyrhynchum* was synonymised with the older name *Ctenopoma weeksii*.

The species grows up to 10 cm (4 in) in length and has long been one of the frequently maintained bushfishes. The sexes cannot be distinguished by differences in coloration or fin form, but only by the areas of spinous scales behind the eyes and on the caudal peduncle in large specimens, which are more strongly developed in males.

The natural habitat was given in the original description as Yembe River, Ubanghi region, in DR Congo (Congo-Kinshasa). Apparently no additional information is available on biotope details and/or water chemistry and temperatures. Maintenance in the aquarium is problem-free. These fishes are peaceful aquarium-dwellers and can readily be housed with other fishes as long as the latter aren't too small. The appearance of this species exhibits affinities with *Ct. acutirostre* and *Ct. ocellatum*, but its behaviour is more lively. For optimal maintenance the aquarium should be densely planted in places and provide hiding-places in the form or wood and caves. A number of floating plants will protect the fishes from view and hence increase their sense of security. Efficient filtration of the water creating slight water movement is advisable. The water temperature should be around 25 °C (77 °F).

If fed a robust and varied diet *Ct. weeksii* will rapidly grow into strong, healthy specimens. Breeding appears not to be difficult and has frequently been achieved in recent years. A pair that are not too young should be used for breeding – on the basis of practical experience the fishes should be around two years old. The breeding tank can be an aquarium set up as for normal maintenance. The females should be ready to spawn and have a round, "full" belly. These fishes do not construct a nest and spawn during a nuptial embrace near the bottom. The eggs rise to the surface of the water and remain suspended among the plants. The parents do not practise any brood care and should be removed from the breeding tank after spawning. Care is advised when doing this – as with all other bushfish species – as when they are netted there is a serious danger of injury through their becoming tangled in the material of the net and this can result in not inconsiderable injuries. Hence capture by hand or with a plastic container or fish trap is advisable.

The larvae hatch after around 24 hours and the fry swim free after a further 70 hours. The number of offspring can be rather high, with 700 to 900 fry being nothing unusual. Soft, mineral-poor, slightly acid water is advisable for optimal development of the eggs and fry. Rearing the young is not difficult and should take place as described for *M. nanum*.

Ctenopoma weeksii

The genus *Microctenopoma*

Microctenopoma ansorgii

Microctenopoma damasi

Microctenopoma fasciolatum

Microctenopoma nanum

Microctenopoma sp.

Orange or Ornate Bushfish.

This species can be described as the most beautiful of the bushfishes by virtue of the coloration it sometimes displays – attractive orange-brown, with partially antique-green flanks and six dark brown crossbands, plus white fin edgings (unfortunately seen only in males). It is one of the peaceful bushfish species and is very good for keeping together with numerous other fishes.

The natural habitat lies mainly in the large areas of rainforest extending from Gabon across Congo-Brazzaville to Central Africa and the Shiloango river system in DR Congo (Congo-Kinshasa). It is also purportedly found in the tributaries of the Ntem and Ogowe river systems to the south of Djoum. These fishes live mainly among dense vegetation in the bank zones of narrow watercourses, where they are caught only fairly rarely. Their occurrence is very variable and apparently restricted to a small number of locations spread over a wide area. The water parameters in the areas where they are collected lie in the mineral-depleted, acid range.

Other bushfishes that are collected nearby or sympatric include *M. nanum*, which is similar in appearance, and *Ct. maculatum*, which has a deeper body form. The water temperature measures around 25 °C (77 °F). These fishes are usually to be found in heavily shaded wooded areas with low light levels. Their maintenance in the aquarium is problem-free. If half-grown specimens can be obtained then they can be easily transferred and acclimatised to the home aquarium. But adult *ansorgii* do not acclimatise well to the aquarium microbiotope and often die after a few months, despite good maintenance. However, acclimatised tank-bred stocks are often available in the aquarium trade.

Microctenopoma ansorgii belongs to the group of species that practise brood care. For breeding they should be provided with a tank that is densely planted in places and only dimly illuminated. The water should be very soft and slightly acid, and filtration should be installed. The temperature should be set to 26 °C (79 °F).

The male constructs his bubblenest beneath floating vegetation, and the pair then spawn beneath it during a nuptial embrace. The brood is tended by the male. Depending on the behaviour of the fishes in question, it is often a good idea to remove the parents after a day in order to ensure that all the young grow on. The fry swim free after around four days and then require small pond foods and/or infusorians. After around four days more they can be fed with freshly-hatched *Artemia* nauplii.

At an age of around 25 days the little Orange Bushfishes start to exhibit their vertical stripes, and after only around six weeks they start to engage in miniature mock battles.

Microctenopoma ansorgii male

Green-Blue Bushfish.

Males attain a total length of around 7 cm (2.75 in), while females are full-grown at somewhat more than 6 cm (2.25 in). Normal coloration is a faint grey-brown in both sexes. Dominant males colour up to a bright green-blue on a dark body, and brilliant blue and black during courtship and mating. This blue-black consists of numerous tiny iridescent spots on the black body, while at this time females merely accentuate their normal coloration with a light grey longitudinal band.

The natural habitat of this species is the area around Lake Edward in East Africa. They are found on both the Congo and the Ugandan sides of the lake. *M. damasi* lives mainly in the usually exposed, heavily vegetated accumulations of water in inundation zones and swamps, and in small shallow pools whose surface is often completely covered by the floating plant *Pistia stratiotes*. It is to be assumed that the temperature in these bodies of water rises dramatically in the heat of the sun and may reach values of around 34 °C (93 °F), sometimes with an accompanying serious shortage of oxygen. The geographically widespread bushfish species *Ctenopoma muriei* is also found in these waters.

The maintenance of this bushfish presents no problems. Because M. *damasi* is relatively uncommon, it is best to keep these fishes in a species tank with no tankmates, or with a small shoal of tetras or a few platies if excessive timidity is a problem. *M. damasi* practise brood care. The male builds a small, compact bubblenest among or beneath floating plants, and the pair spawn beneath it during a nuptial embrace. The eggs are transparent and rise to the surface of the water. The male tends the brood alone. At a water temperature of 28 °C (82.5 °F) the larvae hatch after about a day and a half and the fry swim free after a further 48 hours. The fry hover close to the water's surface and should be fed small quantities of the tiniest pond foods or powdered flake food scattered on the surface. Using additional gentle aeration will ensure good distribution of the food. The fry grow very quickly given good care.

Microctenopoma damasi female

Microctenopoma damasi male

Banded Bushfish

The Banded Bushfish is a peaceful species that grows to only 8 cm (3 in) long, and again is a bubblenest-builder that practises brood care. When fully grown, males exhibit greatly prolonged dorsal and anal fins, unlike the females, who have a dorsal and anal fin that is pointed but not prolonged. Additional differences between the sexes include the intense, almost blue-black zig-zag stripes in males (these are not as boldly expressed in females) and a light grey longitudinal band along the centre of the body when in courtship mood. These fishes are somewhat timid, but in a well-planted aquarium with several tankmates they are active swimmers and good objects of study. Banded Bushfishes are not aggressive among themselves or towards other fishes. Their maintenance is problem-free, and they can also readily be kept with smaller species.

Nevertheless the aquarium should not be too small and be a minimum length of 80 cm (32 in). There should be good planting with stemmed plants, dense in places. Here too sand with a grain size of 1 to 2 mm in diameter will serve as substrate. A few floating plants such as *Pistia stratiotes* and some *Riccia fluitans* will serve to round off the décor. The temperature should be around 25 °C (77 °F). The menu should be varied and include robust foods as well as healthy, vitamin-rich vegetable foods.

In recent decades *Microctenopoma fasciolatum* has often been confused with *Microctenopoma nanum*. Even today the resulting uncertainty still afflicts various authors and aquarists, although the original descriptions of these two species leave their identities in no doubt. *M. nanum* exhibits great similarity of body form to *M. ansorgii* and *M. damasi*. By contrast, at first glance the Banded Bushfish can be seen to be much deeper in body form and its appearance is completely different by virtue of the 10 to 12 zig-zag cross-bands.

The natural habitat of *M. fasciolatum* is the DR Congo region, from the upper Congo at Mosembe to its mouth at Boma. Water analyses in these areas reveal mineral-poor and slightly acid conditions, and accordingly the water chemistry should correspond to these values in any aquarium set up for breeding. The temperature should be raised to 28 °C (82.5 °F), and the tank sited in a quiet place. Dim lighting and low-current filtration are also recommended. Often the patience of the would-be breeder will be severely put to the test before the male begins to build a bubblenest, usually a very loose construction among floating plants at the water's surface, and sometimes including pieces of plants. Although this is followed by dummy spawning runs, it may be many days more before spawning actually takes place. With a little patience and a bit of luck the fishes will then mate beneath the bubblenest, with the male embracing his partner from below while she remains in the normal swimming position. The eggs rise slowly to the water's surface.

Large specimens can be very productive and up to 1000 eggs are not uncommon. It is also not unusual for the pair to spawn over several consecutive days. Because the parents represent no danger to either eggs or fry, the young can develop in safety even with the adults present. The larvae hatch around 18 hours after spawning, and the fry swim free a good two days later and hunt for the tiniest of foods among the plants at the water's surface. They should now be fed for the first few days on the finest live foods (eg infusorians) as well as additional powdered dry food. Given this maintenance, the little Banded Bushfishes will grow on rapidly and attain a length of 1.5 cm (0.5 in) after only eight weeks. At this size they already exhibit the zig-zag patterning of their parents. Unfortunately the fry grow on unevenly, and removing the larger specimens is beneficial for the growth and development of those that remain.

Microctenopoma fasciolatum

Dwarf Bushfish

This is one of the small bushfish species that are easy to maintain. When adult they attain a total length of around 7 cm (2.75 in). The differences between the sexes are easy to see in larger specimens. Males exhibit darker body colours, often ranging from dark grey to blue-black. The dorsal and anal fins are somewhat more pointed than those of females and bordered with light blue fin edgings. Females remain lighter in their coloration overall and usually exhibit a broad light grey longitudinal band extending from the centre of the body to the caudal peduncle. Depending on mood, there are seven to ten dark crossbands in both sexes, more prominent in females in courtship mood.

Because of the wide distribution region there are different colour variants and body forms. The natural habitat of *M. nanum* extends from southern Cameroon to Gabon, Rio Muni, and the lower part of the Congo. I have often caught this species in the south-eastern tributaries of the River Kienke as well as in the affluents of the Ntem and Dja rivers in southern Cameroon. This species has also been recorded in the outlying streams of the Ogowe river system that extend into this region. To my eye the fishes from the Dja affluents look somewhat more red-brownish in their coloration, so that their appearance often exhibits parallels to the synonym species *Ctenopoma brunneum* (AHL, 1927), which lives in this region according to the original description.

M. nanum lives in small watercourses, tributaries of the larger rivers, often up to 3 metres (10 feet) wide and usually only 50 cm (20 in) deep. Favourite haunts are the overgrown bank regions and adjacent inundation zones. These fishes were also found among leaves that had fallen into the water and exposed roots of trees. The water in all these locations was invariably very soft and acid, and almost always flowing. The habitats were always forest streams that were seldom exposed to the sun and where the temperature is always around 25 °C (77 °F), no matter what the time of the year. *M. nanum* is problem-free in its maintenance, but nevertheless a few points need to be borne in mind when setting up the aquarium. Even though also the species doesn't grow very large, it should not be kept in too small a tank. Dense planting in places and occasional groups of floating plants are strongly recommended. Efficient filtration should be used to produce pure, clear water with a gentle current. The temperature should be 24 °C (75 °F).

Even though the locations for *M. nanum* are mainly in tropical Africa, its native waters are usually shaded. Fishes of this species are very susceptible to disease if maintained at too high a temperature in the aquarium.

For breeding, only a single pair should be placed in the breeding tank, if possible individuals that are ready to spawn and have already shown an inclination to mate in the normal aquarium. The male will soon construct a small bubblenest among the floating plants. After a lively courtship the female will follow him beneath the bubblenest, where the pair embrace and the eggs are released and fertilised. They then rise into the bubblenest, where, when spawning is over, they are tended and guarded by the male alone until the fry become free-swimming. The species is fairly productive, and large specimens can release 100 eggs.

For successful breeding it is recommended that soft acid water should be used to match the water parameters of the natural habitat. The larvae hatch after around 24 hours, and the fry swim free after a further 70 hours. At this point the parent fishes should be removed from the breeding tank. The tiniest pond foods or infusorians, along with powdered flake, are recommended as first foods. After a few days the fry can be fed additionally, and after a further three days exclusively, on freshly-hatched *Artemia* nauplii.

Microctenopoma nanum

This species, which is apparently not yet scientifically described, exhibits affinities to the Dwarf Bushfish, *Microctenopoma nanum*, which remains equally small. While *Microctenopoma nanum* exhibits a predominantly grey to brown body coloration, this new species is predominantly light brown to clay-yellow. The stripes on the body also exhibit a very different pattern. These fishes undoubtedly belong to the bubblenesting group. On the basis of observations to date, *Microctenopoma* sp. attains a total length round 6 to 8 cm (2.25 to 3.0 in).

The species was caught in the company of *Ctenopoma pellegrini* in the area near Kindu in DR Congo. The fishes were living in a small stream that flows from the south into the River Congo at Lualaba; the water parameters were general and carbonate hardness only slightly above 1 degree (German), an electrical conductivity of 65 to 75 µS/cm at water temperature of 29 °C (84 °F), and pH of 4.8.

Although several specimens of this species were imported, unfortunately there were no offspring.

Microctenopoma sp.

Displaying *Microctenopoma* sp.

Afterword

The final species description above concludes my portrayal of the labyrinthfishes, in which I have tried to document – especially with photos – all the forms that have been seen alive in the aquarium to date. However, this book makes no claims to completeness. It is only with the greatest of effort that I have caught many species myself in Africa and Asia and brought them back for study in the aquarium. Various rare species have been bred and thus achieved a modest distribution in the hobby. My intention in writing this book has been to produce an overview of the diversity of the labyrinthfishes and in so doing provide the newcomer to the aquarium hobby, and perhaps also the seasoned aquarist, with a little information and food for thought. And it may be that this book also contains illustrations of species that in the near future will no longer have any natural habitat as the result of environmental change and unfortunately have also died out in captivity.

Hence if these lines lead to just one labyrinthfish species or another being maintained or studied in the aquarium, with the result that the species is able to survive, then this book will have served its purpose, as labyrinthfishes are among the most beautiful and interesting aquarium fishes of all. I hope that this book will also follow in the footsteps of the European Anabantoid Club with the Arbeitskreis Labyrinthfische im VDA e.V. (EAC/AKL - founded in 1992) and the Internationale Gemeinschaft für Labyrinthfische (IGL – founded in 1979) and contribute to the wider distribution of these colourful and remarkable fishes.

Rainforest biotope in southern Cameroon

Ctenopoma kingsleyae biotope in Sierra Leone

Woodland watercourse in Ghana

Sand track in southern Cameroon

Microctenopoma nanum biotope in southern Cameroon

Crinum plants in southern Cameroon

Steppe region in northern Nigeria

General Bibliography:

- BLEEKER, P. 1877-1879. Atlas ichthyologique des Indes orientales néêrlandaises. Tomes IX. Percoïdes III.Muller, Amsterdam, 80 pp., pls.355-420

- FREYHOF, J. & HERDER,F., 2002: Review of the fishes of the genus *Macropodus* in Vietnam,

  with description of two new species from Vietnam and southern China

  (Perciformes: Osphronemidae)

  Ichthyol. Explor. Freshwaters, Vol.13,No. 2, pp. 147-167 18 figs., 2 tabs.

- GECK, J. und KOPIC, G. 2005. Prachtguramis , Aquaristik aktuell Sonderheft 1/2005: 30-36

- GOLDSTEIN, R.2004: The betta Handbook Barron's Educational Series, New York 11788

- HALLMANN, M. & SCHMIDT, J. 2007: Kaum bekannt, aber vom Aussterben bedroht.

- Datz,7, Seite32-35, DATZ,8, Seite 56-59

- INGER, F. R. & CHIN, P. K.1962: The fresh-water fishes of north Borneo,

  Chicago Natural History Museum, 3/1962

- KOTTELAT, M. und NG, P.K.L. 2005. Diagnoses of six new species of *Parosphromenus* (Teleostei: Ospronemidae) from Malay Peninsula and Borneo, with Notes on other species. [The Raffles Bulletin of Zoology 2005] Supplement No. 13: 101-113

- KOTTELAT, M. & NG, P.K.L. 1994: Diagnosis of five new species of fighting fishes from Bangka and Borneo (Telostei: Belontiidae). Ichthyol. Explor. Freshwaters 5(1), 65-78

- KOTTELAT, M. & WHITTEN,A.J. & KARTIKASARI,S.N. & WIRJOATMODJO,S.,1993: Freshwater Fishes of Western Indonesia and Sulawesi. Periplus Editions Limited (HK) Ltd. 293 pp.

- KOKOSCHA, M. 1989: Zwei afrikanische Labyrinthfische. *Ctenopoma acutirostre* und *Ctenopoma ocellatum*. DATZ 42(3), 140-142.

- KOKOSCHA, M. 1989: Jungfischpigmentierung als Hinweis auf die systematische Stellung verschiedener Buschfischarten. Der Makropode 11(6), 91-95.

- KOWASUPAT, C. et.al.,2012: *Betta mahachaiensis*, a new species of bubble-nesting fighting fish (Teleostei: Osphronemidae) from Samut Sakhon Province, Thailand, ZOOTAXA 3522:49-60

- KOWASUPAT, C. et.al.,2012: *Betta siamorientalis*, a new species of bubble-nesting fighting fish (Teleostei: Osphronemidae) from eastern Thailand, Vertebrate Zoology 62(3) 2012, 387-397

- KUBOTA, K., MAJIMA, M., MIKI, K. & YAMAZAKI, K. 1996: Anabantoids. Tropical Fish Collection 5, PISCES Publishers Co., Ltd., Minato-ku, 108 Tokio, Japan.

- LINKE, H. 1991: Labyrinth fish. The bubble-nest-builders. Their identification, care and breeding. Tetra Press, Melle, Germany, 174 pp.

- LINKE, H. 1998: Labyrinthfische-Farbe im Aquarium. Tetra Verlag GmbH, Münster. 203 Seiten

- NEUGEBAUER, N. 2005: Neue Fundorte vom Krabi-Kampffisch *Betta simplex*,1994

  Aquarium life, 2, Seite 48-51. Gong Verlag – Ismaning

- NGUYEN, V.H. 2005: Ca Nuoc Ngot Viet Nam. Tap III. (Fresh Water fish of Vietnam)v. 3:[629-646

  Contain new species descriptions]

  NHA XUAT BAN NONG NGHIEP

- PAEPKE, H.-J., 1994: Die Paradiesfische, Die neue Brehm-Bücherei, Band 616, Westarp Wissenschaften, Magdeburg.

- PINTER, H. 1984: Labyrinthfische, Hechtköpfe und Schlangenkopffische, Eugen Ulmer Verlag, Stuttgart

- RAINBOTH, W. J. 1996: Fishes of the Cambodian Mekong – FAO SPECIES IDENTIFICATION FIELD GUIDE FOR FISHERY PURPOSES – FOOD AND AGRICULTURE ORGANIZATION OF THE UNITED NATIONS – p.265

- REGAN, C. T.,1910: The Asiatic fishes of the family Anabantidae.

  Proc. Zool. Soc. London 1909 (pt. 4) 767-787

- RICHTER, H.- J.1979: Das Buch der Labyrinthfische, Neumann Verlag Leipzig-Radebeul.

- SCHÄFER, F. 1997: Aqualog all Labyrinths Verlag A.C.S.GMBH, Mörfelden Walldorf

- SCHÄFER, F. 2009: *Colisa –Trichogaster-Trichopodus*

  Betta News 1/2009, Journal des European Anabantoid Club/ Arbeitskreis

  Labyrinthfische: S. 6-8

- SCHMIDT, J. 1996: Vergleichende Untersuchungen zum Fortpflanzungsverhalten der *Betta* – Arten (Belontiidae, Anabantoidei) VNW, Verlag Natur und Wissen, Solingen.

- SCHINDLER, I. & SCHMIDT, J. 2006; Review of the mouthbrooding *Betta* (Teleostei; Osphronemidae) from Thailand, with descriptions oft two new species. Zeitschrift für Fischkunde

- SMITS, H. M., 1945: The freshwater fishes of Siam, or Thailand.

  Bull. US. Nat Mus., 188: 1-622

- SCHINDLER, I & LINKE, H.,2012: Two new species of the genus *Parosphromenus* (Teleostei: Osphronemidae) from Sumatra, Vertebrate Zoology 62(3) 2012, 399-406

- SCHINDLER, I. & van der VOORT, S., 2011: Re-description of *Betta rubra* Perugia, 1893 (Teleostei: Osphronemidae), an enigmatic fighting fish from Sumatra, Bulletin of Fish Biology, Vol.13, 1/2, 21-32

- TAN, H. H. & NG, P. K. L.2004. Two new species of freshwater fish (Telestei: Balitoridae, Osphronemidae) from southern Sarawak. In YONG, H.S.,F.S.P.NG & E.E.L.YEN (eds.) Sarawak Bau Limestone Biodiversity. Sarawak Museum Journal, Vol. LIX, No. 80 (New Series); Special Issue No. 8: 267-284

- TAN, H. H. und NG, P. K. L. 2005:The fighting fishes (Teleostei: Osphronenmidae: Genus *Betta*) of Singapore, Malaysia and Brunei. [The Raffles Bulletin of Zoology 2005] Supplement No. 13: 43-99